Stochastic Mechanics
Random Media
Signal Processing and Image Synthesis
Mathematical Economics and Finance
Stochastic Optimization
Stochastic Control
Stochastic Models in Life Sciences

Applications of Mathematics

Stochastic Modelling And Applied Probability

45

T0203051

Springer
New York
Berlin
Heidelberg
Hong Kong
London
Milan
Paris
Tokyo

Applications of Mathematics

1 Fleming/Rishel, **Deterministic and Stochastic Optimal Control** (1975)
2 Marchuk, **Methods of Numerical Mathematics,** Second Ed. (1982)
3 Balakrishnan, **Applied Functional Analysis,** Second Ed. (1981)
4 Borovkov, **Stochastic Processes in Queueing Theory** (1976)
5 Liptser/Shiryayev, **Statistics of Random Processes I: General Theory,** Second Ed. (1977)
6 Liptser/Shiryayev, **Statistics of Random Processes II: Applications,** Second Ed. (1978)
7 Vorob'ev, **Game Theory: Lectures for Economists and Systems Scientists** (1977)
8 Shiryayev, **Optimal Stopping Rules** (1978)
9 Ibragimov/Rozanov, **Gaussian Random Processes** (1978)
10 Wonham, **Linear Multivariable Control: A Geometric Approach,** Third Ed. (1985)
11 Hida, **Brownian Motion** (1980)
12 Hestenes, **Conjugate Direction Methods in Optimization** (1980)
13 Kallianpur, **Stochastic Filtering Theory** (1980)
14 Krylov, **Controlled Diffusion Processes** (1980)
15 Prabhu, **Stochastic Storage Processes: Queues, Insurance Risk, Dams, and Data Communication,** Second Ed. (1998)
16 Ibragimov/Has'minskii, **Statistical Estimation: Asymptotic Theory** (1981)
17 Cesari, **Optimization: Theory and Applications** (1982)
18 Elliott, **Stochastic Calculus and Applications** (1982)
19 Marchuk/Shaidourov, **Difference Methods and Their Extrapolations** (1983)
20 Hijab, **Stabilization of Control Systems** (1986)
21 Protter, **Stochastic Integration and Differential Equations** (1990)
22 Benveniste/Métivier/Priouret, **Adaptive Algorithms and Stochastic Approximations** (1990)
23 Kloeden/Platen, **Numerical Solution of Stochastic Differential Equations** (1992)
24 Kushner/Dupuis, **Numerical Methods for Stochastic Control Problems in Continuous Time,** Second Ed. (2001)
25 Fleming/Soner, **Controlled Markov Processes and Viscosity Solutions** (1993)
26 Baccelli/Brémaud, **Elements of Queueing Theory** (1994)
27 Winkler, **Image Analysis, Random Fields, and Dynamic Monte Carlo Methods: An Introduction to Mathematical Aspects** (1994)
28 Kalpazidou, **Cycle Representations of Markov Processes** (1995)
29 Elliott/Aggoun/Moore, **Hidden Markov Models: Estimation and Control** (1995)
30 Hernández-Lerma/Lasserre, **Discrete-Time Markov Control Processes: Basic Optimality Criteria** (1996)
31 Devroye/Györfi/Lugosi, **A Probabilistic Theory of Pattern Recognition** (1996)
32 Maitra/Sudderth, **Discrete Gambling and Stochastic Games** (1996)
33 Embrechts/Klüppelberg/Mikosch, **Modelling Extremal Events** (1997)
34 Duflo, **Random Iterative Models** (1997)

(continued after index)

J. Michael Steele

Stochastic Calculus and Financial Applications

Springer

J. Michael Steele
The Wharton School
Department of Statistics
University of Pennsylvania
3000 Steinberg Hall–Dietrich Hall
Philadelphia, PA 19104-6302, USA

With 3 figures.

Mathematics Subject Classification (2000): 60G44, 60H05, 91B28, 60G42

Library of Congress Cataloging-in-Publication Data
Steele, J. Michael
Stochastic calculus and financial applications / J. Michael Steele.
p. cm. — (Applications of mathematics ; 45)
Includes bibliographical references and index.

1. Stochastic analysis. 2. Business mathematics. I. Title. II. Series.
QA274.2 .S74 2000
519.2—dc21 00-025890

ISBN 978-1-4419-2862-7

Printed in the United States of America. (EB)

9 8 7 6 5 4

Springer-Verlag is a part of *Springer Science+Business Media*

springeronline.com

Preface

This book is designed for students who want to develop professional skill in stochastic calculus and its application to problems in finance. The Wharton School course that forms the basis for this book is designed for energetic students who have had some experience with probability and statistics but have not had advanced courses in stochastic processes. Although the course assumes only a modest background, it moves quickly, and in the end, students can expect to have tools that are deep enough and rich enough to be relied on throughout their professional careers.

The course begins with simple random walk and the analysis of gambling games. This material is used to motivate the theory of martingales, and, after reaching a decent level of confidence with discrete processes, the course takes up the more demanding development of continuous-time stochastic processes, especially Brownian motion. The construction of Brownian motion is given in detail, and enough material on the subtle nature of Brownian paths is developed for the student to evolve a good sense of when intuition can be trusted and when it cannot. The course then takes up the Itô integral in earnest. The development of stochastic integration aims to be careful and complete without being pedantic.

With the Itô integral in hand, the course focuses more on models. Stochastic processes of importance in finance and economics are developed in concert with the tools of stochastic calculus that are needed to solve problems of practical importance. The financial notion of replication is developed, and the Black-Scholes PDE is derived by three different methods. The course then introduces enough of the theory of the diffusion equation to be able to solve the Black–Scholes partial differential equation and prove the uniqueness of the solution. The foundations for the martingale theory of arbitrage pricing are then prefaced by a well-motivated development of the martingale representation theorems and Girsanov theory. Arbitrage pricing is then revisited, and the notions of admissibility and completeness are developed in order to give a clear and professional view of the fundamental formula for the pricing of contingent claims.

This is a text with an attitude, and it is designed to reflect, wherever possible and appropriate, a prejudice for the concrete over the abstract. Given good general skill, many people can penetrate most deeply into a mathematical theory by focusing their energy on the mastery of well-chosen examples. This does not deny that good abstractions are at the heart of all mathematical subjects. Certainly, stochastic calculus has no shortage of important abstractions that have stood the test of time. These abstractions are to be cherished and nurtured. Still, as a matter of principle, each abstraction that entered the text had to clear a high hurdle.

Many people have had the experience of learning a subject in 'spirals.' After penetrating a topic to some depth, one makes a brief retreat and revisits earlier topics with the benefit of fresh insights. This text builds on the spiral model in several ways. For example, there is no shyness about exploring a special case before discussing a general result. There also are some problems that are solved in several different ways, each way illustrating the strength or weakness of a new technique.

Any text must be more formal than a lecture, but here the lecture style is followed as much as possible. There is also more concern with 'pedagogic' issues than is common in advanced texts, and the text aims for a coaching voice. In particular, readers are encouraged to use ideas such as George Pólya's "Looking Back" technique, numerical calculation to build intuition, and the art of guessing before proving. The main goal of the text is to provide a professional view of a body of knowledge, but along the way there are even more valuable skills one can learn, such as general problem-solving skills and general approaches to the invention of new problems.

This book is not designed for experts in probability theory, but there are a few spots where experts will find something new. Changes of substance are far fewer than the changes in style, but some points that might catch the expert eye are the explicit use of wavelets in the construction of Brownian motion, the use of linear algebra (and dyads) in the development of Skorohod's embedding, the use of martingales to achieve the approximation steps needed to define the Itô integral, and a few more.

Many people have helped with the development of this text, and it certainly would have gone unwritten except for the interest and energy of more than eight years of Wharton Ph.D. students. My fear of omissions prevents me from trying to list all the students who have gone out of their way to help with this project. My appreciation for their years of involvement knows no bounds.

Of the colleagues who have helped personally in one way or another with my education in the matters of this text, I am pleased to thank Erhan Çinlar, Kai Lai Chung, Darrell Duffie, David Freedman, J. Michael Harrison, Michael Phelan, Yannis Karatzas, Wenbo Li, Andy Lo, Larry Shepp, Steve Shreve, and John Walsh. I especially thank Jim Pitman, Hristo Sendov, Ruth Williams, and Marc Yor for their comments on earlier versions of this text. They saved me from some grave errors, and they could save me from more if time permitted. Finally, I would like to thank Vladimir Pozdnyakov for hundreds of hours of conversation on this material. His suggestions were especially influential on the last five chapters.

J. Michael Steele
Philadelphia, PA

Contents

CHAPTER 1

Random Walk and First Step Analysis

The fountainhead of the theory of stochastic processes is simple random walk. Already rich in unexpected and elegant phenomena, random walk also leads one inexorably to the development of Brownian motion, the theory of diffusions, the Itô calculus, and myriad important applications in finance, economics, and physical science.

Simple random walk provides a model of the wealth process of a person who makes a living by flipping a fair coin and making fair bets. We will see it is a hard living, but first we need some notation. We let $\{X_i : 1 \le i < \infty\}$ denote a sequence of independent random variables with the probability distribution given by

$$P(X_i = 1) = P(X_i = -1) = \frac{1}{2}.$$

Next, we let S_0 denote an arbitrary integer that we view as our gambler's initial wealth, and for $1 \le n < \infty$ we let S_n denote S_0 plus the partial sum of the X_i:

$$S_n = S_0 + X_1 + X_2 + \cdots + X_n.$$

If we think of $S_n - S_0$ as the net winnings after n fair wagers of one dollar each, we almost have to inquire about the probability of the gambler winning A dollars before losing B dollars. To put this question into useful notation, we do well to consider the first time τ at which the partial sum S_n reaches level A or level $-B$:

$$\tau = \min\{n \ge 0 : S_n = A \text{ or } S_n = -B\}.$$

At the random time τ, we have $S_\tau = A$ or $S_\tau = -B$, so our basic problem is to determine $P(S_\tau = A \mid S_0 = 0)$. Here, of course, we permit the wealth of the idealized gambler to become negative — not an unrealistic situation.

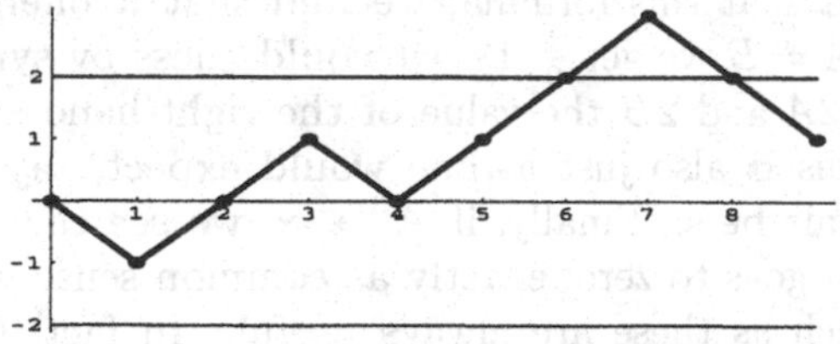

FIGURE 1.1. HITTING TIME OF LEVEL ± 2 IS 6

1.1. First Step Analysis

The solution of this problem can be obtained in several ways, but perhaps the most general method is *first step analysis*. One benefit of this method is that it is completely elementary in the sense that it does not require any advanced mathematics. Still, from our perspective, the main benefit of first step analysis is that it provides a benchmark by which to gauge more sophisticated methods.

For our immediate problem, first step analysis suggests that we consider the gambler's situation after one round of the game. We see that his wealth has either increased by one dollar or decreased by one dollar. We then face a problem that replicates our original problem except that the "initial" wealth has changed. This observation suggests that we look for a recursion relation for the function

$$f(k) = P(S_\tau = A \mid S_0 = k), \quad \text{where } -B \le k \le A.$$

In this notation, $f(0)$ is precisely the desired probability of winning A dollars before losing B dollars.

If we look at what happens as a consequence of the first step, we immediately find the desired recursion for $f(k)$,

$$f(k) = \frac{1}{2}f(k-1) + \frac{1}{2}f(k+1) \text{ for } -B < k < A, \tag{1.1}$$

and this recursion will uniquely determine f when it is joined with the boundary conditions

$$f(A) = 1 \text{ and } f(-B) = 0.$$

The solution turns out to be a snap. For example, if we let $f(-B+1) = \alpha$ and substitute the values of $f(-B)$ and $f(-B+1)$ into equation (1.1), we find that $f(-B+2) = 2\alpha$. If we then substitute the values of $f(-B+1)$ and $f(-B+2)$ into equation (1.1) we find $f(-B+3) = 3\alpha$, whence it is no great leap to guess that we have $f(-B+k) = k\alpha$ for all $0 \le k \le A+B$.

Naturally, we verify the guess simply by substitution into equation (1.1). Finally, we determine that $\alpha = 1/(A+B)$ from the right boundary condition $f(A) = 1$ and the fact that for $k = A+B$ our conjectured formula for f requires $f(A) = (A+B)\alpha$. In the end, we arrive at a formula of remarkable simplicity and grace:

$$P(S_n \text{ reaches } A \text{ before } -B \mid S_0 = 0) = \frac{B}{A+B}. \tag{1.2}$$

Looking Back

When we look back at this formula, we find that it offers several reassuring checks. First, when $A = B$ we get $\frac{1}{2}$, as we would guess by symmetry. Also, if we replace A and B by $2A$ and $2B$ the value of the right-hand side of formula (1.2) does not change. This is also just as one would expect, say by considering the outcome of pairs of fair bets. Finally, if $A \to \infty$ we see the gambler's chance of reaching A before $-B$ goes to zero, exactly as common sense would tell us.

Simple checks such as these are always useful. In fact, George Pólya made "Looking Back" one of the key tenets of his lovely book *How to Solve It*, a volume that may teach as much about doing mathematics as any ever written. From time to time, we will take advantage of further advice that Pólya offered about looking back and other aspects of problem solving.

1.2. Time and Infinity

Our derivation of the hitting probability formula (1.2) would satisfy the building standards of all but the fussiest communities, but when we check the argument we find that there is a logical gap; we have tacitly assumed that τ is finite. How do we know for sure that the gambler's net winnings will eventually reach A or $-B$? This important fact requires proof, and we will call on a technique that exploits a general principle: if something is possible — and there are infinitely many "serious" attempts — then it will happen.

Consider the possibility that the gambler wins $A+B$ times in a row. If the gambler's fortune has not already hit $-B$, then a streak of $A+B$ wins is guaranteed to boost his fortune above A. Such a run of luck is unlikely, but it has positive probability—in fact, probability $p = 2^{-A-B}$. Now, if we let E_k denote the event that the gambler wins on each turn in the time interval $[k(A+B), (k+1)(A+B)-1]$, then the E_k are independent events, and $\tau > n(A+B)$ implies that all of the E_k with $0 \le k \le n$ fail to occur. Thus, we find

$$P(\tau > n(A+B) \mid S_0 = 0) \le P(\cap_{k=0}^{n-1} E_k^c) = (1-p)^n. \tag{1.3}$$

Since $P(\tau = \infty \mid S_0 = 0) \le P(\tau > n(A+B) \mid S_0 = 0)$ for all n, we see from equation (1.3) that $P(\tau = \infty \mid S_0 = 0) = 0$, just as we needed to show to justify our earlier assumption.

By a small variation on this technique, we can even deduce from equation (1.3) that τ has moments of all orders. As a warm-up, first note that if $1(A)$ denotes the indicator function of the event A, then for any integer-valued nonnegative random variable Z we have the identity

$$Z = \sum_{k=1}^{\infty} 1(Z \ge k). \tag{1.4}$$

If we take expectations on both sides of the identity (1.4), we find a handy formula that textbooks sometimes prove by a tedious summation by parts:

$$E(Z) = \sum_{k=1}^{\infty} P(Z \ge k). \tag{1.5}$$

We will use equations (1.4) and (1.5) on many occasions, but much of the time we do not need an exact representation. In order to prove that $E(\tau^d) < \infty$ we can get along just as well with rough bounds. For example, if we sum the crude estimate

$$\tau^d 1[(k-1)(A+B) < \tau \le k(A+B)] \le k^d (A+B)^d 1[(k-1)(A+B) < \tau],$$

over k, then we have

$$\tau^d \le \sum_{k=1}^{\infty} k^d (A+B)^d 1[(A+B)(k-1) < \tau]. \tag{1.6}$$

We can then take expectations on both sides of the inequality (1.6) and apply the tail estimate (1.3). The ratio test finally provides the convergence of the bounding sum:

$$E(\tau^d) \le \sum_{k=1}^{\infty} k^d (A+B)^d (1-p)^{k-1} < \infty.$$

A Second First Step

Once we know that τ has a finite expectation, we are almost immediately drawn to the problem of determining the value of that expectation. Often, such ambitious questions yield only partial answers, but this time the answer could not be more complete or more beautiful.

Again, we use first step analysis, although now we are interested in the function defined by

$$g(k) = E(\tau \mid S_0 = k).$$

After one turn of the game, two things will have happened: the gambler's fortune will have changed, and a unit of time will have passed. The recurrence equation that we obtain differs from the one found earlier only in the appearance of an additional constant term:

$$g(k) = \frac{1}{2}g(k-1) + \frac{1}{2}g(k+1) + 1 \text{ for } -B < k < A. \tag{1.7}$$

Also, since the time to reach A or $-B$ is zero if S_0 already equals A or $-B$, we have new boundary conditions:

$$g(-B) = 0 \text{ and } g(A) = 0.$$

This time our equation is not so trivial that we can guess the answer just by calculating a couple of terms. Here, our guess is best aided by finding an appropriate analogy. To set up the analogy, we introduce the forward difference operator defined by

$$\Delta g(k-1) = g(k) - g(k-1),$$

and we note that applying the operator twice gives

$$\Delta^2 g(k-1) = g(k+1) - 2g(k) + g(k-1).$$

The recurrence equation (1.7) can now be written rather elegantly as a second order difference equation:

$$\frac{1}{2}\Delta^2 g(k-1) = -1 \text{ for } -B < k < A. \tag{1.8}$$

The best feature of this reformulation is that it suggests an immediate analogy. The integer function $g\colon \mathbb{N} \to \mathbb{R}$ has a constant second *difference*, and the real functions with a constant second *derivative* are just quadratic polynomials, so one is naturally led to look for a solution to equation (1.7) that is a quadratic over the integers. By the same analogy, equation (1.8) further suggests that the coefficient of k^2 in the quadratic should be -1. Finally, the two boundary conditions tell us that the quadratic must vanish at $-B$ and A, so we are left with only one reasonable guess,

$$g(k) = -(k-A)(k+B). \tag{1.9}$$

To verify that this guess is indeed an honest solution only requires substitution into equation (1.7). This time we are lucky. The solution does check, and our analogies have provided a reliable guide.

Finally, we note that when we specialize our formula to $k = 0$, we come to a result that could not be more striking:

$$E(\tau \mid S_0 = 0) = AB. \tag{1.10}$$

This formula is a marvel of simplicity — no better answer could even be imagined. Moreover, when we look back on equation (1.10), we find several interesting deductions.

For example, if we let $\tau' = \min\{n \geq 0 : S_n = -1\}$ and set

$$\tau'' = \min\{n \geq 0 : S_n = -1 \text{ or } S_n = A\},$$

then we see that $\tau'' \leq \tau'$. But equation (1.10) tells us $E(\tau'') = A$ so we find that $E(\tau') \geq A$ for all A. The bottom line is that $E(\tau') = \infty$, or, in other words, the expected time until the gambler gets behind by even one dollar is infinite.

This remarkable fact might give the gambler some cause for celebration, except for the sad symmetrical fact that the expected time for the gambler to get ahead by one dollar is also infinite. Strangely, one of these two events must happen on the very first bet; thus we face one of the many paradoxical features of the fair coin game.

There are several further checks that we might apply to formula (1.10), but we will pursue just one more. If we consider the symmetric interval $[-A, A]$, is there some way that we might have guessed that the expected time until the first exit should be a quadratic function of A? One natural approach to this question is to consider the expected size of $|S_n|$. The central limit theorem and a bit of additional work will tell us that $E(|S_n|) \sim \sqrt{2n/\pi}$, so when both n and A are large we see that $E(|S_n|)$ will first leave the interval $[-A, A]$ when $n \sim \pi A^2/2$. This observation does not perfectly parallel our exit-time formula (1.10), but it does suggest that a quadratic growth rate is in the cards.

1.3. Tossing an Unfair Coin

It is often remarked that life is not fair, and, be that as it may, there is no doubt that many gambling games are not even-handed. Considerable insight into the difficulties that face a player of an unfair game can be found by analysis of the simplest model — the biased random walk defined by $S_n = S_0 + X_1 + X_2 + \cdots + X_n$, where

$$P(X_i = 1) = p \text{ and } P(X_i = -1) = 1 - p = q \text{ where } p \neq q.$$

To solve the ruin problem for biased random walk, we take $f(k)$ and τ as before and note that first step analysis leads us to

$$f(k) = pf(k+1) + qf(k-1).$$

This is another equation that is most easily understood if it is written in terms of the difference operator. First, we note that since $p + q = 1$ the equation can be rearranged to give

$$0 = p\{f(k+1) - f(k)\} - q\{f(k) - f(k-1)\},$$

from which we find a simple recursion for $\Delta f(k)$:

$$\Delta f(k) = (q/p)\Delta f(k-1). \tag{1.11}$$

Now, we simply iterate equation (1.11) to find

$$\Delta f(k+j) = (q/p)^j \Delta f(k),$$

so, if we set $\alpha = \Delta f(-B)$, we can exploit the fact that $f(-B) = 0$ and successive cancellations to find

$$f(k) = \sum_{j=0}^{k+B-1} \Delta f(j - B) = \alpha \sum_{j=0}^{k+B-1} (q/p)^j = \alpha \frac{(q/p)^{k+B} - 1}{(q/p) - 1}. \tag{1.12}$$

We can then eliminate α from equation (1.12) if we let $k = A$ and invoke our second boundary condition:

$$1 = f(A) = \alpha \frac{(q/p)^{A+B} - 1}{(q/p) - 1}.$$

After determining α, we return to equation (1.12) and take $k = 0$ to get to the bottom line; for biased random walk, we have a simple and explicit formula for the ruin probability:

$$P(S_n \text{ hits } A \text{ before } -B \mid S_0 = 0) = \frac{(q/p)^B - 1}{(q/p)^{A+B} - 1}. \tag{1.13}$$

This formula would transform the behavior of millions of the world's gamblers, if they could only take it to heart. Such a conversion is unlikely, though perhaps a few might be moved to change their ways if they would work out the implications of equation (1.13) for some typical casino games.

TIME AND TIME AGAIN

The expected time until the biased random walk hits either level A or $-B$ can also be found by first step analysis. If $g(k)$ denotes the expected time until the random walk hits A or $-B$ when we start at k, then the equation given by first step analysis is just

$$g(k) = pg(k+1) + qg(k-1) + 1.$$

As before, this equation is better viewed in difference form

$$\Delta g(k) = (q/p)\Delta g(k-1) - 1/p, \tag{1.14}$$

where the boundary conditions are the same as those we found for the unbiased walk

$$g(-B) = 0 \text{ and } g(A) = 0.$$

To solve equation (1.14), we first note that if we try a solution of the form ck then we find that $g_0(k) = k/(q-p)$ is one solution of the inhomogeneous equation (1.14). From our earlier work we also know that $\alpha + \beta(q/p)^k$ is a solution of the homogeneous equation (1.11), so to obtain a solution that handles the boundary conditions we consider solutions of the form

$$g(k) = \frac{k}{q-p} + \alpha + \beta(q/p)^k.$$

The two boundary conditions give us a pair of equations that we can solve to determine α and β in order to complete the determination of $g(k)$. Finally, when we specialize to $g(0)$, we find the desired formula for the expected hitting time of $-B$ or A for the biased random walk:

$$E(\tau \mid S_0 = 0) = \frac{B}{q-p} - \frac{A+B}{q-p}\,\frac{1-(q/p)^B}{1-(q/p)^{A+B}}. \tag{1.15}$$

The formulas for the hitting probabilities (1.13) and the expected hitting time (1.15) are more complicated than their cousins for unbiased walk, but they answer more complex questions. When we look back on these formulas, we naturally want to verify that they contain the results that were found earlier, but one cannot recapture the simpler formulas just by setting $p = q = \frac{1}{2}$. Nevertheless, formulas (1.13) and (1.15) are consistent with the results that were obtained for unbiased walks. If we let $p = \frac{1}{2} + \epsilon$ and $q = \frac{1}{2} - \epsilon$ in equations (1.13) and (1.15), we find that as $\epsilon \to 0$ equations (1.13) and (1.15) reduce to $B/(A+B)$ and AB, as one would expect.

1.4. Numerical Calculation and Intuition

The formulas for the ruin probabilities and expected hitting times are straightforward, but for someone interested in building serious streetwise intuition there is nothing that beats numerical computation.

- We now know that in a fair game of coin tosses and \$1 wagers the expected time until one of the players gets ahead by \$100 is 10,000 tosses, a much larger number than many people might expect.
- If our gambler takes up a game with probability $p = 0.49$ of winning on each round, he has less than a 2% chance of winning \$100 before losing \$200. This offers a stark contrast to the fair game, where the gambler would have a 2/3 probability of winning \$100 before losing \$200. The cost of even a small bias can be surprisingly high.

In the table that follows, we compute the probability of winning \$100 before losing \$100 in some games with odds that are typical of the world's casinos. The table assumes a constant bet size of \$1 on all rounds of the game.

TABLE 1.1. STREETWISE BENCHMARKS.

Chance on one round	0.500	0.495	0.490	0.480	0.470
Chance to win \$100	0.500	0.1191	0.0179	0.0003	6×10^{-6}
Duration of the game	10,000	7,616	4,820	2,498	1,667

One of the lessons we can extract from this table is that the traditional movie character who chooses to wager everything on a single round of roulette is not so foolish; there is wisdom to back up the bravado. In a game with a 0.47 chance to win on each bet, you are about 78,000 times more likely to win \$100 by betting \$100 on a single round than by playing just \$1 per round. Does this add something to your intuition that goes beyond the simple formula for the ruin probability?

1.5. First Steps with Generating Functions

We have obtained compelling results for the most natural problems of gambling in either fair or unfair games, and these results make a sincere contribution to our understanding of the real world. It would be perfectly reasonable to move to other problems before bothering to press any harder on these simple models. Nevertheless, the first step method is far from exhausted, and, if one has the time and interest, much more detailed information can be obtained with just a little more work.

For example, suppose we go back to simple random walk and consider the problem of determining the probability *distribution* of the first hitting time of level 1 given that the walk starts at zero. Our interest is no longer confined to a single number, so we need a tool that lets us put all of the information of a discrete distribution into a package that is simple enough to crack with first step analysis.

If we let τ denote this hitting time, then the appropriate package turns out to be the probability generating function:

$$\phi(z) = E(z^{\tau} \mid S_0 = 0) = \sum_{k=0}^{\infty} P(\tau = k \mid S_0 = 0) z^k. \tag{1.16}$$

If we can find a formula for $\phi(z)$ and can compute the Taylor expansion of $\phi(z)$ from that formula, then by identifying the corresponding coefficients we will have found $P(\tau = k \mid S_0 = 0)$ for all k. Here, one should also note that once we understand τ we also understand the distribution of the first time to go up k levels; the probability generating function in that case is given by $\phi(z)^k$ because the probability generating function of a sum of independent random variables is simply the product of the probability generating functions.

Now, although we want to determine a function, first step analysis proceeds much as before. When we take our first step, two things happen. First, there is the passage of one unit of time; and, second, we will have moved from zero to either -1 or 1. We therefore find on a moment's reflection that

$$\phi(z) = \frac{1}{2}E(z^{\tau+1} \mid S_0 = -1) + \frac{1}{2}E(z^{\tau+1} \mid S_0 = 1). \tag{1.17}$$

Now, $E(z^\tau \mid S_0 = -1)$ is the same as the probability generating function of the first time to reach level 2 starting at 0, and we noted earlier that this is exactly $\phi(z)^2$. We also have $E(z^\tau \mid S_0 = 1) = 1$, so equation (1.17) yields a quadratic equation for $\phi(z)$:

$$\phi(z) = \frac{1}{2}z\phi(z)^2 + \frac{1}{2}z. \tag{1.18}$$

In principle $\phi(z)$ is now determined, but we can get a thoroughly satisfying answer only if we exercise some discrete mathematics muscle. When we first apply the quadratic formula to solve equation (1.18) for $\phi(z)$ we find two candidate solutions. Since $\tau \geq 1$, the definition of $\phi(z)$ tells us that $\phi(0) = 0$, and only one of the solutions of equation (1.18) evaluates to zero when $z = 0$, so we can deduce that

$$\phi(z) = \frac{1 - \sqrt{1 - z^2}}{z}. \tag{1.19}$$

The issue now boils down to finding the coefficients in the Taylor expansion of $\phi(z)$. To get these coefficients by successive differentiation is terribly boring, but we can get them all rather easily if we recall Newton's generalization of the binomial theorem. This result tells us that for any exponent $\alpha \in \mathbb{R}$, we have

$$(1 + y)^\alpha = \sum_{k=0}^{\infty} \binom{\alpha}{k} y^k, \tag{1.20}$$

where the binomial coefficient is defined to be 1 for $k = 0$ and is defined by

$$\binom{\alpha}{k} = \frac{\alpha(\alpha - 1)\cdots(\alpha - k + 1)}{k!} \tag{1.21}$$

for $k > 0$. Here, we should note that if α is equal to a nonnegative integer m, then the Newton coefficients (1.21) reduce to the usual binomial coefficients, and Newton's series reduces to the usual binomial formula.

When we apply Newton's formula to $(1 - z^2)^{\frac{1}{2}}$, we quickly find the Taylor expansion for ϕ:

$$\phi(z) = \frac{1 - \sqrt{1 - z^2}}{z} = \sum_{k=1}^{\infty} \binom{1/2}{k} (-1)^{k+1} z^{2k-1},$$

and when we compare this expansion with the definition of $\phi(z)$ given by equation (1.16), we can identify the corresponding coefficients to find

$$P(\tau = 2k-1 \mid S_0 = 0) = \binom{1/2}{k}(-1)^{k+1}. \tag{1.22}$$

The last expression is completely explicit, but it can be written a bit more comfortably. If we expand Newton's coefficient and rearrange terms, we quickly find a formula with only conventional binomials:

$$P(\tau = 2k-1 \mid S_0 = 0) = \frac{1}{2k-1}\binom{2k}{k}2^{-2k}. \tag{1.23}$$

This formula and a little arithmetic will answer any question one might have about the distribution of τ. For example, it not only tells us that the probability that our gambler's winnings go positive for the first time on the fifth round is $1/16$, but it also resolves more theoretical questions such as showing

$$E(\tau^\alpha) < \infty \text{ for all } \alpha < 1/2,$$

even though we have

$$E(\tau^\alpha) = \infty \text{ for all } \alpha \geq 1/2.$$

1.6. Exercises

The first exercise suggests how results on biased random walks can be worked into more realistic models. Exercise 1.2 then develops the fundamental recurrence property of simple random walk. Finally, Exercise 1.3 provides a mind-stretching result that may seem unbelievable at first.

EXERCISE 1.1 (Complex Models from Simple Ones). Consider a naive model for a stock that has a support level of \$20/share because of a corporate buy-back program. Suppose also that the stock price moves randomly with a downward bias when the price is above \$20 and randomly with an upward bias when the price is below \$20. To make the problem concrete, we let Y_n denote the stock price at time n, and we express our support hypothesis by the assumption that

$$P(Y_{n+1} = 21 \mid Y_n = 20) = 0.9, \text{ and } P(Y_{n+1} = 19 \mid Y_n = 20) = 0.1.$$

We then reflect the downward bias at price levels above \$20 by requiring for $k > 20$ that

$$P(Y_{n+1} = k+1 \mid Y_n = k) = 1/3 \text{ and } P(Y_{n+1} = k-1 \mid Y_n = k) = 2/3.$$

The upward bias at price levels below \$20 is expressed by assuming for $k < 20$ that

$$P(Y_{n+1} = k+1 \mid Y_n = k) = 2/3 \text{ and } P(Y_{n+1} = k-1 \mid Y_n = k) = 1/3.$$

Calculate the expected time for the stock price to fall from \$25 through the support level of \$20 all the way down to \$18.

EXERCISE 1.2 (Recurrence of SRW). If S_n denotes simple random walk with $S_0 = 0$, then the usual binomial theorem immediately gives us the probability that we are back at 0 at time $2k$:

$$P(S_{2k} = 0 \mid S_0 = 0) = \binom{2k}{k} 2^{-2k}. \tag{1.24}$$

(a) First use Stirling's formula $k! \sim \sqrt{2\pi k}\, k^k e^{-k}$ to justify the approximation

$$P(S_{2k} = 0) \sim (\pi k)^{-\frac{1}{2}},$$

and use this fact to show that if N_n denotes the number of visits made by S_k to 0 up to time n, then $E(N_n) \to \infty$ as $n \to \infty$.

(b) Finally, prove that we have

$$P(S_n = 0 \text{ for infinitely many } n) = 1.$$

This is called the *recurrence property* of random walk; with probability one simple random walk returns to the origin infinitely many times. Anyone who wants a hint might consider the plan of calculating the expected value of

$$N = \sum_{n=1}^{\infty} 1(S_n = 0)$$

in two different ways. The direct method using $P(S_n = 0)$ should then lead without difficulty to $E(N) = \infty$. The second method is to let

$$r = P(S_n = 0 \text{ for some } n \geq 1 | S_0 = 0)$$

and to argue that

$$E(N) = \frac{r}{1-r}.$$

To reconcile this expectation with the calculation that $E(N) = \infty$ then requires $r = 1$, as we wanted to show.

(c) Let $\tau_0 = \min\{n \geq 1 \colon S_n = 0\}$ and use first step analysis together with the first-passage time probability (1.23) to show that we also have

$$P(\tau_0 = 2k) = \frac{1}{2k-1}\binom{2k}{k} 2^{-2k}. \tag{1.25}$$

Use Stirling's formula $n! \sim n^n e^{-n}\sqrt{2\pi n}$ to show that $P(\tau_0 = 2k)$ is bounded above and below by a constant multiple of $k^{-3/2}$, and use these bounds to conclude that $E(\tau_0^{\alpha}) < \infty$ for all $\alpha < \frac{1}{2}$ yet $E(\tau_0^{\frac{1}{2}}) = \infty$.

EXERCISE 1.3. Consider simple random walk beginning at 0 and show that for any $k \neq 0$ the expected number of visits to level k before returning to 0 is exactly 1. Anyone who wants a hint might consider the number N_k of visits to level k before the first return to 0. We have $N_0 = 1$ and can use the results on hitting probabilities to show that for all $k \geq 1$ we have

$$P(N_k > 0) = \frac{1}{2}\frac{1}{k} \text{ and } P(N_k > j+1 \mid N_k > j) = \frac{1}{2} + \frac{1}{2}\frac{k-1}{k}.$$

CHAPTER 2

First Martingale Steps

The theory of martingales began life with the aim of providing insight into the apparent impossibility of making money by placing bets on fair games. The success of the theory has far outstripped its origins, and martingale theory is now one of the main tools in the study of random processes. The aim of this chapter is to introduce the most intuitive features of martingales while minimizing formalities and technical details. A few definitions given here will be refined later, but the redundancy is modest, and the future abstractions should go down more easily with the knowledge that they serve an honest purpose.

We say that a sequence of random variables $\{M_n : 0 \leq n < \infty\}$ is a *martingale* with respect to the sequence of random variables $\{X_n : 1 \leq n < \infty\}$, provided that the sequence $\{M_n\}$ has two basic properties. The first property is that for each $n \geq 1$ there is a function $f_n : \mathbb{R}^n \mapsto \mathbb{R}$ such that $M_n = f_n(X_1, X_2, \ldots, X_n)$, and the second property is that the sequence $\{M_n\}$ satisfies the fundamental *martingale identity*:

$$E(M_n \mid X_1, X_2, \ldots, X_{n-1}) = M_{n-1} \text{ for all } n \geq 1. \tag{2.1}$$

To round out this definition, we will also require that M_n have a finite expectation for each $n \geq 1$, and, for a while at least, we will require that M_0 simply be a constant.

The intuition behind this definition is easy to explain. We can think of the X_i as telling us the ith outcome of some gambling process, say the head or tail that one would observe on a coin flip. We can also think of M_n as the fortune of a gambler who places fair bets in varying amounts on the results of the coin tosses. Formula (2.1) tells us that the *expected* value of the gambler's fortune at time n given all the information in the first $n-1$ flips of the coin is simply M_{n-1}, the *actual* value of the gambler's fortune before the nth round of the coin flip game.

The martingale property (2.1) leads to a theory that brilliantly illuminates the fact that a gambler in a fair game cannot expect to make money, however cleverly he varies his bets. Nevertheless, the reason for studying martingales is not that they provide such wonderful models for gambling games. The compelling reason for studying martingales is that they pop up like mushrooms all over probability theory.

2.1. Classic Examples

To develop some intuition about martingales and their basic properties, we begin with three classic examples. We will rely on these examples throughout the text, and we will find that in each case there are interesting analogs for Brownian motion as well as many other processes.

Example 1

If the X_n are independent random variables with $E(X_n) = 0$ for all $n \geq 1$, then the partial sum process given by taking $S_0 = 0$ and $S_n = X_1 + X_2 + \cdots + X_n$ for $n \geq 1$ is a martingale with respect to the sequence $\{X_n : 1 \leq n < \infty\}$.

Example 2

If the X_n are independent random variables with $E(X_n) = 0$ and $\mathrm{Var}(X_n) = \sigma^2$ for all $n \geq 1$, then setting $M_0 = 0$ and $M_n = S_n^2 - n\sigma^2$ for $n \geq 1$ gives us a martingale with respect to the sequence $\{X_n : 1 \leq n < \infty\}$.

One can verify the martingale property in the first example almost without thought, so we focus on the second example. Often, the first step one takes in order to check the martingale property is to separate the conditioned and unconditioned parts of the process:

$$E(M_n \mid X_1, X_2, \ldots, X_{n-1}) = E(S_{n-1}^2 + 2S_{n-1}X_n + X_n^2 - n\sigma^2 \mid X_1, X_2, \ldots, X_{n-1}).$$

Now, since S_{n-1}^2 is a function of $\{X_1, X_2, \ldots, X_{n-1}\}$, its conditional expectation given $\{X_1, X_2, \ldots, X_{n-1}\}$ is just S_{n-1}^2. When we consider the second summand, we note that when we calculate the conditional expectation given $\{X_1, X_2, \ldots, X_{n-1}\}$ the sum S_{n-1} can be brought outside of the expectation

$$E(S_{n-1}X_n \mid X_1, X_2, \ldots, X_{n-1}) = S_{n-1}E(X_n \mid X_1, X_2, \ldots, X_{n-1}).$$

Next, we note that $E(X_n \mid X_1, X_2, \ldots, X_{n-1}) = E(X_n) = 0$ since X_n is independent of $X_1, X_2, \ldots, X_{n-1}$; by parallel reasoning, we also find

$$E(X_n^2 \mid X_1, X_2, \ldots, X_{n-1}) = \sigma^2.$$

When we reassemble the pieces, the verification of the martingale property for $M_n = S_n^2 - n\sigma^2$ is complete.

Example 3

For the third example, we consider independent random variables X_n such that $X_n \geq 0$ and $E(X_n) = 1$ for all $n \geq 1$. We then let $M_0 = 1$ and set

$$M_n = X_1 \cdot X_2 \cdots X_n \text{ for } n \geq 1.$$

One can easily check that M_n is a martingale. To be sure, it is an obvious multiplicative analog to our first example. Nevertheless, this third martingale offers some useful twists that will help us solve some interesting problems.

For example, if the independent identically distributed random variables Y_n have a moment generating function

$$\phi(\lambda) = E(\exp(\lambda Y_n)) < \infty,$$

then the independent random variables $X_n = \exp(\lambda Y_n)/\phi(\lambda)$ have mean one so their product leads us to a whole *parametric family* of martingales indexed by λ:

$$M_n = \exp(\lambda \sum_{i=1}^{n} Y_i)/\phi(\lambda)^n.$$

Now, if there exists a $\lambda_0 \neq 0$ such that $\phi(\lambda_0) = 1$, then there is an especially useful member of this family. In this case, when we set $S_n = \sum_{i=1}^n Y_i$ we find that

$$M_n = e^{\lambda_0 S_n}$$

is a martingale. As we will see shortly, the fact that this martingale is an explicit function of S_n makes it a particularly handy tool for study of the partial sums S_n.

SHORTHAND NOTATION

Formulas that involve conditioning on $X_1, X_2, \ldots, X_n$ can be expressed more tidily if we introduce some shorthand. First, we will write $E(Z \mid \mathcal{F}_n)$ in place of $E(Z \mid X_1, X_2, \ldots, X_n)$, and when $\{M_n : 1 \leq n < \infty\}$ is a martingale with respect to $\{X_n : 1 \leq n < \infty\}$ we will just call $\{M_n\}$ a martingale with respect to $\{\mathcal{F}_n\}$. Finally, we use the notation $Y \in \mathcal{F}_n$ to mean that Y can be written as $Y = f(X_1, X_2, \ldots, X_n)$ for some function f, and in particular if A is an event in our probability space, we will write $A \in \mathcal{F}_n$ provided that the indicator function of A is a function of the variables $\{X_1, X_2, \ldots, X_n\}$. The idea that unifies this shorthand is that we think of $\mathcal{F}_n$ as a representation of the information in the set of observations $\{X_1, X_2, \ldots, X_n\}$. A little later, we will provide this shorthand with a richer interpretation and some technical polish.

2.2. New Martingales from Old

Our intuition about gambling tells us that a gambler cannot turn a fair game into an advantageous one by periodically deciding to double the bet or by cleverly choosing the time to quit playing. This intuition will lead us to a simple theorem that has many important implications. As a necessary first step, we need a definition that comes to grips with the fact that the gambler's life would be easy if future information could be used to guide present actions.

DEFINITION 2.1. *A sequence of random variables* $\{A_n : 1 \leq n < \infty\}$ *is called* nonanticipating *with respect to the sequence* $\{\mathcal{F}_n\}$ *if for all* $1 \leq n < \infty$, *we have*

$$A_n \in \mathcal{F}_{n-1}.$$

In the gambling context, a nonanticipating A_n is simply a function that depends only on the information $\mathcal{F}_{n-1}$ that is known before placing a bet on the nth round of the game. This restriction on A_n makes it feasible for the gambler to permit A_n to influence the size of the nth bet, say by doubling the bet that would have been made otherwise. In fact, if we think of A_n itself as the bet multiplier, then $A_n(M_n - M_{n-1})$ would be the change in the gambler's fortune that is caused by the nth round of play. The idea of a bet size multiplier leads us to a concept that is known in more scholarly circles as the martingale transform.

DEFINITION 2.2. *The process* $\{\widetilde{M}_n : 0 \leq n < \infty\}$ *defined by setting* $\widetilde{M}_0 = M_0$ *and by setting*

$$\widetilde{M}_n = M_0 + A_1(M_1 - M_0) + A_2(M_2 - M_1) + \cdots + A_n(M_n - M_{n-1})$$

for $n \geq 1$ *is called the* martingale transform *of* $\{M_n\}$ *by* $\{A_n\}$.

The martingale transform gives us a general method for building new martingales out of old ones. Under a variety of mild conditions, the transform of a martingale is again a martingale. The next theorem illustrates this principle in its most useful instance.

THEOREM 2.1 (Martingale Transform Theorem). *If* $\{M_n\}$ *is a martingale with respect to the sequence* $\{\mathcal{F}_n\}$, *and if* $\{A_n : 1 \le n < \infty\}$ *is a sequence of* bounded *random variables that are nonanticipating with respect to* $\{\mathcal{F}_n\}$, *then the sequence of martingale transforms* $\{\widetilde{M}_n\}$ *is itself a martingale with respect to* $\{\mathcal{F}_n\}$.

PROOF. We obviously have $\widetilde{M}_n \in \mathcal{F}_n$, and the boundedness of the A_k guarantees that $\widetilde{M}_n$ has a finite expectation for all n. Finally, the martingale property follows from a simple calculation. We simply note that

$$E\left(\widetilde{M}_n - \widetilde{M}_{n-1} \mid \mathcal{F}_{n-1}\right) = E(A_n(M_n - M_{n-1}) \mid \mathcal{F}_{n-1}) = A_n E(M_n - M_{n-1} \mid \mathcal{F}_{n-1}) = 0, \tag{2.2}$$

and the martingale identity

$$E(\widetilde{M}_n | \mathcal{F}_{n-1}) = \widetilde{M}_{n-1}$$

is equivalent to equation (2.2). □

STOPPING TIMES PROVIDE MARTINGALE TRANSFORMS

One of the notions that lies at the heart of martingale theory is that of a *stopping time.* Intuitively, a stopping time is a random variable that describes a rule that we could use to decide to stop playing a gambling game. Obviously, such a rule cannot depend on the outcome of a round of the game that has yet to be played. This intuition is captured in the following definition.

DEFINITION 2.3. *A random variable* τ *that takes values in* $\{0, 1, 2, \dots\} \cup \{\infty\}$ *is called a* stopping time *for the sequence* $\{\mathcal{F}_n\}$ *if*

$$\{\tau \le n\} \in \mathcal{F}_n \quad \textit{for all} \quad 0 \le n < \infty.$$

In many circumstances, we are interested in the behavior of a random process, say Y_n, precisely at the stopping time τ. If $\tau < \infty$ with probability one, then we can define the *stopped process* Y_τ by setting

$$Y_\tau = \sum_{k=0}^{\infty} 1(\tau = k) Y_k.$$

The fact that we define Y_τ only when we have $P(\tau < \infty) = 1$ should underscore that our definition of a stopping time permits the possibility that $\tau = \infty$, and it should also highlight the benefit of finding stopping times that are finite with probability one. Nevertheless, we always have truncation at our disposal; if we let $n \wedge \tau = \min\{n, \tau\}$, then $n \wedge \tau$ is a bounded stopping time, and for any sequence of random variables Y_n the stopped process $Y_{n\wedge\tau}$ is well defined. Also, the truncated stopping times $n \wedge \tau$ lead to an important class of martingales that we will use on many future occasions.

THEOREM 2.2 (Stopping Time Theorem). *If* $\{M_n\}$ *is a martingale with respect to* $\{\mathcal{F}_n\}$, *then the stopped process* $\{M_{n\wedge\tau}\}$ *is also a martingale with respect to* $\{\mathcal{F}_n\}$.

PROOF. First, we note that there is no loss of generality in assuming $M_0 = 0$ since otherwise we can introduce the martingale $M'_n = M_n - M_0$. Next, we note that the bounded random variables A_k defined by

$$A_k = 1(\tau \geq k) = 1 - 1(\tau \leq k - 1)$$

are nonanticipating since τ is a stopping time. Finally,

$$\sum_{k=1}^{n} A_k\{M_k - M_{k-1}\} = M_\tau 1(\tau \leq n-1) + M_n 1(\tau \geq n) = M_{n\wedge\tau},$$

so we that see $\{M_{n\wedge\tau}\}$ is the martingale transform of M_n by the process $\{A_n\}$ which is bounded and nonanticipating. Theorem 2.1 then confirms that $\{M_{n\wedge\tau}\}$ is a martingale. □

2.3. Revisiting the Old Ruins

The stopping time theorem provides a new perspective on our earlier calculation of the probability that a random walk S_n starting at 0 has a probability $B/(A+B)$ of hitting level A before hitting level $-B$. If we let

$$\tau = \min\{n : S_n = A \text{ or } S_n = -B\},$$

then the stopping time theorem and the fact that S_n is a martingale combine to tell us that $S_{n\wedge\tau}$ is also a martingale, so we have

$$E[S_{n\wedge\tau}] = E[S_{0\wedge\tau}] = 0 \quad \text{for all} \quad n \geq 0. \tag{2.3}$$

Now, we checked earlier that τ is finite with probability one, so we also have

$$\lim_{n\to\infty} S_{n\wedge\tau} = S_\tau \text{ with probability one.}$$

The random variables $|S_{n\wedge\tau}|$ are bounded by $\max(A, B)$ so the dominated convergence theorem[1] tells us

$$\lim_{n\to\infty} E[S_{n\wedge\tau}] = E[S_\tau],$$

so equation (2.3) tells us

$$0 = E[S_\tau]. \tag{2.4}$$

Remarkably, we have a second way to calculate $E[S_\tau]$. We have the random variable representation

$$S_\tau = A1(S_\tau = A) - B1(S_\tau = -B),$$

so if we take expectations, we find

$$E[S_\tau] = P(S_\tau = A) \cdot A - (1 - P(S_\tau = A)) \cdot B. \tag{2.5}$$

From equations (2.5) and (2.4), we therefore find that

$$0 = E[S_\tau] = P(S_\tau = A) \cdot A - (1 - P(S_\tau = A)) \cdot B,$$

and we can solve this equation to find the classical formula:

$$P(S_\tau = A) = \frac{B}{A+B}.$$

[1]This is the first time we have used one of the three great tools of integration theory: the dominated convergence theorem, the monotone convergence theorem, and Fatou's lemma. A discussion of these results and a quick review of the Lebesgue integral can be found in the Appendix on Mathematical Tools.

Once More Quickly

The martingale method for calculating the ruin probability may seem long-winded compared to first step analysis, but the impression is due to the detail with which the calculations were given. With some experience behind us, we can pick up the pace considerably. For example, if we now calculate $E[\tau]$, the expected hitting time of A or $-B$ for an unbiased random walk, we can see that the martingale method is actually very quick.

For unbiased simple random walk $S_n = X_1 + X_2 + \cdots + X_n$, where the X_i are independent symmetric Bernoulli random variables, we have $\text{Var}(X_i) = 1$, so we know from the first calculation in this section that $M_n = S_n^2 - n$ is a martingale. Next, we note that

$$|M_{n\wedge\tau}| \leq \max(A^2, B^2) + \tau,$$

and, since we showed earlier that $E[\tau] < \infty$, we see that for all $n \geq 1$ the random variables $M_{n\wedge\tau}$ are dominated by an integrable random variable. The martingale property of $M_{n\wedge\tau}$ gives us $E[M_{n\wedge\tau}] = 0$ for all $n \geq 0$, and $M_{n\wedge\tau}$ converges to M_τ with probability one by the finiteness of τ, so the dominated convergence theorem finally gives us

$$E[M_\tau] = \lim_{n\to\infty} E[M_{n\wedge\tau}] = 0.$$

What makes this fact useful is that again we have a second way to calculate $E[M_\tau]$. If we let $\alpha = P(S_\tau = A)$ then we have $P(S_\tau = -B) = 1 - \alpha$, so we can calculate $E[M_\tau]$ directly from the elementary definition of expectation and our earlier discovery that $\alpha = B/(A + B)$ to find

$$E[M_\tau] = E[S_\tau^2] - E[\tau] = \alpha A^2 + (1 - \alpha)B^2 - E[\tau] = AB - E[\tau].$$

Because our martingale argument already established that $E[M_\tau] = 0$, we again find the lovely formula $E[\tau] = AB$.

Now With Bias

How about the ruin probabilities for biased random walk? This case is more interesting since we will make use of a new martingale. To be precise, if X_i are independent random variables with $P(X_i = 1) = p$ and $P(X_i = -1) = q$ where $q = 1 - p$ and $p \neq \frac{1}{2}$, then we can define a new martingale by setting $M_0 = 1$ and setting

$$M_n = (q/p)^{S_n} \quad \text{for all} \quad n \geq 1.$$

One easily verifies that M_n is indeed a martingale, so we can go directly to the calculation of $P(S_\tau = A)$. By our usual argument, $P(\tau < \infty) = 1$, so as $n \to \infty$ we see that $M_{n\wedge\tau}$ converges with probability one to M_τ. Because $M_{n\wedge\tau}$ is a martingale, we have $E[M_{n\wedge\tau}] = 1$ for all $n \geq 0$, and the random variables $M_{n\wedge\tau}$ are bounded, so the dominated convergence theorem tells us that

$$E[M_\tau] = \lim_{n\to\infty} E[M_{n\wedge\tau}] = 1.$$

We also have a bare-handed calculation of $E[M_\tau]$,

$$E[M_\tau] = P(S_\tau = A) \cdot (q/p)^A + (1 - P(S_\tau = A)) \cdot (q/p)^{-B},$$

so from the fact that $E[M_\tau] = 1$ we find an equation for $P(S_\tau = A)$. When we solve this equation, we find

$$P(S_\tau = A) = \frac{(q/p)^B - 1}{(q/p)^{A+B} - 1},$$

exactly as we found by first step analysis.

Some Perspective

The martingale $M_n = (q/p)^{S_n}$ may seem to have popped out of thin air, but it is actually an old friend. The more principled (and less magical) way of coming to this martingale is to use the parametric family of martingales that we built using the moment generating function. For each step X_i of a biased random walk, the moment generating function is given by

$$\phi(\lambda) = E(\exp(\lambda X_i)) = pe^{\lambda} + qe^{-\lambda}, \tag{2.6}$$

so by our earlier calculations we know that the process defined by

$$M_n = e^{\lambda S_n}/(pe^{\lambda} + qe^{-\lambda})^n$$

is a martingale for all λ.

Now, if we can find λ_0 such that

$$\phi(\lambda_0) = pe^{\lambda_0} + qe^{-\lambda_0} = 1, \tag{2.7}$$

then we see that the simple process $M_n = e^{\lambda_0 S_n}$ is a martingale. To make the last martingale completely explicit, we only need to find e^{λ_0}. To do this, we multiply equation (2.7) by $x = e^{\lambda_0}$ to get a quadratic equation in x, and then we solve that equation to find two solutions: $x = 1$ and $x = q/p$. The solution $x = 1$ gives us the trivial martingale $M_n \equiv 1$, but when we take the second choice we find the martingale $M_n = (q/p)^{S_n}$ that we found to be so handy in the solution of the ruin problem for biased random walk.

2.4. Submartingales

The applicability of martingale theory can be extended greatly if we relax the martingale identity to an analogous inequality. This wider class of processes retains many of the good features of martingales, yet it is far more flexible and robust.

Definition 2.4. *If the integrable random variables $M_n \in \mathcal{F}_n$ satisfy the inequality*

$$M_{n-1} \le E(M_n \mid \mathcal{F}_{n-1}) \quad \textit{for all} \quad n \ge 1,$$

we say $\{M_n\}$ is a submartingale *adapted to $\{\mathcal{F}_n\}$.*

Submartingales are handy because many natural operations on submartingales (or martingales) lead us directly to another submartingale. For example, if $\{M_n\}$ is a martingale then $\{|M_n|\}$ is a submartingale, and if $p \ge 1$, then $\{|M_n|^p\}$ is also a submartingale, provided that $E(|M_n|^p) < \infty$ for all $n \ge 0$. As we will see shortly, these results are best understood as corollaries of a general inequality for the conditional expectation of a convex function of a random variable.

Jensen's Inequality for Conditional Expectations

To develop the required inequality, we first recall that one of the most important features of a convex function $\phi(x)$ is that its graph $\{(x, \phi(x)) : x \in \mathbb{R}\}$ is equal to the upper envelope of the set of linear functions that lie below the graph. Expressed in a formula, this says:

$$\phi(x) = \sup_{L \in \mathcal{L}} L(x), \tag{2.8}$$

where

$$\mathcal{L} = \{L : L(u) = au + b \leq \phi(u) \quad \text{for all} \quad -\infty < u < \infty\}.$$

One of the most important consequences of convexity in probability theory is Jensen's inequality.

THEOREM 2.3 (Jensen's Inequality). *For any convex function ϕ, we have*

$$\phi(E(X \mid \mathcal{F})) \leq E(\phi(X) \mid \mathcal{F}). \tag{2.9}$$

PROOF. All we need to do is to use the envelope characterization of convexity and the linearity of the conditional expectation:

$$\begin{aligned} E(\phi(X) \mid \mathcal{F}) &= E(\sup_{L \in \mathcal{L}} L(X) \mid \mathcal{F}) \\ &\geq \sup_{L \in \mathcal{L}} E(L(X) \mid \mathcal{F}) = \sup_{L \in \mathcal{L}} L(E(X \mid \mathcal{F})) \\ &= \phi(E(X \mid \mathcal{F})). \end{aligned}$$

□

Here, we should note how remarkably little we needed to know about the conditional expectation in order to prove Jensen's inequality. We only needed two facts: (1) $E(\cdot \mid \mathcal{F})$ is monotone in the sense that $X \leq Y$ implies that $E(X \mid \mathcal{F}) \leq E(Y \mid \mathcal{F})$ and (2) $E(\cdot \mid \mathcal{F})$ linear in the sense that for any constants a and b we have $E(aX + b \mid \mathcal{F}) = aE(X \mid \mathcal{F}) + b$.

Before we go too far, we should also note that Jensen's inequality is often applied in its unconditional form, which corresponds to taking $\mathcal{F}$ to be the trivial σ-field. For such a choice, the conditional inequality (2.9) boils down to the simpler bound:

$$\phi(E(X)) \leq E(\phi(X)). \tag{2.10}$$

L^p Spaces and Jensen's Inequality

We often need to work with random variables that satisfy some size constraint, and the most common way to express such constraints is in terms of function spaces, especially the space L^p that consists of all X such that $E[|X|^p] < \infty$. We will not call on any detailed information about these spaces, but there are a few facts that one should know.

For any $1 \leq p \leq \infty$, we can consider L^p as a normed linear space where for finite $p \geq 1$ the norm is given by $||X||_p = (E[|X|^p])^{\frac{1}{p}}$ and for infinite p we take $||X||_\infty = \inf\{t\colon P(|X| \leq t) = 1\}$.

For us, the most important L^p spaces are those with $p = 1, p = 2$, and $p = \infty$. The first of these is just the set of random variables with finite mean, and the second is just the set of random variables with finite variance. The third space is the set of random variables that are bounded on a set of probability one. The most basic relation between these spaces is that

$$\|X\|_1 \leq \|X\|_2 \leq \|X\|_\infty \text{ and } L^\infty \subset L^2 \subset L^1. \tag{2.11}$$

The only one of these relations that might require proof is the bound $\|X\|_1 \leq \|X\|_2$, but this is also easy if one recalls Cauchy's inequality $E[|XY|] \leq \|X\|_2\|Y\|_2$ and makes the special choice $Y = 1$.

The bounds given in relation (2.11) suggest that there is a monotonicity relation for the L^p norms, and Jensen's inequality turns out to be the perfect tool to confirm this suggestion. If we take the convex function defined by $\phi(x) = x^\lambda$ for some $\lambda \geq 1$, then Jensen's inequality applied to $Y = |X|^p$ tells us that

$$(E[|X|^p])^\lambda \leq E(|X|^{p\lambda}),$$

so, if we then let $p' = p\lambda$ and take the $p\lambda$th root of the last inequality, we find exactly what we suspected:

$$\|X\|_p \leq \|X\|_{p'} \text{ for all } 1 \leq p \leq p' < \infty. \tag{2.12}$$

It even turns out that one has the natural relation

$$\lim_{p\to\infty} \|X\|_p = \|X\|_\infty,$$

but this limit relation does not come into play nearly as often as the important monotonicity result (2.12).

2.5. Doob's Inequalities

Many applications of probability theory depend on qualitative inferences rather than the explicit calculation of probabilities. The main sources of such qualitative inferences are the limit theorems that tell us that some event has probability one, and at the heart of such "strong laws" one almost always finds a *maximal inequality*. In this section, we develop two important martingale maximal inequalities that will serve as key tools in the development of stochastic calculus.

DEFINITION 2.5. *If* $\{M_n: 0 \leq n < \infty\}$ *is any sequence of random variables, the sequence defined by*

$$M_n^* = \sup_{0\leq m\leq n} M_m$$

is called the maximal sequence *associated with* M_n.

THEOREM 2.4 (Doob's Maximal Inequality). *If* $\{M_n\}$ *is a nonnegative submartingale and* $\lambda > 0$, *then*

$$\lambda P(M_n^* \geq \lambda) \leq E[M_n 1(M_n^* \geq \lambda)] \leq E[M_n]. \tag{2.13}$$

PROOF. The proof of this result is not difficult, but it does bring us to two observations that are often useful. The first of these observations is the simple *index shifting inequality*, which tells us that for any submartingale $\{M_n\}$ we have

$$E[M_m 1_A] \leq E[M_n 1_A] \text{ for all } A \in \mathcal{F}_m \text{ and all } n \geq m. \tag{2.14}$$

To prove the bound (2.14), we first note by the definition of a submartingale that

$$M_m \leq E(M_{m+1}|\mathcal{F}_m) \text{ and } M_{m+1} \leq E(M_{m+2}|\mathcal{F}_{m+1})$$

so when we substitute the second inequality into the first we find

$$M_m \leq E(E(M_{m+2}|\mathcal{F}_{m+1})|\mathcal{F}_m) = E(M_{m+2}|\mathcal{F}_m).$$

We can repeat this computation as many times as we like, so in general we find

$$M_m \le E(M_n|\mathcal{F}_m) \text{ for all } n \ge m.$$

Finally, when we multiply the last equation by $1_A \in \mathcal{F}_m$ and take expectations, the defining property of conditional expectation gives us inequality (2.14).

The second useful observation is that many random variables are best understood when written in terms of the values of an associated stopping time. Here, we will use the stopping time

$$\tau = \min\{m : M_m \ge \lambda\},$$

which has the immediate benefit of providing an alternative expression for the tail probability of the maximal sequence:

$$P(M_n^* \ge \lambda) = P\Big(\sup_{0\le m\le n} M_m \ge \lambda\Big) = P(\tau \le n).$$

The further benefit gained by introducing τ is that on the set $\{\tau \le n\}$ we have $M_\tau \ge \lambda$, so we have the basic decomposition

$$\lambda 1(\tau \le n) \le M_\tau 1(\tau \le n) = \sum_{0\le m\le n} M_m 1(\tau = m). \tag{2.15}$$

Now, to prove the desired maximal inequality, we only need to do honest arithmetic when we calculate the expected value of both sides of equation (2.15).

Since the event $\{\tau = m\}$ is measurable with respect to $\mathcal{F}_m$, the index shifting inequality (2.14) tells us that

$$E(M_m 1(\tau = m)) \le E(M_n 1(\tau = m)) \quad \text{for all} \quad m \le n. \tag{2.16}$$

Finally, when we go back to equation (2.15) to take expectations, we can apply the bound (2.16) to find

$$\lambda P\Big(\sup_{0\le m\le n} M_m \ge \lambda\Big) \le E\Big[\sum_{0\le m\le n} M_n 1(\tau = m)\Big] = E[M_n 1(M_n^* \ge \lambda)] \le E[M_n],$$

which completes the proof. □

In most applications of Doob's maximal inequality, one only needs the simple outside bounds of inequality (2.13), but there is one important application where the critical edge is given by the first half of the double inequality (2.13). This sharper bound is essential in the proof of Doob's L^p inequality, a result that will serve as a basic tool in our study of the continuity of stochastic integrals.

Self-Improving Inequalities

Some inequalities have the interesting feature that they are self-improving in the sense that they imply inequalities that are apparently stronger versions of themselves. For example, from the inequality (2.13), we can immediately deduce the apparently stronger fact that for any nonnegative submartingale and any $p \ge 1$ we have

$$\lambda^p P(M_n^* \ge \lambda) \le E[M_n^p]. \tag{2.17}$$

In this case, the proof is easy. We merely note that $M_n' = M_n^p$ is also a nonnegative submartingale, so when inequality (2.13) is applied to M_n' we immediately find inequality (2.17).

DOOB'S L^p INEQUALITY

One of the central facts about L^p spaces that we have not reviewed before is Hölder's inequality, which tells us that for all $1 \le p \le \infty$ we have

$$||XY||_1 \le ||X||_p ||Y||_q, \quad \text{provided } \frac{1}{p} + \frac{1}{q} = 1.$$

We will not prove Hölder's inequality at this point, although we will give a proof later in order to illustrate an interesting probabilistic technique. Our immediate goal is to prove a lemma that shows how Hölder's inequality can be used to squeeze the juice out of an inequality like that given by the inside bound of inequality (2.13).

LEMMA 2.1. *If X and Y are nonnegative random variables for which we have the bounds $E[Y^p] < \infty$ for some $p > 1$ and*

$$\lambda P(X \ge \lambda) \le E[Y 1(X \ge \lambda)] \quad \text{for all} \quad \lambda \ge 0, \tag{2.18}$$

then

$$||X||_p \le \frac{p}{p-1} ||Y||_p. \tag{2.19}$$

PROOF. An important conclusion of the lemma is that $E[X^p]$ is finite, so we must be careful to avoid making the tacit assumption that $E[X^p] < \infty$. If we let $X_n = \min(X, n)$, then X_n is a bounded random variable, and if we can prove the bound (2.19) for X_n the corresponding inequality for X will follow by Fatou's lemma. A nice feature of this truncation is that we can immediately check that X_n again satisfies the bound (2.18).

Now, for any $z \ge 0$, we have

$$z^p = p \int_0^z x^{p-1} dx = p \int_0^\infty x^{p-1} 1(z \ge x) dx,$$

so assigning $z \mapsto X_n$ and taking expectations, we find:

$$\begin{aligned} E(X_n^p) &= p \int_0^\infty x^{p-1} P(X_n \ge x) dx \\ &\le p \int_0^\infty x^{p-2} E(Y 1(X_n \ge x)) dx \\ &= p\, E(Y \int_0^\infty x^{p-2} 1(X_n \ge x) dx) \\ &= \frac{p}{p-1} E(Y \cdot (X_n)^{p-1}). \end{aligned}$$

We can now apply Hölder's inequality to decouple Y and X_n. Hölder permits us to raise Y to the power p and to raise X_n^{p-1} to the conjugate power $q = p/(p-1)$ to find

$$\begin{aligned} ||X_n||_p^p = E(X^p) &\le \frac{p}{p-1} E(Y (X_n)^{p-1}) \\ &\le \frac{p}{p-1} ||Y||_p ||X_n||_p^{p-1}. \end{aligned}$$

The inequality of the lemma follows when we divide both sides by $||X_n||_p^{p-1}$ and then use Fatou's lemma as planned. □

In view of Lemma 2.1 and Doob's maximal inequality, we immediately find an inequality that tells us that the L^p norm of the maximal function of a submartingale is almost as well behaved as the original submartingale.

THEOREM 2.5 (Doob's L^p Inequality). *If $\{M_n\}$ is a nonnegative submartingale, then for all $p > 1$ and all $n \geq 0$, we have*

$$||M_n^*||_p \leq \frac{p}{p-1}||M_n||_p. \tag{2.20}$$

2.6. Martingale Convergence

Martingales have the grandest habit of converging. If a martingale is positive, then it converges with probability one. If it is L^1-bounded — that is to say, $E[|M_n|] \leq B < \infty$ for all $n \geq 1$ — then it converges with probability one to a limit X_∞ that is an element of L^1. If $p > 1$ and the martingale M_n is L^p-bounded (so that $E[|M_n|^p] \leq B < \infty$ for all $n \geq 1$), then M_n converges with probability one and, moreover, converges as an L^p sequence.

For the theory of stochastic calculus, the most important martingale convergence theorem is probably that for L^2-bounded martingales. We will prove this result by a method that illustrates one of the fundamental ways one can use Doob's maximal inequality.

THEOREM 2.6 (L^2-Bounded Martingale Convergence Theorem). *If $\{M_n\}$ is a martingale that satisfies $E[M_n^2] \leq B < \infty$ for all $n \geq 0$, then there exists a random variable M_∞ with $E[M_\infty^2] \leq B$ such that*

$$P\Big(\lim_{n\to\infty} M_n = M_\infty\Big) = 1 \text{ and } \lim_{n\to\infty} ||M_n - M_\infty||_2 = 0. \tag{2.21}$$

PROOF. For any pair of rational numbers $a < b$, we let

$$A_{ab} = \Big\{\omega\colon \liminf_{n\to\infty} M_n(\omega) \leq a < b \leq \limsup_{n\to\infty} M_n(\omega)\Big\},$$

and we then note that to show M_n converges with probability one, we only have to show that $P(A_{ab}) = 0$. To make the connection to the theory of maximal inequalities, we note that for any $\epsilon > (b-a)/2$ we have

$$A_{ab} \subset \Big\{\omega\colon \sup_{m\leq k<\infty} |M_k - M_m| \geq \epsilon\Big\} \quad \text{for all} \quad m \geq 0,$$

so to prove the first conclusion of the theorem we only have to bound the last probability.

A key property of square integrable martingales is that the differences $d_k = M_k - M_{k-1}$ are orthogonal: $E[d_k d_j] = 0$ if $k \neq j$. Now, there is no loss of generality in assuming that $M_0 = 0$, so the telescoping of the summed differences gives us the representation $M_n = \sum_{k=1}^n d_k$. The orthogonality of the d_k and the L^2-boundedness of M_n then give us that

$$E[M_n^2] = E\Big[\Big(\sum_{k=1}^n d_k\Big)^2\Big] = \sum_{k=1}^n E[d_k^2].$$

Our hypothesis tells us that $E[M_n^2] \leq B$ for all n, so when we let $n \to \infty$ we arrive at the fundamental observation that the $E[d_k^2]$ have a convergent sum:

$$\sum_{k=1}^{\infty} E[d_k^2] \leq B. \tag{2.22}$$

Finally, by Doob's maximal inequality applied to the submartingale defined by $M'_k = (M_{k+m} - M_m)^2$, we find that

$$P\left(\sup_{k\geq m} |M_k - M_m| \geq \epsilon\right) = P\left(\sup_{k\geq m} (M_k - M_m)^2 \geq \epsilon^2\right) \leq \frac{1}{\epsilon^2} \sum_{k=m+1}^{\infty} E[d_k^2].$$

The bottom line is that

$$P(A_{ab}) \leq \frac{1}{\epsilon^2} \sum_{k=m+1}^{\infty} E[d_k^2],$$

so letting $m \to \infty$ shows that $P(A_{ab}) = 0$. The arbitrariness of the rational pair (a, b) then establishes the convergence of M_n on a set of probability one.

To prove the second conclusion of the theorem, we simply let M_∞ denote the almost sure limit of the M_n, and note that we have

$$M_\infty - M_n = \sum_{k=n+1}^{\infty} d_k \text{ and } E[(M_\infty - M_n)^2] = \sum_{k=n+1}^{\infty} E[d_k^2].$$

By the convergence of the sum given in inequality (2.22), we know that the tail sums converge to zero, and, as a consequence, we see $||M_\infty - M_n||_2 \to 0$, as required for the second conclusion of the theorem. □

The Localization Idea

The *localization method* is a remarkably general technique that can often be used to apply L^2 results in problems that do not begin with any direct connection to L^2. This technique will soon become one of our daily companions, and it may be fairly considered one of the handiest technical tools in the theory of martingales. Although the next result is one that we will shortly surpass, it still nicely illustrates how localization works in the simplest possible context.

THEOREM 2.7 (Convergence Theorem Illustrating Localization). *If $\{M_n\}$ is a martingale and there is a constant B such that*

$$E[|M_n|] \leq B < \infty \text{ and } |M_{n+1} - M_n| \leq B < \infty \tag{2.23}$$

for all $n \geq 0$, then there exists a random variable M_∞ with $E[|M_\infty|] \leq B$ such that

$$P\left(\lim_{n\to\infty} M_n = M_\infty\right) = 1. \tag{2.24}$$

PROOF. We let $\tau = \inf\{n\colon |M_n| \geq \lambda\}$, and we write

$$M_n = M_{n\wedge\tau} + R_n,$$

where $R_n = M_n - M_{n\wedge\tau}$. By Doob's stopping time theorem, $M_{n\wedge\tau}$ is a martingale, and $|M_{n\wedge\tau}| \leq \lambda + B$ so $E[|M_{n\wedge\tau}|^2]$ is a bounded sequence. Theorem 2.6 then tells us that the martingale $M_{n\wedge\tau}$ will converge with probability one.

Now, since $R_n(\omega) = 0$ for $\omega \in \{\tau = \infty\}$, we see that the set S of all ω where $M_n(\omega)$ does not converge satisfies

$$S \subset \{\tau < \infty\} = \left\{ \sup_{0 \le n < \infty} |M_n| \ge \lambda \right\}.$$

When we apply Doob's maximal inequality to the submartingale $|M_n|$, we have for all N that

$$P\left(\sup_{0 \le n \le N} |M_n| \ge \lambda \right) \le E[|M_N|]/\lambda \le B/\lambda,$$

so letting $N \to \infty$ we have

$$P(S) \le P\left(\sup_{0 \le n < \infty} |M_n| \ge \lambda \right) \le B/\lambda. \tag{2.25}$$

Since $\lambda \ge 0$ is arbitrary, we see that $P(S) = 0$ and the proof of the convergence is complete. Finally, the bound $E[|M_\infty|] \le B$ follows from Fatou's lemma and the convergence of $\{M_n\}$. □

L^1-Bounded Martingales

The proof of Theorem 2.7 is a fine illustration of localization, but, as luck would have it, the result that we obtain is not the best possible. The second hypothesis of the pair (2.23) is unnecessary. We seldom press for the strongest possible results, but in this case the light is definitely worth the candle. The stronger theorem is one of the most important results in probability theory, and its proof will lead us to several ideas that are useful in their own right.

THEOREM 2.8 (L^1-Bounded Martingale Convergence Theorem). *If $\{M_n\}$ is a martingale and there is a constant B such that*

$$E[|M_n|] \le B < \infty \tag{2.26}$$

for all $n \ge 0$, then there exists a random variable M_∞ with $E[|M_\infty|] \le B$ such that

$$P(\lim_{n \to \infty} M_n = M_\infty) = 1. \tag{2.27}$$

Good Games and the Up-Crossing Inequality

Our first step toward the proof of Theorem 2.8 is to recall an intuitive fact from gambling theory: we should never sit out a round when we are lucky enough to be playing a favorable game.

PROPOSITION 2.1 (Sticking with a Good Game). *If M_n is a nonnegative submartingale and if*

$$\widetilde{M_n} = A_n(M_n - M_{n-1}) + A_{n-1}(M_{n-1} - M_{n-2}) + \cdots + A_1(M_1 - M_0) + M_0$$

is the martingale transform of M_n by a nonanticipating sequence A_n of random variables that take only the values 0 and 1, then

$$E[\widetilde{M_n}] \le E[M_n].$$

PROOF. From the fact that $A_k \in \mathcal{F}_{k-1}$, $A_k \in \{0,1\}$, and the submartingale property of $\{M_n\}$ we have

$$E[A_k(M_k - M_{k-1}) \mid \mathcal{F}_{k-1}] = A_k E[(M_k - M_{k-1}) \mid \mathcal{F}_{k-1}]$$
$$\leq E[M_k - M_{k-1} | \mathcal{F}_{k-1}].$$

Now, when we take the expectation of this inequality, we find

$$E[A_k(M_k - M_{k-1})] \leq E[M_k - M_{k-1}],$$

and when we sum this inequality over $1 \leq k \leq n$, we find

$$E[\widetilde{M_n}] - E[M_0] \leq E[M_n] - E[M_0]. \qquad \square$$

UP-CROSSINGS AND THE MEANING OF CONVERGENCE

We now need to think a bit about the meaning of convergence. If a sequence of real numbers converges (to a finite or infinite limit), then for every pair $a < b$ of rational numbers there are only a finite number of times that the sequence can pass from a (or below) on up to b (or above). Our plan is to use the gambling proposition to show that one cannot expect many such passages from a martingale. The next definition helps make this plan precise.

DEFINITION 2.6. *If $S = \{c_i : 0 \leq i \leq n\}$ is a sequence of real numbers, we define the* number of up-crossings *of the interval $[a, b]$ by S to be the largest k such that there exist $2k$ integers $0 \leq i_1 < j_1 < i_2 < j_2 < \cdots < i_k < j_k \leq n$ for which all of the elements of $\{c_{i_s} : 1 \leq s \leq k\}$ are less than or equal to a and all of the elements of $\{c_{j_s} : 1 \leq s \leq k\}$ are greater than or equal to b.*

Finally, we consider a submartigale $\{M_n\}$ and define the random variable $N_n(a, b)$ to be the number of up-crossings of the interval $[a, b]$ that are made by the sequence $\{M_k(\omega) : 0 \leq k \leq n\}$. The key fact about the $N_n(a, b)$ is that its expectation can be bounded in terms of the expectation of $(M_n - a)_+ = \max(M_n - a, 0)$.

THEOREM 2.9 (Up-crossing Inequality). *If $\{M_n\}$ is a submartingale, then for any $a < b$ we have*

$$(b - a)E[N_n(a, b)] \leq E[(M_n - a)_+]. \tag{2.28}$$

PROOF. The first observation is that $(M_n - a)_+$ is also a nonnegative submartingale since the function $f(x) = (x - a)_+$ is convex and monotone increasing. Moreover, the number of up-crossings of $[a, b]$ by the sequence $\{M_k(\omega) : 0 \leq k \leq n\}$ is equal to the number of up-crossings of $[0, b - a]$ by the sequence $\{(M_k(\omega) - a)_+ : 0 \leq k \leq n\}$.

Now we put on our gambling hats. Suppose we make bets on the sequence $\{(M_n - a)_+\}$ by following the rule of putting on the bet when the sequence first hits zero and then taking off the bet when it first gets up to $b - a$ (or higher). We then continue to play by repeating our rule: put the bet back on when we get back down to zero and take it off when we get back up to $b - a$ (or higher). This betting scheme is a transform $\widetilde{M_n}$ of the submartingale $(M_n - a)_+$ by a nonanticipating sequence $\{A_n\}$ that takes on the values 0 and 1, so by Proposition 2.1 we have

$$E[\widetilde{M_n}] \leq E[(M_n - a)_+]. \tag{2.29}$$

The final observation is that if we use the betting strategy that gives us $\widetilde{M_n}$ then we are guaranteed to win at least $(b-a)N_n(a,b)$ units; that is, $(b-a)N_n(a,b) \leq \widetilde{M_n}$. When we take expectations in this inequality and apply the bound (2.29), the proof of the up-crossing inequality is complete. □

Proof of the Convergence Theorem

The proof of Theorem 2.8 is now a piece of cake. To prove the almost sure convergence of the sequence $\{M_n\}$ to a (possibly infinite) limit M_∞, all we must show is that for each rational pair $a < b$ we have $P(A_{ab}) = 0$, where A_{ab} is the event defined by

$$A_{ab} = \left\{\omega\colon \liminf_{n\to\infty} M_n(\omega) \leq a < b \leq \limsup_{n\to\infty} M_n(\omega)\right\}.$$

Next we note that $N_n(a,b)$ is a monotone sequence of random variables that converges with probability one to a (possibly infinite) random variable. If we denote this limit by $N_\infty(a,b)$, we have

$$A_{ab} \subset \{\omega : N_\infty(a,b) = \infty\},$$

so all we must do to prove $P(A_{ab}) = 0$ is to show that $E[N_\infty(a,b)] < \infty$. But this last bound is easy, since the up-crossing inequality tells us that for each n we have

$$E[N_n(a,b)] \leq \frac{1}{b-a} \sup_m E[(M_m - a)_+] \leq \frac{1}{b-a}(|a| + B),$$

so the bound $E[N_\infty(a,b)] < \infty$ follows from Fatou's lemma.

Our arguments have so far established that M_n converges with probability one to a (possibly infinite) random variable. If we denote this limit by M_∞, then the fact that $E[|M_n|] \leq B$ for all n teams up with Fatou's lemma to tell us $E[|M_\infty|] \leq B$. This is more than we need to conclude that M_∞ is finite with probability one so the proof of Theorem 2.8 is complete.

Amusing Convergence Side Effects

The martingale convergence theorems sometimes surprise us with their consequences — mere convergence can have interesting side effects. To give a quick example, we take simple random walk S_n and consider the problem of showing that the stopping time $\tau = \min\{n : S_n = A \text{ or } S_n = -B\}$ is finite with probability one. The martingale $M_n = S_{n\wedge\tau}$ cannot converge for any $\omega \in \{\omega : \tau(\omega) = \infty\}$ since for such ω we have $|M_n(\omega) - M_{n+1}(\omega)| = |S_n(\omega) - S_{n+1}(\omega)| = 1$, but M_n is a bounded martingale, so Theorem 2.6 tells us M_n does indeed converge with probability one. As a consequence, we must have $P(\tau < \infty) = 1$. This slick proof certainly offers an interesting contrast to our first proof that $P(\tau < \infty) = 1$, although, as usual, the slick proof is more specialized.

2.7. Exercises

The first exercise offers experience in the solution of ruin problems for walks with steps governed by general discrete-valued random variables. The exercise also shows that the martingale $M_n = (q/p)^{S_n}$ that we used in the analysis of biased random walk S_n is not as special as it might seem at first. Exercise 2.2 then gives an intuitive formula for the expected time of the kth occurrence of an arbitrary sequence of independent events. Finally, the last three exercises provide experience with more general techniques that are important for the theory and applications of martingales.

EXERCISE 2.1 (Finding a Martingale). Consider a gambling game with multiple payouts: the player loses \$1 with probability α, wins \$1 with probability β, and wins \$2 with probability γ. Specifically, we assume that $\alpha = 0.52$, $\beta = 0.45$, and $\gamma = 0.03$, so the expected value of each round of the game is only \$$-0.01$.

(a) Suppose the gambler bets one dollar on each round of the game and that $\{X_k\}$ is the amount won or lost on the k'th round. Find a real number x such that $M_n = x^{S_n}$ is a martingale where $S_n = X_1 + X_2 + \cdots + X_n$ tallies the gambler's winnings. Note: You will need to find the numerical solutions to a cubic equation, but $x = 1$ is one solution so the cubic can be reduced to a quadratic.

(b) Let p denote the probability of winning \$100 or more before losing \$100. Give numerical bounds p_0 and p_1 such that $p_0 \le p \le p_1$ and $p_1 - p_0 \le 3 \times 10^{-3}$. You should be sure to take proper account of the fact that the gambler's fortune may skip over \$100 if a win of \$2 takes place when the gambler's fortune is \$99. This issue of "overshoot" is one that comes up in many problems where a process can skip over intervening states.

EXERCISE 2.2 (Expected Time to kth Occurrence).
Suppose that the sequence of independent events $\{A_i\}$ satisfies

$$\phi(n) = \sum_{i=1}^{n} P(A_i) \to \infty \text{ as } n \to \infty,$$

and let $\tau_k = \min\{n\colon \sum_{i=1}^{n} 1(A_i) = k\}$. Note that by the Borel–Cantelli Lemma[2] that we have $P(\tau_k < \infty) = 1$. By applying Doob's stopping time Theorem to an appropriate martingale, prove the more quantitative result that

$$E[\phi(\tau_k)] = k \text{ for all } k \ge 1.$$

EXERCISE 2.3. Use Jensen's inequality, the L^1 martingale convergence theorem, and Doob's L^p maximal inequality to prove the L^p-bounded martingale convergence theorem. Specifically, show that if $\{M_n\}$ is a martingale that satisfies $E[|M_n|^p] \le B < \infty$ for some $p > 1$ and for all $n \ge 0$, then there exists a random variable M_∞ with $E[|M_\infty|^p] \le B$ such that

$$P\Big(\lim_{n\to\infty} M_n = M_\infty\Big) = 1 \text{ and } \lim_{n\to\infty} \|M_n - M_\infty\|_p = 0. \tag{2.30}$$

[2]The Borel–Cantelli lemma is one of the basic facts of probability theory that is discussed in the Appendix on Mathematical Tools.

EXERCISE 2.4 (Doob's Decomposition). If $\{M_n, \mathcal{F}_n\}$ is a martingale with

$$E[M_n^2] < \infty \text{ for all } n,$$

show that we can write

$$M_n^2 = N_n + A_n,$$

where (1) $\{N_n, \mathcal{F}_n\}$ is a martingale; (2) A_n is monotone (so, $A_n \geq A_{n-1}$); and (3) A_n is nonanticipating (so $A_n \in \mathcal{F}_{n-1}$). Those who want a hint might try defining the A_n by first taking $A_0 = 0$ and then setting

$$A_{n+1} = A_n + E\big[(M_{n+1} - M_n)^2 \,|\, \mathcal{F}_n\big] \text{ for } n \geq 1.$$

EXERCISE 2.5. Suppose that $\{M_n\}$ is a submartingale and that ν and τ are *bounded* stopping times such that $\nu \leq \tau$. Prove that $E[M_\nu] \leq E[M_\tau]$.

CHAPTER 3

Brownian Motion

Brownian motion is the most important stochastic process. As a practical tool, it has had profound impact on almost every branch of physical science, as well as several branches of social science. As a creation of pure mathematics, it is an entity of uncommon beauty. It reflects a perfection that seems closer to a law of nature than to a human invention.

The first important applications of Brownian motion were made by L. Bachelier and A. Einstein. Bachelier is the acknowledged father of quantitative methods in finance, and Einstein is ... well ... Einstein. Bachelier's aim was to provide a model for financial markets, and Einstein wanted to model the movement of a particle suspended in a liquid. Bachelier hoped that Brownian motion would lead to a model for security prices that would provide a sound basis for the pricing of options, a hope that was vindicated after some modifications. Einstein's aim was to provide a means of measuring Avagadro's number, the number of molecules in a mole of gas, and experiments suggested by Einstein proved to be consistent with his predictions.

Both of these early investigators were willing to set aside mathematical nuances, since for them a financial market or a physics laboratory could provide a test of validity with more sway than any formal justification offered by logical rigor. Still, good foundations make for good houses, and both theory and practice were profoundly served when N. Wiener directed his attention to the *mathematics* of Brownian motion. Among Wiener's many contributions is the first proof that Brownian motion *exists* as a rigorously defined mathematical object, rather than as a physical phenomenon for which one might pose a variety of models. One way that Wiener's contribution is recognized today is that in many parts of the scientific world a Brownian motion is equally likely to be called a Wiener process.

DEFINITION 3.1. *A continuous-time stochastic process* $\{B_t : 0 \leq t < T\}$ *is called a* Standard Brownian Motion *on* $[0, T)$ *if it has the following four properties:*

(i) $B_0 = 0$.

(ii) The increments of B_t *are independent; that is, for any finite set of times* $0 \leq t_1 < t_2 < \cdots < t_n < T$ *the random variables*

$$B_{t_2} - B_{t_1}, B_{t_3} - B_{t_2}, \ldots, B_{t_n} - B_{t_{n-1}}$$

are independent.

(iii) For any $0 \leq s \leq t < T$ *the increment* $B_t - B_s$ *has the Gaussian distribution with mean* 0 *and variance* $t - s$.

(iv) For all ω *in a set of probability one,* $B_t(\omega)$ *is a continuous function of* t.

The main goal of this chapter is to show that one can represent Brownian motion as a random sum of integrals of orthogonal functions. This representation satisfies the theoretician's need to prove the existence of a process with the four defining properties of Brownian motion, but it also serves more concrete demands. In particular, the series representation can be used to derive almost all of the most important analytical properties of Brownian motion. It even gives us a powerful numerical method for generating the Brownian motion paths that are required in computer simulations.

3.1. Covariances and Characteristic Functions

In order to confirm that our construction of Brownian motion is honest, we will need to use some basic facts about the multivariate Gaussian distribution. In particular, we will need tools that help us check that the increments of our process are Gaussian and that those increments are independent.

Multivariate Gaussians

If V is any d-dimensional random vector, then we define the mean vector of

$$V = \begin{bmatrix} V_1 \\ V_2 \\ \cdot \\ \cdot \\ \cdot \\ V_d \end{bmatrix} \quad \text{to be the vector} \quad \mu = E[V] = \begin{bmatrix} \mu_1 \\ \mu_2 \\ \cdot \\ \cdot \\ \cdot \\ \mu_d \end{bmatrix}, \tag{3.1}$$

and we define the covariance matrix of V to be the matrix

$$\Sigma = \begin{bmatrix} \sigma_{11} & \sigma_{12} & \cdots & \sigma_{1d} \\ \sigma_{21} & \sigma_{22} & \cdots & \sigma_{2d} \\ \cdot & \cdot & \cdots & \cdot \\ \cdot & \cdot & \cdots & \cdot \\ \cdot & \cdot & \cdots & \cdot \\ \sigma_{d1} & \sigma_{d2} & \cdots & \sigma_{dd} \end{bmatrix}, \quad \text{where } \sigma_{ij} = E[(V_i - \mu_i)(V_j - \mu_j)]. \tag{3.2}$$

Perhaps the most important fact about multivariate Gaussians is that their joint density is completely determined by the mean vector and the covariance matrix. At first glance, that density may look awkward, but, after some experience, many people come to see the Gaussian density as a genuine marvel of parsimony and elegance.

Definition 3.2 (Multivariate Gaussian). *A d-dimensional random vector V is said to have the d-dimensional multivariate Gaussian distribution with mean μ and covariance Σ if the density of V is given by*

$$(2\pi)^{-\frac{d}{2}}(\det\Sigma)^{-\frac{1}{2}}\exp\left(-\frac{1}{2}(x-\mu)^T\Sigma^{-1}(x-\mu)\right) \quad \textit{for all } x \in \mathbb{R}^d. \tag{3.3}$$

Gaussian Miracle

When we look at the simple case of a bivariate Gaussian $V = (X, Y)$, we easily see that if $\operatorname{Cov}(X, Y) = 0$ then Σ is a diagonal matrix. The importance of this observation is that the special form of the density (3.3) then implies that we can factor the joint density of $V = (X, Y)$ as a product of the density of X and the density of Y. This factorization then yields the miraculous conclusion that X and Y are independent random variables.

This rare circumstance deserves to be savored. Where else does a measure as crude as the vanishing of a single integral suffice to guarantee the independence of a pair of random variables? The theory of Brownian motion often exploits this fact, and it is of particular importance that the same argument provides a natural analog for sets of more than two variables. The coordinates $\{V_i : 1 \leq i \leq d\}$ of a multivariate Gaussian vector are independent if and only if the covariance matrix Σ is a diagonal matrix.

Characteristic Functions

If X is any random variable, the characteristic function of X is defined to be the function $\phi : \mathbb{R} \to \mathbb{C}$ given by $\phi(t) = E(e^{itX})$. Similarly, if V is a d-dimensional vector of random variables and $\theta = (\theta_1, \theta_2, \ldots, \theta_d)$ is a vector of real numbers, then the multivariate characteristic function is defined by $\phi(\theta) = E[\exp(i\theta^T V)]$. There are inversion formulas for both the univariate and multivariate cases that permit one to calculate the underlying distribution function from the characteristic function, but for our purposes it suffices simply to note that the characteristic function of a random variable (or random vector) will uniquely determine its distribution.

For a Gaussian random variable X with mean μ and variance σ^2, the characteristic function is readily computed:

$$E[\exp(itX)] = \frac{1}{\sigma\sqrt{2\pi}} \int_{-\infty}^{\infty} e^{itx} e^{(x-\mu)^2/2\sigma^2}\, dx = e^{it\mu} e^{-t^2\sigma^2/2}. \tag{3.4}$$

With a bit more work, one can also use direct integration and the definition (3.3) to find that the characteristic function of a multivariate Gaussian with mean μ and covariance matrix Σ is given by

$$E[\exp(i\theta^T V)] = \exp\left(i\theta^T \mu - \frac{1}{2}\theta^T \Sigma \theta\right). \tag{3.5}$$

One of the most valuable features of this formula is that it provides us with a handy characterization of multivariate Gaussians.

PROPOSITION 3.1 (Gaussian Characterization). *The d-dimensional random vector V is a multivariate Gaussian if and only if each of the linear combinations $\theta V_1 + \theta_2 V_2 + \cdots + \theta_d V_d$ is a univariate Gaussian.*

PROOF. To prove the more interesting implication, we first suppose that for each θ the linear combination defined by

$$Z_\theta = \theta_1 V_1 + \theta_2 V_2 + \cdots + \theta_d V_d$$

is a univariate Gaussian. If μ and Σ are the mean vector and covariance matrix for V, then we can easily check that

$$E[Z_\theta] = \theta^T \mu \text{ and } \mathrm{Var} Z_\theta = \theta^T \Sigma \theta.$$

In view of our assumption that Z_θ is a univariate Gaussian, we can use our expressions for $E(Z_\theta)$ and $\mathrm{Var}(Z_\theta)$ in the characteristic function formula (3.4), and set

$t = 1$ to find

$$E[\exp(i\theta^T V)] = E(e^{iZ_\theta}) = \exp\left(i\theta^T\mu - \frac{1}{2}\theta^T\Sigma\theta\right).$$

Finally, by formula (3.5), this is exactly what one needs in order to show that V is multivariate Gaussian.

There are several ways to show the more elementary fact that the linear combination $Z_\theta = \theta_1 V_1 + \theta_2 V_2 + \cdots + \theta_d V_d$ is a univariate Gaussian whenever the d-dimensional random vector V is a multivariate Gaussian. A direct (but tedious) method is to calculate the distribution function of Z_θ by integrating the density (3.3) over a half-space. The less direct (but simpler) method is to reverse the characteristic function argument that was just given for the first implication. □

The purpose of reviewing these facts about covariances and characteristic functions is that they have important consequences for Brownian motion. The covariance properties of Brownian motion are intimately tied to the independent increment property, and a covariance calculation will be our key tool for establishing the independent increments property for the process that we construct.

Covariance Functions and Gaussian Processes

If a stochastic process $\{X_t : 0 \le t < \infty\}$ has the property that the vector $(X_{t_1}, X_{t_2}, \ldots, X_{t_n})$ has the multivariate Gaussian distribution for any finite sequence $0 \le t_1 < t_2 < \cdots < t_n$, then $\{X_t\}$ is called a *Gaussian process.* Brownian motion is the Gaussian process *par excellence*, and many of the most important models in applied probability also lead us to Gaussian processes.

Gaussian processes are more amenable to exact calculation than almost any other processes because the joint distributions of a Gaussian process are completely determined by the mean function $\mu(t) = E[X_t]$ and the covariance function

$$f(s,t) = \mathrm{Cov}(X_s, X_t).$$

We can easily calculate the covariance function for Brownian motion just by noting that for any $s \le t$ we have $\mathrm{Cov}(B_s, B_t) = E[(B_t - B_s + B_s)B_s] = E[B_s^2] = s$, so in general we have

$$\text{(3.6)} \qquad \mathrm{Cov}(B_s, B_t) = \min(s,t) \quad 0 \le s,t < \infty.$$

Many basic properties of Brownian motion are revealed by this covariance formula, and central among these is the fact that any Gaussian process with a covariance function given by equation (3.6) must have independent increments. This fact is quite easy to prove but is so important it should be recorded as a lemma.

LEMMA 3.1. *If a Gaussian process $\{X_t : 0 \le t < T\}$ has $E(X_t) = 0$ for all $0 \le t < T$ and if*

$$\mathrm{Cov}(X_s, X_t) = \min(s,t) \text{ for all } 0 \le s,t < T,$$

then the process $\{X_t\}$ has independent increments. Moreover, if the process has continuous paths and $X_0 = 0$, then it is a standard Brownian motion on $[0,T)$.

PROOF. For the first claim, we just need to show that the random vector of process increments

$$(X_{t_2} - X_{t_1}, X_{t_3} - X_{t_2}, \ldots, X_{t_n} - X_{t_{n-1}})$$

has a diagonal covariance matrix. We can confirm the vanishing of the off-diagonal terms just by noting that for $i < j$ we can expand $E[(X_{t_i} - X_{t_{i-1}})(X_{t_j} - X_{t_{j-1}})]$ to find

$$E[X_{t_i}X_{t_j}] - E[X_{t_i}X_{t_{j-1}}] - E[X_{t_{i-1}}X_{t_j}] + E[X_{t_{i-1}}X_{t_{j-1}}]$$

$$= t_i - t_i - t_{i-1} + t_{i-1} = 0.$$

The second claim is now immediate from the definition of Brownian motion, so the proof of the lemma is complete. □

3.2. Visions of a Series Approximation

We will establish the existence of Brownian motion by providing an explicit series expansion. The calculations we make with this series are quite basic, but we still need to spell out some facts about function spaces. First, we recall that for any $p \geq 1$, $L^p[0,1]$ denotes the set of functions $f : [0,1] \to \mathbb{R}$ such that

$$\int_0^1 |f(x)|^p \, dx < \infty,$$

and $L^p[0,1]$ is viewed as a normed linear space with norm

$$||f||_p = \left(\int_0^1 |f(x)|^p \, dx \right)^{1/p}.$$

In the special case when $p = 2$, we can also view $L^2[0,1]$ as an inner product space with inner product defined by

$$\langle f, g \rangle = \int_0^1 f(x)g(x) \, dx < \infty.$$

Finally, if the functions in the set $\{\phi_n \in L^2[0,1] : 0 \leq n < \infty\}$ satisfy $\langle \phi_n, \phi_n \rangle = 1$ for all $n \geq 0$ and $\langle \phi_n, \phi_m \rangle = 0$ for all $n \neq m$, then $\{\phi_n\}$ is called an *orthonormal sequence.* If the finite linear combinations of the ϕ_n also form a dense set in $L^2[0,1]$, then $\{\phi_n\}$ is called a *complete* orthonormal sequence. [1]

There are two facts that make complete orthonormal sequences of particular importance for us. The first is that for any $f \in L^2[0,1]$ we have the representation

$$f = \sum_{n=0}^{\infty} \langle f, \phi_n \rangle \phi_n,$$

where the precise meaning of the convergence of the sum is that

$$||f - \sum_{n=0}^{N} \langle f, \phi_n \rangle \phi_n||_2 \to 0 \text{ as } N \to \infty.$$

[1] The Appendix on Mathematical Tools provides additional background on complete orthonormal sequences, including a proof of Parseval's identity.

The second fact is that we can compute the inner product of f and g in two different ways by means of *Parseval's identity*:

$$\int_0^1 f(x)g(x)\,dx = \sum_{n=0}^{\infty} \langle f, \phi_n\rangle\langle g, \phi_n\rangle. \tag{3.7}$$

If we glimpse over the horizon just a bit, we can see how Parseval's identity might be useful in the study of Brownian motion. The connection comes from specializing f and g to be the indicator functions of the intervals $[0, s]$ and $[0, t]$, that is

$$f(x) = 1_{[0,s]}(x) \text{ and } g(x) = 1_{[0,t]}(x).$$

In this case, Parseval's identity simplifies somewhat and tells us

$$\min(s, t) = \sum_{n=0}^{\infty} \int_0^s \phi_n(x)\,dx \int_0^t \phi_n(x)\,dx. \tag{3.8}$$

The expansion on the right-hand side may seem like a long-winded way of writing $\min(s, t)$, but, long-winded or not, the formula turns out to be exceptionally fruitful.

THE MIN(s, t) CONNECTION

The benefit of the Parseval expansion (3.8) comes from the fact that Brownian motion has covariance $\text{Cov}(B_s, B_t) = \min(s, t)$, so the identity (3.8) creates a connection between Brownian motion and the integrals of a complete orthonormal sequence. There are several ways one might explore this link, but perhaps the most natural idea is to consider the stochastic process defined by

$$X_t = \sum_{n=0}^{\infty} Z_n \int_0^t \phi_n(x)\,dx, \tag{3.9}$$

where $\{Z_n : 0 \leq n < \infty\}$ is a sequence of independent Gaussian random variables with mean zero and variance one.

For the moment, we will not consider the issue of the convergence of the sum (3.9), but we will make an exploratory calculation to see how X_t might relate to Brownian motion. Formally, we have $E(X_t) = 0$, and since X_t is a linear combination of Gaussian variables it is reasonable to compute the covariance:

$$\begin{aligned}
E(X_s X_t) &= E\left[\sum_{n=0}^{\infty} Z_n \int_0^s \phi_n(x)\,dx \sum_{m=0}^{\infty} Z_m \int_0^t \phi_m(x)\,dx\right] \\
&= \sum_{n=0}^{\infty}\sum_{m=0}^{\infty} E\left[Z_n Z_m \int_0^s \phi_n(x)\,dx \int_0^t \phi_m(x)\,dx\right] \\
&= \sum_{n=0}^{\infty} \int_0^s \phi_n(x)\,dx \int_0^t \phi_n(x)\,dx \\
&= \min(s, t).
\end{aligned}$$

This is a very promising calculation! To be sure, much remains to be justified; so far we have not even seen that the series representation of X_t is convergent.

Nevertheless, if we make a reasonable choice of the orthonormal series, the convergence of the sum (3.9) seems inevitable. Moreover, if we make a shrewd choice, the representation should also give us a process that has continuous paths with probability one.

3.3. Two Wavelets

In our construction of Brownian motion, we will use two sequences of functions that have been studied for many years but nowadays are seen in a new light. Both sequences are examples of *wavelets*, a topic of great current activity. To define these functions, we first consider a function that can serve as a "mother wavelet":

$$H(t) = \begin{cases} 1 & \text{for } 0 \le t < \frac{1}{2} \\ -1 & \text{for } \frac{1}{2} \le t \le 1 \\ 0 & \text{otherwise.} \end{cases}$$

Next, for $n \ge 1$ we define a sequence of functions by scaling and translating the mother wavelet $H(\cdot)$. Specifically, we note that any $n \ge 1$ can be written uniquely in the form $n = 2^j + k$, where $0 \le k < 2^j$ and $j \ge 0$, and we exploit this connection to write

$$H_n(t) = 2^{j/2} H(2^j t - k) \text{ for } n = 2^j + k \text{ where } j \ge 0 \text{ and } 0 \le k < 2^j.$$

If we also set $H_0(t) = 1$, we can easily check that $\{H_n : 0 \le n < \infty\}$ is a complete orthonormal sequence for $L^2[0,1]$. Orthonormality follows from simple integrations, and to check completeness we note that — up to sets of measure zero — the indicator function of any binary interval $[k \cdot 2^{-j}, (k+1) \cdot 2^{-j}]$ is in the linear span of the $\{H_n\}$. Since the linear combinations of such indicators can approximate any $f \in L^2[0,1]$ as closely as we like, we conclude that the linear span of $\{H_n\}$ is dense.

Here, we should note that by writing n as $2^j + k$ where $j \ge 0$ and $0 \le k < 2^j$, we get a second view of the sequence $\{H_n\}$ as a doubly indexed array where the rows are indexed by j, as we see in the more natural layout:

$$\begin{array}{c|cccccccc} j=0 & H_1 \\ j=1 & H_2 & H_3 \\ j=2 & H_4 & H_5 & H_6 & H_7 \\ j=3 & H_8 & H_9 & H_{10} & H_{11} & H_{12} & H_{13} & H_{14} & H_{15} \end{array}$$

This way of looking at the sequence is quite handy, since it makes several aspects of the geometry of the sequence $\{H_n\}$ much more evident. In particular, one should note that within any row all of the H_n are simply translations of the first element of the row. Also, within any row the H_n have disjoint support, so for each $x \in [0,1]$ we have $H_n(x) \ne 0$ for at most one n in each row. This last observation is of considerable importance when we want to estimate the size of a linear combination of the H_n.

The next step of the plan is to obtain a good representation for the integrals of the $\{H_n\}$ — the key elements of our intended representation for Brownian motion. These integrals turn out to have an elegant expression in terms of another wavelet sequence that is generated by another mother wavelet, which this time is given by the triangle function:

$$\Delta(t) = \begin{cases} 2t & \text{for } 0 \le t < \frac{1}{2} \\ 2(1-t) & \text{for } \frac{1}{2} \le t \le 1 \\ 0 & \text{otherwise.} \end{cases}$$

This function is just twice the integral of the mother wavelet $H(\cdot)$ taken from 0 to t.

Next, for $n \ge 1$, we use internal scaling and translating of the mother wavelet to define the sequence

$$\Delta_n(t) = \Delta(2^j t - k) \text{ for } n = 2^j + k, \quad \text{where } j \ge 0 \text{ and } 0 \le k < 2^j,$$

and for $n = 0$ we simply take $\Delta_0(t) = t$. The functions Δ_n, $0 \le n < \infty$, will serve as the fundamental building blocks in our representation of Brownian motion. Because we have

$$0 \le \Delta_n(t) \le 1 \text{ for all } t \in [0,1] \text{ and for all } n \ge 0,$$

we will be able to estimate the supremum norm of series of the $\Delta_n(t)$ just by studying the coefficients of the $\Delta_n(t)$. Also, since the mother wavelet $\Delta(t)$ is the integral of the mother wavelet $H(t)$, we naturally expect a close connection between the integrals of the H_n and the Δ_n. In fact, one can check that

$$\int_0^t H_n(u)\,du = \lambda_n \Delta_n(t),$$

where $\lambda_0 = 1$ and for $n \ge 1$ we have

$$\lambda_n = \frac{1}{2} \cdot 2^{-j/2} \text{ where } n \ge 1 \text{ and } n = 2^j + k \text{ with } 0 \le k < 2^j.$$

One should draw the first 32 or so of the $\Delta_n(t)$ just to see how the little wave runs along the unit interval until the last moment when — like Disney's Sorcerer's Apprentice — the wave splits in two, doubles its height, and jumps back to the beginning of the interval. The bottom line is that our candidate for Brownian motion has an explicit, even simple, representation in terms of the triangular wavelets Δ_n.

3.4. Wavelet Representation of Brownian Motion

THEOREM 3.1. *If $\{Z_n : 0 \le n < \infty\}$ is a sequence of independent Gaussian random variables with mean 0 and variance 1, then the series defined by*

$$X_t = \sum_{n=0}^{\infty} \lambda_n Z_n \Delta_n(t) \tag{3.10}$$

converges uniformly on $[0,1]$ with probability one. Moreover, the process $\{X_t\}$ defined by the limit is a standard Brownian motion for $0 \le t \le 1$.

Before proving this result, we need to gather some useful information about the behavior of a sequence of independent Gaussian random variables.

LEMMA 3.2. *If* $\{Z_n\}$ *is a sequence of independent Gaussian random variables with mean zero and variance one, then there is a random variable* C *such that*

$$|Z_n| \le C\sqrt{\log n} \text{ for all } n \ge 2$$

and

$$P(C < \infty) = 1.$$

PROOF. For $x \ge 1$ we have

$$P(|Z_n| \ge x) = \frac{2}{\sqrt{2\pi}} \int_x^\infty e^{-u^2/2}\,du$$

$$\le \sqrt{\frac{2}{\pi}} \int_x^\infty u e^{-u^2/2}\,du = e^{-x^2/2}\sqrt{\frac{2}{\pi}}, \tag{3.11}$$

and hence for any $\alpha > 1$ we find

$$P(|Z_n| \ge \sqrt{2\alpha \log n}) \le \exp(-\alpha \log n)\sqrt{\frac{2}{\pi}} = n^{-\alpha}\sqrt{\frac{2}{\pi}}. \tag{3.12}$$

Now, for $\alpha > 1$ the last bound is summable, so the Borel–Cantelli lemma tells us that

$$P(|Z_n| \ge \sqrt{2\alpha \log n} \quad \text{for infinitely many } n) = 0. \tag{3.13}$$

By equation (3.13), we then see that the random variable defined by

$$\sup_{2 \le n < \infty} \frac{|Z_n|}{\sqrt{\log n}} = C$$

is finite with probability one. □

UNIFORM CONVERGENCE WITH PROBABILITY ONE

We can now provide the proof of Theorem 3.1. The first item of business is to establish the required convergence of the series representation (3.10). For any $n \in [2^j, 2^{j+1})$, we have $\log n < j + 1$, and for any $0 \le x \le 1$ we have $\Delta_n(x) = 0$ for all but one value of n in the interval $[2^j, 2^{j+1})$; so, when we bring in the bound from Lemma 3.2, we find that for any $M \ge 2^J$ we have

$$\begin{aligned}
\sum_{n=M}^\infty \lambda_n |Z_n| \Delta_n(t) &\le C \sum_{n=M}^\infty \lambda_n \sqrt{\log n}\Delta_n(t) \\
&\le C \sum_{j=J}^\infty \sum_{k=0}^{2^j-1} \frac{1}{2} \cdot 2^{-j/2}\sqrt{j+1}\Delta_{2^j+k}(t) \\
&\le C \sum_{j=J}^\infty \frac{1}{2} \cdot 2^{-j/2}\sqrt{j+1}.
\end{aligned}$$

Since the last term goes to zero as $J \to \infty$, we see that with probability one the series (3.10) that defines X_t is uniformly and absolutely convergent. We therefore find that the paths of the process $\{X_t : 0 \le t \le 1\}$ are continuous with probability one.

Calculation of the Covariance Function

The next step in the proof of Theorem 3.1 is to show that

$$\mathrm{Cov}(X_s, X_t) = \min(s,t) \text{ for all } 0 \le s, t < \infty.$$

Because of the absolute convergence of the series defining $\{X_t\}$, we can rearrange terms any way we like. In particular, we can return to our earlier heuristic calculation and check that all the steps leading to the application of Parseval's identity are rigorously justified:

$$\begin{aligned} E(X_s X_t) &= E\Big[\sum_{n=0}^{\infty} \lambda_n Z_n \Delta_n(s) \sum_{m=0}^{\infty} \lambda_m Z_m \Delta_m(t)\Big] \\ &= \sum_{n=0}^{\infty} \lambda_n^2 \Delta_n(s)\Delta_n(t) \\ &= \sum_{n=0}^{\infty} \int_0^s H_n(u)\,du \int_0^t H_n(u)\,du \\ &= \min(s,t). \end{aligned} \tag{3.14}$$

With the covariance function in hand, it is tempting to declare victory, since the process $\{X_t\}$ is "obviously" Gaussian "since it is the sum of independent Gaussian random variables." Such arguments usually provide sound guidance, but an honest proof requires us to look deeper.

The Process is Gaussian

The final step will be to show that for any finite sequence $t_1 < t_2 < \cdots < t_m$ the vector $(X_{t_1}, X_{t_2}, \ldots, X_{t_m})$ has the multivariate Gaussian distribution. Since we have already determined that the covariance function of the process $\{X_t\}$ is given by $\min(s,t)$, Lemma 3.1 will then confirm that $\{X_t\}$ has the independent increment property.

In order to show that $(X_{t_1}, X_{t_2}, \ldots, X_{t_m})$ is a multivariate Gaussian, we will simply compute its multivariate characteristic function. We begin with the definition of X_t and then exploit the independence of the Z_n:

$$\begin{aligned} E\Big[\exp\Big(i\sum_{j=1}^{m} \theta_j X_{t_j}\Big)\Big] &= E\Big[\exp\Big(i\sum_{j=1}^{m} \theta_j \sum_{n=0}^{\infty} \lambda_n Z_n \Delta_n(t_j)\Big)\Big] \\ &= \prod_{n=0}^{\infty} E\Big[\exp\Big(i\lambda_n Z_n \sum_{j=1}^{m} \theta_j \Delta_n(t_j)\Big)\Big]. \end{aligned} \tag{3.15}$$

Next, we calculate the expectations inside the product by the formula for the one-dimensional Gaussian characteristic function and then we collect terms into a single exponential:

$$\prod_{n=0}^{\infty} \exp\Big(-\frac{1}{2}\lambda_n^2\Big(\sum_{j=1}^{m}\theta_j\Delta_n(t_j)\Big)^2\Big) = \exp\Big(-\frac{1}{2}\sum_{n=0}^{\infty}\lambda_n^2\Big(\sum_{j=1}^{m}\theta_j\Delta_n(t_j)\Big)^2\Big).$$

If we expand the exponent in the last expression and recall the special case of Parseval's identity given by equation (3.14), we see that

$$\sum_{n=0}^{\infty} \lambda_n^2 \sum_{j=1}^{m}\sum_{k=1}^{m} \theta_j\theta_k \Delta_n(t_j)\Delta_n(t_k) = \sum_{j=1}^{m}\sum_{k=1}^{m} \theta_j\theta_k \min(t_j, t_k),$$

and, when we carry this identity back to equation (3.15), we find

$$E\left[\exp\left(i\sum_{j=1}^{m}\theta_j X_{t_j}\right)\right] = \exp\left(-\frac{1}{2}\sum_{j=1}^{m}\sum_{k=1}^{m}\theta_j\theta_k \min(t_j, t_k)\right). \tag{3.16}$$

The last expression is precisely the characteristic function of a multivariate Gaussian with mean zero and covariance matrix $\Sigma = (\min(t_i, t_j))$, so indeed the vector $(X_{t_1}, X_{t_2}, \ldots, X_{t_m})$ is a multivariate Gaussian. This is the last fact that we needed in order to complete the proof of Theorem 3.1.

Examples of the Approximation

This construction of Brownian motion has very pleasant analytical properties, which we will soon exploit with abandon, but first we should note that the construction gives us a practical method for computer simulation of the paths of a Brownian motion. When we use as few as eight summands, then we see from Figure 3.1 that the paths are too crude for most purposes, but one can similarly deduce from Figure 3.2 that a path that uses 128 summands might serve quite well in many circumstances.

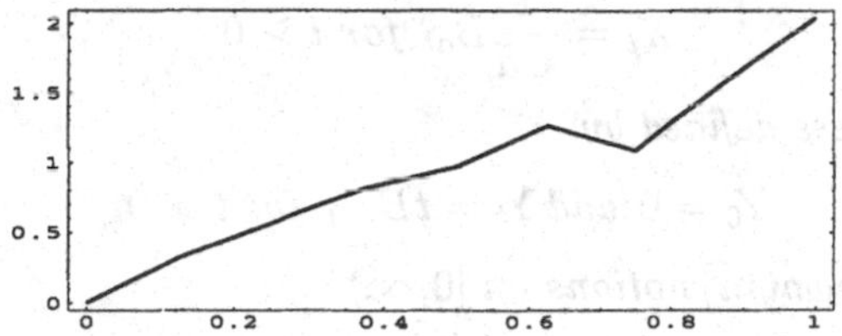

FIGURE 3.1. AN APPROXIMATION BASED ON 2^3 SUMMANDS

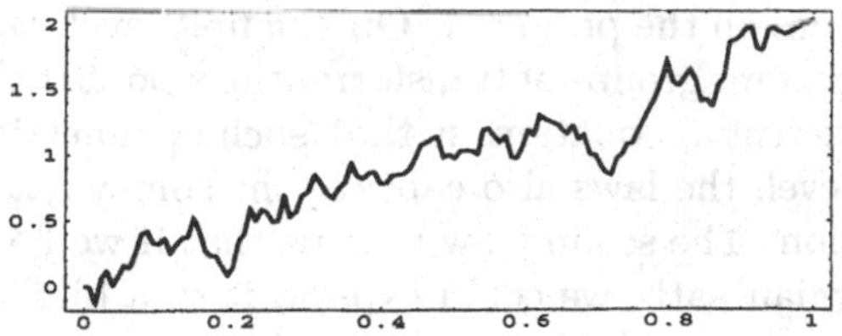

FIGURE 3.2. AN APPROXIMATION BASED ON 2^7 SUMMANDS

3.5. Scaling and Inverting Brownian Motion

Now that we have a standard Brownian motion on $[0,1]$, we can construct a standard Brownian motion on $[0,\infty)$ in a couple of ways. The most straightforward idea is to take a countable number of independent Brownian motions and to string them together by starting each new one where the last one left off. Though notation cannot really make the idea any clearer, we can still root out any ambiguity by casting the construction as a formula. For each $1 \le n < \infty$, we take an independent standard Brownian motion $B_t^{(n)}$ on $[0,1]$, and for any $0 \le t < \infty$ we define B_t by the sum

$$B_t = \sum_{k=1}^{n} B_1^{(k)} + B_{t-n}^{(n+1)} \text{ whenever } t \in [n, n+1);$$

so, for example,

$$B_{3.5} = B_1^{(1)} + B_1^{(2)} + B_1^{(3)} + B_{0.5}^{(4)}.$$

Once one is comfortable with the way the unit interval Brownian motions are pasted together, there is no difficulty in verifying that B_t has all the properties that one requires of standard Brownian motion on $[0,\infty)$.

The interval $[0,\infty)$ is the natural domain for Brownian motion. From the modeler's perspective there is no good reason to confine the process to an artificially finite time interval, but there are even more compelling reasons to prefer to view Brownian motion on an infinite interval. In particular, there are symmetries that are made much more evident when we consider Brownian motion on $[0,\infty)$. These symmetries can be expressed by noting that there are basic transformations of Brownian motion that turn out to be Brownian motion again.

PROPOSITION 3.2 (Scaling and Inversion Laws). *For any $a > 0$, the scaled process defined by*

$$X_t = \frac{1}{\sqrt{a}} B_{at} \text{ for } t \ge 0$$

and the inverted process defined by

$$Y_0 = 0 \text{ and } Y_t = tB_{1/t} \text{ for } t > 0$$

are both standard Brownian motions on $[0,\infty)$.

No one will have any difficulty checking that X_t meets the definition of a standard Brownian motion. A bit later we will see that Y_t is continuous at zero with probability one; and, if we assume this continuity for the moment, we also find it easy to check that Y_t is a standard Brownian motion.

Although there is no great depth to the mathematics of these results, the physics behind them still touches on the profound. On the first level, these laws tie Brownian motion to two important groups of transformations on $[0,\infty)$, and a basic lesson from the theory of differential equations is that such symmetries can be extremely useful. On a second level, the laws also capture the somewhat magical fractal nature of Brownian motion. The scaling law tells us that if we had even one billionth of a second of a Brownian path, we could expand it to a billion years worth of an equally valid Brownian path. The inversion law is perhaps even more impressive. It tells us that the first second of the life of a Brownian path is rich enough to capture the behavior of a Brownian path from the end of the first second until the end of time.

3.6. Exercises

The first exercise shows that our construction of Brownian motion automatically provides a construction of a second important Gaussian process, the Brownian bridge. The remaining exercises review important properties of Gaussian random variables.

EXERCISE 3.1 (Brownian Bridge). In our representation for Brownian motion

$$B_t = \sum_{n=0}^{\infty} \lambda_n Z_n \Delta_n(t),$$

we have $\Delta_0(1) = 1$ and $\Delta_n(1) = 0$ for all $n \geq 1$; so, if we leave off the first term in our representation for Brownian motion to define

$$U_t = \sum_{n=1}^{\infty} \lambda_n Z_n \Delta_n(t),$$

we see that U_t is a continuous process on $[0, 1]$ such that $U_0 = 0$ and $U_1 = 0$. This process is called a *Brownian bridge*, and it is often useful when we need to model a quantity that starts at some level and that must return to a specified level at a specified future time.

(a) Show that we can write $U_t = B_t - tB_1$ for $0 \leq t \leq 1$.

(b) Show that we have $\text{Cov}(U_s, U_t) = s(1-t)$ for $0 \leq s \leq t \leq 1$.

(c) Let $X_t = g(t)B_{h(t)}$, and find functions g and h such that X_t has the same covariance as a Brownian bridge.

(d) Show that the process defined by $Y_t = (1+t)U_{t/(1+t)}$ is a Brownian motion on $[0, \infty)$. Note: Since we have a direct construction of the Brownian bridge U_t, this observation gives us a third way to build Brownian motion on $[0, \infty)$.

EXERCISE 3.2 (Cautionary Tale: Covariance and Independence). The notions of independence and covariance are less closely related than elementary courses sometimes lead one to suspect. Give an example of random variables X and Y with $\text{Cov}(X, Y) = 0$ such that both X and Y are Gaussian yet X and Y are not independent. Naturally, X and Y cannot be jointly Gaussian.

EXERCISE 3.3 (Multivariate Gaussians). Most of the following problems can be done by using the change of variables formula of multivariate calculus or, perhaps more easily, by using characteristic functions.

(a) Let V be a multivariate Gaussian column vector with mean $\mu = E(V)$ and covariance Σ, and let A be an $n \times n$ real matrix. Show that AV is a multivariate Gaussian vector with mean $A\mu$ and covariance $A\Sigma A^T$.

(b) Show that if X and Y are independent Gaussians with mean zero and variance one, then $X - Y$ and $X + Y$ are independent Gaussians with mean zero and variance two.

(c) Prove that if X and Y are jointly Gaussian and $\text{Cov}(X, Y) = 0$ then X and Y are independent.

(d) Prove that if (X, Y) are jointly Gaussian, then the conditional distribution of Y given that $X = x$ is normal with mean

$$\mu_{Y|X=x} = \mu_Y + \frac{\text{Cov}(X,Y)}{\text{Var}X}(x - \mu_X)$$

and variance

$$\sigma^2_{Y|X=x} = \sigma^2_Y - \frac{\text{Cov}^2(X,Y)}{\text{Var}X}.$$

EXERCISE 3.4 (Auxiliary Functions and Moments). For a random variable X, the characteristic function $E(e^{itX})$ always exists since e^{itX} is a bounded random variable (with complex values). The moment generating function $E(e^{tX})$ only exists if X has moments of all orders and those moments do not grow too fast.

(a) If Z is Gaussian with mean 0 and variance 1, find the moment generating function of Z and use it to find the first six moments of Z. In particular, show that $E(Z^4) = 3$, a fact that we will need later. [Hint: Series expansion can be easier than differentiation.]

(b) If $W = Z^4$, then W has moments of all orders, since Z has all moments. Show nevertheless that W does not have a moment generating function.

EXERCISE 3.5 (Gaussian Bounds). The estimation of the Gaussian tail probability used in Lemma 3.2 can be sharpened a bit, and there are complementary inequalities that are also useful. For Gaussian Z with mean zero and variance one, show that

$$\frac{1}{\sqrt{2\pi}} \frac{e^{-x^2/2}}{(x + x^{-1})} \leq P(Z \geq x) \leq \frac{1}{\sqrt{2\pi}} \frac{e^{-x^2/2}}{x} \quad \text{for } x > 0, \tag{3.17}$$

and that

$$\frac{2x}{\sqrt{2\pi e}} \leq P(|Z| \leq x) \leq \frac{2x}{\sqrt{2\pi}} \quad \text{for } 0 < x \leq 1.$$

The second set of inequalities will be useful when we show that the paths of Brownian motion are never differentiable.

Anyone who wants a hint for the proof of the double inequality (3.17) might consider the identity $(u^{-1}e^{-u^2/2})' = -(1 + u^{-2})e^{-u^2/2}$ and the simple fact that for $x \leq u$ one has $e^{-u^2/2} \leq (u/x)e^{-u^2/2}$.

CHAPTER 4

Martingales: The Next Steps

Discrete-time martingales live under a star of grace. They offer inferences of power, purpose, and surprise, yet they impose little in the way of technical nuisance. In continuous time, martingale theory requires more attention to technical foundations — at a minimum, we need to build up some of the basic machinery of probability spaces and to sharpen our view of conditional expectations. Nevertheless, by focusing on martingales with continuous sample paths, we can keep technicalities to a minimum, and, once we have a stopping time theorem, our continuous-time tools will be on a par with those we used in discrete time. We can then examine three classic martingales of Brownian motion that parallel the classic martingales of simple random walk. These martingales quickly capture the most fundamental information about hitting times and hitting probabilities for Brownian motion.

4.1. Foundation Stones

Ever since Kolmogorov first set down a clear mathematical foundation for probability theory in 1933, the tradition has been for probability texts from the most elementary to the most advanced to begin their exposition by defining a *probability space* $(\Omega, \mathcal{F}, P)$. Nowadays there is not much fanfare attached to this seemingly bland formality, but there is no disputing that Kolmogorov's foundations profoundly changed the way that the world views probability theory. Before Kolmogorov's efforts, probability theory and probability modeling were inextricably intertwined, so much so that one might doubt results in probability just as one might doubt the implications of a model. After Kolmogorov's efforts, probability theory offered no more room for doubt than one can find in linear algebra or in the theory of differential equations.

Traditions can be deferred but not denied, so eventually we have to recall some of the basic facts from the foundations of probability theory. First, Ω is only required to be a set, and $\mathcal{F}$ is a collection of subsets of Ω with the property that (i) $\emptyset \in \mathcal{F}$, (ii) $A \in \mathcal{F} \Rightarrow A^c \in \mathcal{F}$, and (iii) for any countable collection of $A_i \in \mathcal{F}$ we have $\cup A_i \in \mathcal{F}$. Such an $\mathcal{F}$ is traditionally called a σ-field, though some authors now use the term *tribe*. The final and most interesting element of the probability triple (or Probabilists' Trinity) is P, the probability measure. We assume nothing more of P than that it be a function from $\mathcal{F}$ into $[0,1]$ such that $P(\Omega) = 1$ and for any disjoint countable collection $\{A_i\}$ of elements of $\mathcal{F}$ one has $P(\cup A_i) = \sum P(A_i)$.

Side Effects of Abstract Foundations

The fundamental benefit of a rigorous mathematical foundation for probability theory is freedom from ambiguity, but there are also side effects that can be both useful and amusing. When we are no longer tied to intuitive probabilistic models, we are free to invent new probability spaces where the methods of probability theory can be used to achieve unexpected ends. In particular, Kolmogorov helped us see that we can use probability theory to prove purely analytical facts.

To give just one example of a useful analytical result that can be proved by introducing an abstract (or artificial) probability space, we will derive Hölder's inequality,

$$E(|XY|) \leq ||X||_p||Y||_q \text{ where } 1 \leq p, q \leq \infty \text{ and } \frac{1}{p} + \frac{1}{q} = 1,$$

by means of the more intuitive Jensen inequality,

$$\phi(E(X)) \leq E(\phi(X)) \text{ for all convex } \phi.$$

The trick is to leverage the abstraction that is implicit in the fact that Jensen's inequality is valid in *any* probability space. Specifically, the idea is to design an artificial probability measure so that when we apply Jensen's inequality to the artificial measure, Hölder's inequality pops out. This strategy may seem odd; but, as we will discover on several occasions, it can be remarkably effective.

To apply the strategy to prove Hölder's inequality, we first note that there is no loss of generality in assuming that $X > 0$ and $Y > 0$. Also, since Hölder's inequality is trivial when $p = 1$, we can assume $1 < p, q < \infty$. Now, to come to the real trick, we introduce a new probability measure Q by taking

$$Q(A) = E(X^p 1_A)/E(X^p). \tag{4.1}$$

One can easily check that Q is an honest probability measure, and if we denote the expectation of Z with respect to the probability measure Q by $E_Q(Z)$, then in terms of the usual expectation taken under P, we have $E_Q(Z) = E(ZX^p)/E(X^p)$. If we now let $\phi(z) = z^q$ and apply Jensen's inequality, we find $(E_Q[Z])^q \leq E_Q[Z^q]$. All that remains is to unwrap the definitions and to make a good choice of Z. First, we note that

$$(E_Q[Z])^q = (E[ZX^p]/E[X^p])^q \text{ and } E_Q[Z^q] = E[Z^q X^p]/E[X^p],$$

so, in longhand, Jensen's inequality tells us that

$$\left(\frac{E[ZX^p]}{E[X^p]}\right)^q \leq \frac{E[Z^q X^p]}{E[X^p]}. \tag{4.2}$$

To get closer to Hölder's inequality, we would like to set $ZX^p = XY$, and, since Z is completely at our disposal, we simply take $Z = Y/X^{p-1}$. This choice also gives us $Z^p X^q = Y^q$ since $(p-1)q = p$, and we then see that the bound (4.2) immediately reduces to Hölder's inequality.

This proof may not be as succinct as the customary derivation via the real variable inequality $xy \leq x^p/p + y^q/q$, but, from the probabilist's point of view, the artificial measure method has a considerable charm. At a minimum, this proof of Hölder's inequality reminds us that each of our probability inequalities may hide treasures that can be brought to light by leveraging the generality that is calmly concealed in the abstract definition of a probability space.

4.2. Conditional Expectations

Since we now have a probability space $(\Omega, \mathcal{F}, P)$, we can take a more sophisticated view of some of our earlier intuitive shorthand notations. In the formalism of Kolmogorov, a random variable is nothing but a function $X : \Omega \mapsto \mathbb{R}$ with the property that $\{\omega\colon X(\omega) \le x\} \in \mathcal{F}$ for all $x \in \mathbb{R}$. This is a perfectly natural definition because such an X meets the minimal requirement one could impose and still be guaranteed to be able to talk about the probability $P(X \le x)$. In the parallel language of measure theory, we say that such an X is *measurable* with respect to $\mathcal{F}$.

We have $A \in \mathcal{F}$ if and only if the indicator function 1_A is measurable with respect to $\mathcal{F}$, so we will overburden the symbol $\in$ by continuing to write $X \in \mathcal{F}$ to indicate that X is measurable with respect to $\mathcal{F}$. Finally, if $\{X_1, X_2, \dots, X_n\}$ is a collection of random variables, then we will often need to consider the smallest σ-field $\mathcal{G}$ such that each of the X_i is measurable with respect to $\mathcal{G}$. This subfield of $\mathcal{F}$ is called the σ-field *generated* by $\{X_1, X_2, \dots, X_n\}$, and it is denoted by $\sigma\{X_1, X_2, \dots, X_n\}$.

Earlier, we used notation such as $E(X \mid \mathcal{G})$ to stand for $E(X \mid X_1, X_2, \dots, X_n)$, and we left the meaning of the second conditional expectation to rest on the understanding that one obtains in elementary courses — an understanding that requires the random variables to be discrete or nicely behaved continuous random variables. Fortunately, there is an interpretation of expectation and conditional expectation that frees us from these restrictions. The random variable X is just a function from Ω to $\mathbb{R}$, and to define the expectation we simply take

$$E(X) = \int_\Omega X(\omega)\, dP(\omega),$$

where the integral sign denotes the Lebesgue integral.[1] The definition of conditional expectation is a little more slippery.

DEFINITION 4.1. *If X is an integrable random variable in the probability space $(\Omega, \mathcal{F}, P)$, and if $\mathcal{G}$ is a subfield of $\mathcal{F}$, we say that Y is a conditional expectation of X with respect to $\mathcal{G}$ if Y is measurable with respect to $\mathcal{G}$ and if*

$$E(X1_A) = E(Y1_A) \text{ for all } A \in \mathcal{G}. \tag{4.3}$$

Also, if Y is any function that is measurable with respect to $\mathcal{G}$ that satisfies (4.3), we write

$$Y = E(X \mid \mathcal{G}).$$

This definition is not easy to love. Fortunately, love is not required. The definition is certainly abstract enough; on first consideration it barely evokes the good uses we have already made of conditional expectation. Nevertheless, we cannot do without this definition. If we continued to rely entirely on our intuitive understanding and the elementary constructions, we would soon be on rocky shoals. We must learn to live with the definition and to learn to benefit from what it has to offer.

The best way to become comfortable with Definition 4.1 is to work through a list of the properties that we have come to expect from conditional expectation and to show that these properties follow from the definition. As a typical example, we first consider the linearity property:

$$E(X + Y \mid \mathcal{G}) = E(X \mid \mathcal{G}) + E(Y \mid \mathcal{G}).$$

To see that linearity follows from Definition 4.1, we let $\alpha = E(X \mid \mathcal{G})$ and $\beta = E(Y \mid \mathcal{G})$, and then we verify that the random variable $\alpha + \beta$ satisfies the properties that are required by the definition of the conditional expectation of $X + Y$.

---

[1] The Appendix on Mathematical Tools gives a quick review of the construction and most important properties of the Lebesgue integral.

Since $\alpha + \beta$ is $\mathcal{G}$-measurable, we only have to show that

$$E[(\alpha + \beta)1_A] = E[(X+Y)1_A] \text{ for all } A \in \mathcal{G}. \tag{4.4}$$

Naturally, this is easy; we just have to keep the logic tight. First we note that

$$E[(\alpha + \beta)1_A] = E[(\alpha 1_A)] + E[(\beta 1_A)]$$

by the linearity of the usual (unconditional) expectation. Next, we note that $E(\alpha 1_A) = E(X1_A)$ and $E(\beta 1_A) = E(Y1_A)$ simply by the definition of α and β as conditional expectations. Finally, by a second application of linearity of (unconditional) expectation, we find $E(X1_A) + E(Y1_A) = E[(X+Y)1_A]$, and, when we string together our chain of equalities, we complete the proof of the required identity (4.4).

This symbol shuffling can look a little silly, but it is part of the price of a general definition of conditional probability. The diligent reader should make sure that the dance is clear, say by showing that $E(X \mid \mathcal{G}) \geq 0$ if $X \geq 0$, or, better yet, by proving the important *Tower Property of Conditional Expectation*: If $\mathcal{H}$ is a subfield of $\mathcal{G}$, then

$$E(E(X \mid \mathcal{G}) \mid \mathcal{H}) = E(X \mid \mathcal{H}). \tag{4.5}$$

Here, we should note that if $\mathcal{H}$ denotes the trivial σ-field consisting of just Ω and $\emptyset$, then the only functions that are $\mathcal{H}$-measurable are the constants. This tells us that $E(X \mid \mathcal{H}) = E(X)$, so the tower property tells us in turn that

$$E(E(X \mid \mathcal{G})) = E(X).$$

The last identity is particularly useful, and it is sometimes called the *law of total probability*.

As a final exercise, one might take on the factorization property: If Y is $\mathcal{G}$-measurable and $|XY|$ and $|X|$ are integrable, then Y can be brought outside the conditional expectation,

$$E(XY \mid \mathcal{G}) = YE(X \mid \mathcal{G}).$$

This is also a fact that we will use on numerous occasions.

Issue of Versions

If $P(Y = Y') = 1$ and Y satisfies the defining equation (4.3) of conditional expectations $E(X \mid \mathcal{G})$, then Y' also satisfies the equation. This circumstance obligates us to live with a certain ambiguity whenever we use conditional expectations. We acknowledge this ambiguity by referring to Y (or Y') as a *version* of the conditional expectation.

This ambiguity usually causes us no trouble, but if we need to make a statement about a whole set of conditional expectations, then there can be delicate moments if the set under consideration has the cardinality of the continuum. We usually dance around this problem by finding a way to do much of our work with countable sets of expectations. Nevertheless, there are times when the passage from the countable set to the continuum can prove challenging, and we will need new tools to help us with the transition.

Issue of Existence

The conditional expectation has now been defined at a level of generality that is appropriate for our work. However, we have not yet considered the *existence* of these conditional expectations. If we are given a random variable X on the probability space $(\Omega, \mathcal{F}, P)$ and a σ-field $\mathcal{G} \subset \mathcal{F}$, how do we know that there exists a random variable Y that meets the definition of $E(X|\mathcal{G})$? We would stray too far from our path if we gave all the details in the construction of $E(X|\mathcal{G})$, but it is easy to give a sketch of the construction. In fact, it is easy to give two such sketches.

The first construction begins with the assumption that there is a sequence of finite partitions $\{G_n\}$ of Ω such that $\{G_{n+1}\}$ is a refinement of $\{G_n\}$ for all n and such that $\mathcal{G}$ is the smallest σ-field that contains all of the G_n. This assumption is actually quite mild, and it is easily checked in all of the probability spaces that one is likely to meet in practice. Next, we let $\mathcal{G}_n$ be the smallest σ-field containing G_n, and, if the partition G_n is given by $\{B_1, B_2, \ldots, B_m\}$, then we can build a candidate for $E(H \mid \mathcal{G}_n)$ by defining a random variable Y_n on Ω by taking

$$Y_n(\omega) = E(X 1_{B_k}) \text{ for all } \omega \in B_k.$$

One then checks that Y_n does indeed satisfy the definition of a conditional expectation of X given $\mathcal{G}_n$. The next step is to show that the sequence $\{Y_n\}$ is an L^1 bounded martingale with respect to the filtration $\{\mathcal{G}_m\}$, so by the discrete martingale convergence theorem Y_n converges in L^1 to a limit $Y \in L^1(\Omega, \mathcal{G}, P)$. This limit is our candidate for $E(X|\mathcal{G})$, and the last step of the construction is to verify that Y meets the definition of a conditional expectation.

For readers who are familiar with projections in L^2, there is a second approach to the construction of $E(X|\mathcal{G})$ that is quicker and more geometric.[2] We first consider X such that $E(X^2) < \infty$, and we then note that the set of functions defined by $S = \{Y \in \mathcal{G} \text{ with } E(Y^2) < \infty\}$ is a closed linear subspace of the space $L^2(\Omega, \mathcal{F}, P)$. Next, we recall the basic fact that for any closed linear space there is a projection mapping $\pi : L^2(\Omega, \mathcal{F}, P) \mapsto S$. In fact, such a map can be defined simply by taking Y to be the element of S that is closest to X in the L^2 distance. One then checks that Y does indeed satisfy the definition of a conditional expectation of X. Finally, after having established the existence of conditional expectations for square integrable random variables, one uses an approximation argument to construct the conditional expectation in the basic case where we only assume that $E(|X|) < \infty$.

4.3. Uniform Integrability

The main goal of this chapter is to develop the continuous-time analogs to the Doob stopping time theorem and the martingale convergence theorems. The most natural way to prove these results is to use what we know about discrete-time martingales and to find appropriate ways to approximate continuous-time processes by discrete-time processes. Before we can begin such an approximation in earnest, we need to develop a few technical tools that help us relate different types of convergence. The first of these is the notion of uniform integrability.

DEFINITION 4.2. *We say that a collection C of random variables is* uniformly integrable *provided that*

$$\rho(x) = \sup_{Z \in C} E(|Z| 1(|Z| > x)) \text{ satisfies } \rho(x) \to 0 \text{ as } x \to \infty. \tag{4.6}$$

The reason that uniform integrability emerges as a useful concept is that it provides us with benefits that parallel L^1 domination, yet it applies in some important situations where L^1 domination is unavailable. For example, the next lemma

[2]A discussion of L^2 and subspace projection is given in the Appendix on Mathematical Tools.

provides an analog of the dominated convergence theorem for sequences that are not necessarily dominated.

CONSEQUENCES OF UNIFORM INTEGRABILITY

LEMMA 4.1 (Uniform Integrability and L^1 Convergence).
If $\{Z_n\}$ is a uniformly integrable sequence and Z_n converges to Z with probability one, then $E(|Z_n - Z|) \to 0$ as $n \to \infty$.

PROOF. We first note that if we apply Fatou's lemma to the definition of $\rho(x)$, then we find that $E\big(|Z|1(|Z| > x)\big) \le \rho(x)$, so, in particular, we see that $Z \in L^1$ and $E(|Z|) \le \rho(x) + x$.

Next, we note that $|Z_n - Z|$ is bounded above by the three-term sum

$$|Z_n - Z|1(|Z_n| \le x) | + |Z|1(|Z_n| > x) + |Z_n|1(|Z_n| > x). \tag{4.7}$$

The first term is bounded by $x + Z \in L^1$, so the dominated convergence theorem tells us

$$\lim_{n\to\infty} E\left(|Z_n - Z|1(|Z_n| \le x) \right) = 0.$$

The second term of the sum (4.7) is dominated by $|Z| \in L^1$, so again we can apply the DCT to find

$$\lim_{n\to\infty} E(|Z|1(|Z_n| > x)) = E(|Z|1(|Z| > x)) \le \rho(x).$$

The expectation of the last term of the sum (4.7) is bounded by $\rho(x)$ for all n, so when we put the three pieces back together we find

$$\limsup_{n\to\infty} E(|Z_n - Z|) \le 2\rho(x).$$

Finally, we can choose x to make $\rho(x)$ as small as we like, so the proof of the lemma is complete. □

The next lemma is trivial to prove, but it isolates an important fact that we will put to immediate use in the proof of Lemma 4.3.

LEMMA 4.2 (Conditional Expectation is a Contraction). *The map that takes $Z \in L^1$ to $E(Z|\mathcal{G})$ is an L^1 contraction; that is, we have*

$$||E(Z|\mathcal{G})||_1 \le ||Z||_1.$$

LEMMA 4.3 (Uniform Integrability and Conditional Expectations). *If Z_n converges to Z with probability one, and if the sequence $\{Z_n\}$ is uniformly integrable, then $E(Z_n|\mathcal{G})$ converges in L^1 and in probability to $E(Z|\mathcal{G})$.*

PROOF. We know from Lemma 4.1 that our hypotheses imply that Z_n converges to Z in L^1, and, since conditional expectation is an L^1 contraction, we immediately deduce that $E(Z_n \mid \mathcal{G})$ converges to $E(Z \mid \mathcal{G})$ in L^1. Finally, Markov's inequality tells us that L^1 convergence implies convergence in probability, so the proof of the lemma is complete. □

CONDITIONS FOR UNIFORM INTEGRABILITY

Our next task is to provide criteria for uniform integrability that can be used more easily than the definition. One useful criterion that follows easily from

Hölder's inequality is that $\{Z_n\}$ is uniformly integrable provided that there is some constant B and some $p > 1$ such that $E(|Z_n|^p) \leq B$ for all $n \geq 1$. The next lemma generalizes this L^p boundedness criterion in a way that we will find particularly handy.

LEMMA 4.4. *If $\phi(x)/x \to \infty$ as $x \to \infty$ and $\mathcal{C}$ is a collection of random variables such that*

$$E[\phi(|Z|)] \leq B < \infty \text{ for all } Z \in \mathcal{C},$$

then $\mathcal{C}$ is uniformly integrable.

To prove the lemma, we set $\psi(x) = \phi(x)/x$ and note that for any $Z \in \mathcal{C}$ we have for all $x > 0$ that

$$E(|Z|1(|Z| \geq x)) = E\left[\frac{\phi(|Z|)}{\psi(|Z|)}1(|Z| \geq x)\right] \leq B/\min\{\psi(y): y \geq x\}.$$

Now, the fact that $\psi(x) \to \infty$ completes the proof.

To get bounds of the type needed in Lemma 4.4, we can sometimes exploit the fact that any integrable random variable can always be viewed as an element of a space that is a bit like L^p for a $p > 1$. The next lemma makes this idea precise.

LEMMA 4.5. *If Z is a random variable with $E(|Z|) < \infty$, there is a convex ϕ such that $\phi(x)/x \to \infty$ as $x \to \infty$ and*

$$E[\phi(|Z|)] < \infty.$$

PROOF. We have $E(|Z|) = \int_0^\infty P(|Z| \geq x)dx < \infty$, and by Exercise 4.8 we can find a nondecreasing function $a(x)$ such that $a(x) \uparrow \infty$ and

$$\int_0^\infty a(x)P(|Z| \geq x)\,dx < \infty.$$

If we then define

$$\phi(x) = \int_0^x a(u)du,$$

then we have

$$E(\phi(|Z|) = E\left[\int_0^{|Z|} a(x)\,dx\right] = E\left[\int_0^\infty a(x)1(x \leq |Z|)\,dx\right]$$

$$= \int_0^\infty a(x)P(|Z| \geq x)\,dx < \infty.$$

We note that $\phi(x)$ is convex by the monotonicity of $\phi'(x) = a(x)$. Finally, we note that $\phi(x)/x \to \infty$ follows from the fact that $a(x) \to \infty$, so the proof of the lemma is complete. □

One of the most useful consequences of Lemma 4.5 is that families of conditional expectations are uniformly integrable. The formal statement of this fact is set out in the next lemma, which will provide a key step in our continuous-time version of Doob's stopping time theorem. The lemma may seem only technical at first, but it captures an important aspect of the geometry of random variables. It has interesting consequences in many parts of probability theory.

LEMMA 4.6. *If Z is an $\mathcal{F}$-measurable random variable with $E(|Z|) < \infty$, then the collection of random variables given by*

$$\{Y : Y = E(Z|\mathcal{G}) \text{ for some } \sigma\text{-field } \mathcal{G} \subset \mathcal{F}\} \tag{4.8}$$

is uniformly integrable.

PROOF. The proof is almost immediate. By Lemma 4.5 we can choose a convex ϕ such that $\phi(x)/x \to \infty$ as $x \to \infty$ and

$$E[\phi(|Z|)] < \infty;$$

so, from Jensen's inequality, we have

$$E[\phi(|E(Z|\mathcal{G})|)] \leq E((E(\phi(|Z|)|\mathcal{G})) \leq E(\phi(|Z|)) < \infty.$$

Now, by Lemma 4.4, the last bound is all we need to establish the required uniform integrability. □

4.4. Martingales in Continuous Time

We now have the tools that we need in order to take a more general view of martingales. If a collection $\{\mathcal{F}_t : 0 \leq t < \infty\}$ of sub-σ-fields of $\mathcal{F}$ has the property that $s \leq t$ implies $\mathcal{F}_s \subset \mathcal{F}_t$, then the collection is called a *filtration*; and, if the random variables $\{X_t : 0 \leq t < \infty\}$ are such that each X_t is $\mathcal{F}_t$ measurable, then we say that X_t is *adapted* to the filtration. Finally, if $\{X_t\}$ is adapted to $\{\mathcal{F}_t\}$, we say $\{X_t\}$ is a *martingale* if we have two additional properties:

(1) $E(|X_t|) < \infty$ for all $0 \leq t < \infty$ and
(2) $E(X_t \mid \mathcal{F}_s) = X_s$ for all $0 \leq s \leq t < \infty$.

For us, the most important continuous-time martingales are those $\{X_t\}$ for which there is an $\Omega_0 \subset \Omega$ with $P(\Omega_0) = 1$ such that for all $\omega \in \Omega_0$ the function on $[0, \infty)$ defined by $t \mapsto X_t(\omega)$ is continuous. Naturally, these will be called *continuous martingales*. We define the *continuous submartingales* by a similar extension of our definition of discrete-time submartingales.

THE STANDARD BROWNIAN FILTRATION

For us, the most important filtrations are those associated with Brownian motion. The natural filtration of Brownian motion is — naturally — the filtration given by $\sigma\{B_s : s \leq t\}$, but the evolution of the theory of stochastic processes has brought out the interesting fact that life flows more smoothly if we work with a slightly different filtration.

This new filtration is called the *augmented filtration*, and it is given by a construction that is widely used. To define this filtration for Brownian motion on $[0, T]$ where $T \leq \infty$, we first consider the collection $\mathcal{C}$ of all of the sets of probability zero in the σ-field $\sigma\{B_s : s \leq T\}$. Next, we consider the collection $\mathcal{N}$ of all A such that $A \subset B$ for some $B \in \mathcal{C}$. The collection $\mathcal{N}$ is called the set of null sets, and we assume that the probability measure P is extended so that $P(A) = 0$ for all $A \in \mathcal{N}$. Finally, we define the augmented filtration for Brownian motion to be the filtration given by $\{\mathcal{F}_t\}$, where for each $t \in [0, T]$ we take $\mathcal{F}_t$ to be the smallest σ-field containing $\sigma\{B_s : s \leq t\}$ and $\mathcal{N}$.

In the case of Brownian motion, the augmented filtration is also called the *standard Brownian filtration*, and one of the consequences of the augmentation construction is that the standard Brownian filtration $\{\mathcal{F}_t\}$ has two basic properties: (1) $\mathcal{F}_0$ contains all of the null sets $\mathcal{N}$, and (2) for all $t \geq 0$ we have

$$\mathcal{F}_t = \bigcap_{s:s>t} \mathcal{F}_s.$$

The second of these conditions is often called right continuity, and we sometimes write $\mathcal{F}_{t+}$ as shorthand for $\mathcal{F}_t = \cap_{s:s>t}\mathcal{F}_s$. In a sweep of intellectual inspiration, conditions (1) and (2) were christened the *usual conditions.* The world still searches for a more suggestive name.

STOPPING TIMES

Martingales and stopping times are like nuts and bolts; they were made to work together. Now that we have martingales in continuous time, we should not waste a second before introducing stopping times. If $\{\mathcal{F}_t\}$ is a filtration, we say that $\tau\colon \Omega \to \mathbb{R} \cup \{\infty\}$ is a *stopping time* with respect to $\{\mathcal{F}_t\}$, provided that

$$\{\omega : \tau(\omega) \leq t\} \subset \mathcal{F}_t \text{ for all } t \geq 0.$$

Also, if X_t is any collection of random variables indexed by $t \in [0, \infty)$, we can define the *stopped variable* X_τ on the set $\{\omega\colon \tau(\omega) < \infty\}$ by taking

$$X_\tau(\omega) = X_t(\omega), \quad \text{provided } \tau(\omega) = t.$$

DOOB'S STOPPING TIME THEOREM

We are now in a position to prove the main result of this chapter and to take our profits on the technological investments that we made in the last section. The ideas and methods of uniform integration provide the key to our proof of the extension of Doob's stopping time theorem from discrete time to continuous time.

THEOREM 4.1 (Doob's Continuous-Time Stopping Theorem).
Suppose $\{M_t\}$ is a continuous martingale with respect to a filtration $\{\mathcal{F}_t\}$ that satisfies the usual conditions. If τ is a stopping time for $\{\mathcal{F}_t\}$, then the process defined by

$$X_t = M_{t\wedge\tau}$$

is also a continuous martingale with respect to $\{\mathcal{F}_t\}$.

PROOF. Our two tasks are to verify the integrability $E(|X_t|) < \infty$ and to prove the martingale identity:

$$E(X_t \mid \mathcal{F}_s) = X_s \text{ for all } s \leq t. \tag{4.9}$$

The natural strategy is to bootstrap ourselves up from the results that we have already obtained for discrete-time martingales.

We begin by introducing a discrete approximation to τ that is always at least as large as τ. This is one of the standard tricks when using discrete time theory to prove continuous-time results. In our particular case, we can also design our approximation τ_n in a way that guarantees a good relationship to s and t. For any $n \geq 1$, we define a random time τ_n with values in the set

$$S(n) = \{s + (t-s)k/2^n\colon -\infty < k < \infty\}$$

by letting $\tau_n(\omega)$ be the smallest element of $S(n)$ that is at least as large as $\tau(\omega)$. With this definition, we can easily check that τ_n is a stopping time and that we have $\tau_n(\omega) \downarrow \tau(\omega)$ for all ω. Moreover, if we let $\{M_u, \mathcal{F}_u\}_{S(n)}$ denote the martingale $\{M_u\}$

restricted to the times in $S(n)$, then $\{M_u, \mathcal{F}_u\}_{S(n)}$ is a discrete-time martingale and $\{|M_u|, \mathcal{F}_u\}_{S(n)}$ is a discrete-time submartingale.

From the second observation and the fact that both τ_n and t are elements of $S(n)$, we find that

$$E[|M_{t\wedge\tau_n}|] \leq E[|M_t|].$$

If we let $n \to \infty$, then the last inequality and Fatou's lemma tell us that we have $E(|M_{t\wedge\tau}|) < \infty$ for all $t \geq 0$. This bound establishes the integrability of the random variables $\{M_{t\wedge\tau}\}$, so all that remains is to prove the martingale identity (4.9).

We first note that s, t, and τ_n are elements of $S(n)$, so the discrete stopping time theorem tells us that

$$E(M_{t\wedge\tau_n} \mid \mathcal{F}_s) = M_{s\wedge\tau_n}. \tag{4.10}$$

To complete the proof of the martingale identity (4.9), we just need to examine the two sides of the identity (4.10) as $n \to \infty$.

By the continuity of the martingale $\{M_t\}$, the righthand side of (4.10) converges for all ω to $M_{s\wedge\tau} = X_s$. In the same way, $M_{t\wedge\tau_n}$ converges for all ω to $M_{t\wedge\tau}$, so the martingale identity (4.9) is established if we can show that we also have L^1 convergence. The proof of (4.9) therefore comes down to establishing uniform integrability of the sequence $M_{t\wedge\tau_n}$.

By the integrability of M_t and Lemma 4.5, we know there is a convex ϕ such that $\phi(x)/x \to \infty$ such that $E[\phi(|M_t|)] < \infty$. The convexity of ϕ and the integrability of $\phi(|M_t|)$ then tell us that $\{\phi(|M_u|), \mathcal{F}_u\}_{S(n)\cap[0,t]}$ is a discrete-time submartingale. The stopping time $t\wedge\tau_n$ takes its values in $S(n)\cap[0,t]$, and t is also an element of $S(n)\cap[0,t]$, so, by the discrete stopping time theorem, we have

$$E[\phi(|M_{t\wedge\tau_n}|)] \leq E[\phi(|M_t|)].$$

Finally, by Lemma 4.4 this bound implies the uniform integrability of the set of variables $\{M_{t\wedge\tau_n}\}$, so we can apply Lemma 4.3 to conclude that $E(M_{t\wedge\tau_n} \mid \mathcal{F}_s)$ converges to $E(M_{t\wedge\tau} \mid \mathcal{F}_s)$ in L^1. The bottom line is that we can take the limits on both sides of equation (4.10) to deduce the martingale identity (4.9) and thus complete the proof of the theorem. (see Exercise 4.2, Part c). □

Doob's Maximal Inequality

The proof of the stopping time theorem required us to build up a substantial collection of new tools, but the continuous-time versions of the maximal inequalities and the convergence theorems will not be so demanding. We simply need to focus our attention on an appropriately chosen subset of times where we can apply our discrete-time results.

Theorem 4.2 (Doob's Maximal Inequalities in Continuous Time). *If $\{M_t\}$ is a continuous nonnegative submartingale and $\lambda > 0$, then for all $p \geq 1$ we have*

$$\lambda^p P\Big(\sup_{\{t:0\leq t\leq T\}} M_t > \lambda\Big) \leq E(M_T^p), \tag{4.11}$$

and, if $M_T \in L^p(dP)$ for some $p > 1$, then we also have

$$\|\sup_{t:0\leq t\leq T} M_t\|_p \leq \frac{p}{p-1}\|M_T\|_p. \tag{4.12}$$

PROOF. If we let $S(n,T) = \{t_i : t_i = iT/2^n, 0 \le i \le 2^n\}$, then the continuity of M_t tells us that for all ω we have

$$\lim_{n\to\infty} \sup_{t\in S(n,T)} M_t = \sup_{0\le t\le T} M_t \tag{4.13}$$

and

$$\lim_{n\to\infty} 1\Big(\sup_{t\in S(n,T)} M_t > \lambda\Big) = 1\Big(\sup_{0\le t\le T} M_t > \lambda\Big). \tag{4.14}$$

Also, by Doob's discrete-time maximal inequality (2.17), we have

$$\lambda^p P\Big(\sup_{t\in S(n,T)} M_t > \lambda\Big) \le E(M_T^p), \tag{4.15}$$

so when we apply Fatou's lemma to (4.15) we obtain (4.11). Finally, to complete the proof of the theorem, we note that inequality (4.12) follows immediately upon application of Fatou's lemma to the discrete-time L^p inequality

$$\|\sup_{t\in S(n,T)} M_t\|_p \le \frac{p}{p-1}\|M_T\|_p.$$ □

MARTINGALE LIMIT THEOREMS IN CONTINUOUS TIME

THEOREM 4.3 (Martingale Convergence Theorems in Continuous Time). *If a continuous martingale $\{M_t\}$ satisfies $E(|M_t|^p) \le B < \infty$ for some $p > 1$ and all $t \ge 0$, then there exists a random variable M_∞ with $E(|M_\infty|^p) \le B$ such that*

$$P\Big(\lim_{t\to\infty} M_t = M_\infty\Big) = 1 \text{ and } \lim_{t\to\infty} \|M_t - M_\infty\|_p = 0. \tag{4.16}$$

Also, if $\{M_t\}$ is a continuous martingale that satisfies $E(|M_t|) \le B < \infty$ for all $t \ge 0$, then there exists a random variable M_∞ with $E(|M_\infty|) \le B$ such that

$$P\Big(\lim_{t\to\infty} M_t = M_\infty\Big) = 1. \tag{4.17}$$

PROOF. We first note that $\{M_n : n = 0, 1, 2, \ldots\}$ is a discrete time martingale, so there is some M_∞ such that M_n converges to M_∞ with probability one and in $L^p(dP)$. Also, for all integers $m \ge 0$ and all real $t \ge m$, we have the trivial bound

$$|M_t - M_\infty| \le |M_m - M_\infty| + \sup_{\{t: m\le t<\infty\}} |M_t - M_m|, \tag{4.18}$$

and we already know that $|M_m - M_\infty| \to 0$ with probability one, so (4.18) gives us

$$\limsup_{t\to\infty} |M_t - M_\infty| \le \lim_{m\to\infty} \sup_{\{t: m\le t<\infty\}} |M_t - M_m|. \tag{4.19}$$

To estimate the last term, we note that $\{M_t - M_n : n \le t < \infty\}$ is a continuous martingale, so our freshly minted maximal inequality (4.11) tells us that for $\lambda > 0$ we have

$$P\Big(\sup_{\{t: m\le t\le n\}} |M_t - M_m| > \lambda\Big) \le \lambda^{-p} E(|M_n - M_m|^p).$$

Now, since M_n converges to M_∞ in $L^p(dP)$, we just let $n \to \infty$ in the last inequality to find

$$P\Big(\sup_{\{t: m \le t < \infty\}} |M_t - M_m| > \lambda \Big) \le \lambda^{-p} E(|M_\infty - M_m|^p). \tag{4.20}$$

If we let $m \to \infty$ in the bound (4.20), then the L^p convergence of M_m implies that the right-hand side converges to zero and the dominated convergence theorem tells us that we can take the limit inside the probability, so we find

$$P\Big(\lim_{m \to \infty} \sup_{\{t: m \le t < \infty\}} |M_t - M_m| > \lambda \Big) = 0. \tag{4.21}$$

Finally, we see from (4.19) and (4.21) that $M_t \to M_\infty$ with probability one, thus completing the proof of the first assertion of (4.16).

The L^p convergence (4.16) is proved by a similar, but simpler, argument. We first note that for all m we have

$$||M_t - M_\infty||_p \le ||M_t - M_m||_p + ||M_m - M_\infty||_p.$$

Since $S_t = |M_t - M_m|$ is a submartingale, we have for all $n > t$ that $||M_t - M_m||_p \le ||M_n - M_m||_p$, so we find that for all m we have

$$\limsup_{t \to \infty} ||M_t - M_\infty||_p \le ||M_m - M_\infty||_p + \sup_{\{n: n \ge m\}} ||M_n - M_m||_p.$$

Because M_m converges to M_∞ in L^p, the last two terms go to zero as $m \to \infty$, establishing the L^p convergence of M_t.

At last, we come to the L^1 part of the theorem. Here we will use a localization argument that parallels one we introduced in the discrete case, although this time it works without a hitch. First, we let $\tau_n = \inf\{t\colon |M_t| \ge n\}$ and note by Doob's stopping time theorem that $M_{t \wedge \tau_n}$ is a martingale for each n. By the definition of τ_n and the continuity of M_t, we have that $M_{t \wedge \tau_n}$ is bounded, so the first part of the theorem tells us that $M_{t \wedge \tau_n}$ converges with probability one as $t \to \infty$. Since $M_t(\omega) = M_{t \wedge \tau_n}(\omega)$ for all t if $\tau_n(\omega) = \infty$, we see that $M_t(\omega)$ converges for all $\omega \in \{\omega\colon \tau_n(\omega) = \infty\}$, except perhaps for a set of probability zero.

Now, by Doob's maximal inequality (4.12) applied to the submartingale $|M_t|$, we have for all T that

$$P\Big(\sup_{0 \le t \le T} |M_t| \ge \lambda \Big) \le E(|M_T|)/\lambda \le B/\lambda;$$

so, by letting $T \to \infty$, we have

$$P\Big(\sup_{0 \le t < \infty} |M_t| \ge \lambda \Big) \le B/\lambda. \tag{4.22}$$

In terms of τ_n, the last inequality tells us that for all $n \ge 1$ we have

$$P(\tau_n = \infty) \ge 1 - B/n,$$

so, when we take unions, we find

$$P\left(\cup_{n=1}^\infty \{\omega\colon \tau_n = \infty\} \right) = 1.$$

Now, since $M_t(\omega)$ converges for all $\omega \in \cup_{n=1}^\infty \{\omega\colon \tau_n = \infty\}$, we finally conclude that M_t converges with probability one. Moreover, if we let M_∞ denote the value of this limit, we see from Fatou's lemma and the bound $E(|M_t|) \le B$ that we also have $E(|M_\infty|) \le B$. □

4.5. Classic Brownian Motion Martingales

THEOREM 4.4. *Each of the following processes is a continuous martingale with respect to the standard Brownian filtration :*

1. B_t,
2. $B_t^2 - t$,
3. $\exp(\alpha B_t - \alpha^2 t/2)$.

The theorem barely stands in need of proof, although it yields a basket of consequences. The fact that the processes are adapted, continuous, and integrable is immediate; and for the first two processes we can verify the martingale identity just as we did for simple random walk. To check the martingale identity for the last process, we set $X_t = \exp(\alpha B_t - \alpha^2 t/2)$, and note for $t > s$ that

$$E(X_t \mid \mathcal{F}_s) = X_s E\Big(\exp\big(\alpha(B_t - B_s) - \alpha^2(t-s)/2\big) \mid \mathcal{F}_s\Big) = X_s,$$

since the conditional expectation reduces to an ordinary expectation by independence. To begin the harvest of consequences, we first note that we get *exact copies* of our wonderful formulas for the ruin probabilities and expected hitting times for simple random walk.

RUIN PROBABILITIES FOR BROWNIAN MOTION

THEOREM 4.5. *If* $A, B > 0$ *and* $\tau = \min\{t\colon B_t = -B \text{ or } B_t = A\}$, *then we have* $P(\tau < \infty) = 1$ *and*

$$P(B_\tau = A) = \frac{B}{A+B} \text{ and } E(\tau) = AB. \tag{4.23}$$

PROOF. The finiteness of τ is proved even more easily for Brownian motion than for simple random walk. Since $P(|B_{n+1} - B_n| > A + B) = \epsilon > 0$ and since the events $E_n = \{|B_{n+1} - B_n| > A + B\}$ are independent, we have

$$P(\tau > n+1) \le (1-\epsilon)^n,$$

so, as before, we find that τ has finite moments of all orders.

To prove the formula for the ruin probability, we first note that

$$E(B_\tau) = A \cdot P(B_\tau = A) - B \cdot (1 - P(B_\tau = A)).$$

Since $B_{t\wedge\tau}$ is a martingale, we also have $E(B_{t\wedge\tau}) = 0$ for all t. Moreover, since we have the bound $|B_{t\wedge\tau}| \le A + B$, the dominated convergence theorem tells us that

$$E(B_\tau) = \lim_{t\to\infty} E(B_{t\wedge\tau}) = 0.$$

The last two equations can now be solved for $P(B_\tau = A)$ to obtain the first of our two formulas (4.23).

Just as for simple random walk, the proof of the expected hitting-time formula follows immediately from the formula for the ruin probabilities and the identity

$$E(B_\tau^2) = E(\tau). \tag{4.24}$$

To justify the identity (4.24), we use the martingale $M_t = B_t^2 - t$ and the observation that for all $t \ge 0$ we have $|M_{t\wedge\tau}| \le A^2 + B^2 + \tau$. The last bound has finite expectation, so one gets (4.24) from the dominated convergence theorem and the fact that for all $t \ge 0$ we have $E(M_{t\wedge\tau}) = 0$. □

Hitting Time of a Level

In addition to results on the hitting time of a two-sided boundary (4.23), we can also obtain useful information about the hitting time of a one-sided boundary:

$$\tau_a = \inf\{t\colon B_t = a\}.$$

In fact, we will find both the Laplace transform and the density of τ_a.

THEOREM 4.6. *For any real value a, we have $P(\tau_a < \infty) = 1$ and*

$$E[\exp(-\lambda\tau_a)] = \exp(-|a|\sqrt{2\lambda}\,). \tag{4.25}$$

PROOF. By symmetry, it suffices to consider $a \geq 0$. Now, in order to prove that $P(\tau_a < \infty) = 1$, we just note that the first formula of (4.23) tells us that for all $b > 0$ we have

$$P(\tau_a < \infty) \geq P(B_{\tau_a \wedge \tau_{-b}} = a) = \frac{b}{a+b},$$

so the arbitrariness of b guarantees the desired result.

To prove the identity (4.25), we use the continuous-time martingale $M_t = \exp(\alpha B_t - \frac{1}{2}\alpha^2 t)$, and we note that the nonnegative martingale $M_{t\wedge\tau_a}$ is bounded above by $\exp(\alpha a)$. We then find by the continuity of M_t, the stopping time theorem, and the dominated convergence theorem that we have

$$1 = \lim_{t\to\infty} E(M_{\tau_a \wedge t}) = E(M_{\tau_a}).$$

Finally, since $B_{\tau_a} = a$, we have $M_{\tau_a} = e^{\alpha a}\exp(-\frac{1}{2}\alpha^2\tau_a)$, so, if we use the fact that $E[M_{\tau_a}] = 1$, then we see that we can take $\alpha = \sqrt{2\lambda}$ to complete the proof of formula (4.25). □

First Consequences

From the expected hitting-time formula for a two-sided boundary, we can immediately deduce that $E(\tau_a) = \infty$; we just note that for any $b > 0$ we have $\tau_a \geq \tau_a \wedge \tau_{-b}$ so $E(\tau_a) \geq ab$ by (4.23). Simple though this may be, the fact still may strain our intuition, and it is useful to note that the Laplace transform can also be used to show that $E(\tau_a) = \infty$. This time, we just note that

$$E(\tau_a) = -\lim_{\lambda\to 0}\frac{d}{d\lambda}E[\exp(-\lambda\tau_a)] = \lim_{\lambda\to 0}\frac{a}{\sqrt{2\lambda}}\exp(-a\sqrt{2\lambda}) = \infty.$$

In fact, this derivation actually offers us a path to more detailed information. For example, one can use the Laplace transform (4.25) to prove the amusing formula

$$E(\tau_a^{-1}) = a^{-2}. \tag{4.26}$$

This result is a strange reciprocal analog to the familiar fact that the expected time until standard Brownian motion leaves $[-a, a]$ is a^2.

The derivation of the identity (4.26) turns out to be particularly simple if we begin with the easy calculus identity

$$t^{-1} = \int_0^\infty e^{-\lambda t}\,d\lambda.$$

From here, we just replace t by τ_a, take expectations, and use our formula for $E(e^{-\lambda\tau_a})$ to calculate that

$$E(\tau_a^{-1}) = \int_0^\infty E(e^{-\lambda\tau_a})\,d\lambda = \int_0^\infty \exp(-a\sqrt{2\lambda})\,d\lambda = \int_0^\infty e^{-u}u\,\frac{du}{a^2} = \frac{1}{a^2}.$$

This derivation may look a bit magical, but it is more or less typical of the special manipulations that can be used to extract information from a Laplace transform.

Nevertheless, we could have proceeded in a much more systematic way. In fact, the explicit inversion of the Laplace transform (4.25) is not difficult if one has access to a good table of Laplace transforms. Rather remarkably, the inversion gives us an explicit closed form for the density of τ_a:

$$f_{\tau_a}(t) = \frac{|a|}{\sqrt{2\pi t^3}}\exp(-a^2/2t) \text{ for } t \geq 0. \tag{4.27}$$

In a later chapter, we will obtain this density by a more direct method, so for the moment we will be content to note one amusing curiosity — the density of τ_a is the same as that of Z^{-2}, where Z is a Gaussian random variable with mean zero and variance a^{-2}.

Looking Back

The formula for the Laplace transform of τ_a may not seem like one that we could have guessed, but it turns out that we can come very close just by appealing to the scaling properties of Brownian motion. The main observation is that if we let $f(a,\lambda) = E[\exp(-\lambda\tau_a)]$, then we can show that f must satisfy a simple functional equation:

$$f(a+b,\lambda) = f(a,\lambda)\cdot f(b,\lambda) \text{ for all } a \geq 0 \text{ and } b \geq 0. \tag{4.28}$$

It is easy to see why equation (4.28) holds. We simply need to note that τ_{a+b} is equal in distribution to $\tau_a + \tau_b'$, where τ_b' is a random variable that is independent of τ_a and that has the same distribution as τ_b.

The functional equation (4.28) is well known to analysts, and it is not difficult to show that the continuity of the function $f(\cdot,\lambda)$ and the validity of functional equation (4.28) are enough for one to prove that $f(a,\lambda) = \exp(ag(\lambda))$, where g is some unspecified function of λ. Finally, in order to get rid of $g(\lambda)$, we first note that by the space-time scaling property of Brownian motion, we have that τ_{ab} has the same distribution as $b^2\,\tau_a$. In terms of the Laplace transform, this tells us that $f(ab,\lambda) = f(a,\lambda b^2)$, or, in terms that make g explicit:

$$\exp(abg(\lambda)) = \exp(ag(\lambda b^2)).$$

Taking the logarithm, we find $bg(\lambda) = g(\lambda b^2)$, so setting $b = 1/\sqrt{\lambda}$ and $g(1) = c$ we find

$$f(a,\lambda) = \exp(ca\sqrt{\lambda}).$$

Thus, except for showing that the constant c equals $-\sqrt{2}$, we come to the pleasing conclusion that the Laplace transform (4.25) can be obtained by symmetry arguments alone.

4.6. Exercises

The first exercise is intended to help build comfort and skill with the formal notion of conditional expectation. The next three exercises then offer practice with the relationships between various modes of convergence. These are not difficult exercises, but their value should not be underestimated. Much of our future work will count on the reader being able to move easily between all of the different convergence concepts.

The fifth exercise gives an elegant criterion for the finiteness of a stopping time, and Exercise 4.6 shows how one can sometimes calculate the higher moments of stopping times. Finally, Exercise 4.7 provides another important example of the fact that martingale theory can be used to achieve purely analytical results, and it also anticipates an approximation that will be used in our development of the Itô integral.

EXERCISE 4.1 (Properties of Conditional Expectation).

(a) Tower Property. Use the definition of conditional expectation to prove the tower property: If $\mathcal{H}$ is a subfield of $\mathcal{G}$, then

$$E(E(X \mid \mathcal{G}) \mid \mathcal{H}) = E(X \mid \mathcal{H}). \tag{4.29}$$

(b) Factorization Property. If $Y \in \mathcal{G} \subset \mathcal{F}$, show that if $E(|X|) < \infty$ and Y is bounded then

$$E(XY|\mathcal{G}) = YE(X|\mathcal{G}).$$

A good part of the challenge of problems like these is the introduction of notation that makes clear that there has been no "begging of the question." Without due care, one can easily make accidental use of the tower property to prove the tower property!

EXERCISE 4.2 (Modes of Convergence). The central result of stochastic calculus is surely Itô's formula, and its proof will use several different modes of convergence. This exercise provides a useful warmup for many future arguments.

(a) Show that if $E(|X_n - X|^\alpha) \to 0$ for some $\alpha > 0$ then X_n converges to X in probability.

(b) Use the Borel–Cantelli lemma to show that if $X_n \to X$ in probability, then there is a sequence of integers $n_1 < n_2 < \cdots$ such that $X_{n_k} \to X$ with probability one as $k \to \infty$.

(c) If $X_n \to X$ in probability and $X_n \to Y$ with probability one, show that $P(X = Y) = 1$. By the way, one should note that this fact was used in the last line of the proof of Theorem 4.1.

(d) Give an example for which $E(|X_n - X|) \to 0$, yet X_n does not converge to X with probability one.

(e) Suppose that $E(|X|) < \infty$ and $Y_n = E(X|\mathcal{F}_n)$ for a sequence of σ-fields $\mathcal{F}_n \subset \mathcal{F}$. Show that if $Y_n \to Y$ with probability one, then $Y_n \to Y$ in L^1.

EXERCISE 4.3 (L^1-Bounded Martingales Need Not Be Uniformly Integrable). Consider $X_t = \exp(B_t - t/2)$ and show that X_t is a continuous martingale with $E(|X_t|) = 1$ for all $t \geq 0$. Next, show that X_t converges with probability one to $X = 0$. Explain why this implies that X_t does not converge in L^1 to X and explain why X_t is not uniformly integrable, despite being L^1-bounded.

EXERCISE 4.4 (Time Inversion of Brownian Motion). We claimed in an earlier chapter that if we define a process $\{Y_t\}$ by the time inversion of Brownian motion,

$$Y_t = \begin{cases} 0 & \text{if } t = 0 \\ tB_{1/t} & \text{if } t > 0, \end{cases}$$

then $\{Y_t\}$ is again a standard Brownian motion. One can easily check that $\{Y_t\}$ is a Gaussian process with covariance function $\min(s,t)$, so the only sticky point to showing that $\{Y_t\}$ is Brownian motion is to prove that it is continuous at zero. Give a verification of this fact by completing the following program:

(a) Show that for all $\epsilon > 0$ the process defined by $\{X_t = Y_t - Y_\epsilon : t \geq \epsilon\}$ is a martingale.

(b) Check that for all $0 < s \leq t$ we have $E[(Y_t - Y_s)^2] = t - s$.

(c) Use Doob's maximal inequality (in continuous time) to give a careful proof of the fact that

$$P\left(\lim_{t\to 0} Y_t = 0\right) = 1.$$

EXERCISE 4.5 (Williams' "Sooner Rather Than Later" Lemma).

Suppose that τ is a stopping time for the filtration $\{\mathcal{F}_n\}$ and that there is a constant N such that for all $n \geq 0$ we have

$$P(\tau \leq n + N \mid \mathcal{F}_n) \geq \epsilon > 0. \tag{4.30}$$

Informally, equation (4.30) tells us that — no matter what has happened so far — there is at least an ϵ chance that we will stop sometime in the next N steps.

Use induction and the trivial relation

$$P(\tau > kN) = P(\tau > kN \text{ and } \tau > (k-1)N)$$

to show that $P(\tau > kN) \leq (1-\epsilon)^k$. Conclude that we have $P(\tau < \infty) = 1$ and that $E[\tau^p] < \infty$ for all $p \geq 1$.

Remark: Here one should note that $P(A|\mathcal{F}_n)$ is interpreted as $E(1_A|\mathcal{F}_n)$, and any honest calculation with $P(A|\mathcal{F}_n)$ must rest on the properties of conditional expectations.

EXERCISE 4.6. Use the martingale

$$X_t = \exp(\alpha B_t - \alpha^2 t/2)$$

to calculate $\phi(\lambda) = E[\exp(-\lambda\tau)]$, where $\tau = \inf\{t\colon B_t = A \text{ or } B_t = -A\}$. Use this result to calculate $E[\tau^2]$. What difficulty do we face if we try to calculate $E[\tau^2]$ when the the boundary is not symmetric?

EXERCISE 4.7 (Approximation by Step Functions). Let $\Omega = [0,1]$, and let $\mathcal{F}$ be the Borel field of $[0,1]$ (which by definition is the smallest σ-field that contains all of the open sets). Also, let P denote the probability measure on $\mathcal{F}$ given by Lebesgue measure, and let $\mathcal{F}_n$ denote the smallest σ-field containing the dyadic intervals $J_i = [i2^{-n}, (i+1)2^{-n}]$ for all $1 \le i < 2^n$. Suppose also that $f\colon [0,1] \mapsto \mathbb{R}$ is a Borel measurable function such that $E(|f|^2) < \infty$.

(a) Write $f_n = E(f \mid \mathcal{F}_n)$ as explicitly as you can. Specifically, argue that it is constant on each of the dyadic interval J_i and find the value of the constant.

(b) Show that $\{f_n\}$ is an L^2-bounded martingale, and conclude that f_n converges to some g in L^2.

(c) Assume that f is continuous, and show that $g(x) = f(x)$ for all x with probability one.

EXERCISE 4.8. Show that the function $a(x)$ used in the proof of Lemma 4.5 can be taken to be

$$a(x) = \left\{ \int_x^\infty P(|Z| \ge u)\, du \right\}^{-\alpha} \quad \text{for any } 0 < \alpha < 1.$$

Hint: If we introduce a sequence $\{x_k\}$ by setting $x_0 = 0$ and by setting

$$x_k = \min\{x : \int_x^\infty P(|Z| \ge u)\, du \le 2^{-k}\} \quad \text{for } k \ge 1,$$

then we can easily estimate the integral

$$\int_0^\infty a(x) P(|Z| \ge x)\, dx = \sum_{k=0}^\infty \int_{x_k}^{x_{k+1}} a(x) P(|Z| \ge x)\, dx.$$

CHAPTER 5

Richness of Paths

One could spend a lifetime exploring the delicate — and fascinating — properties of the paths of Brownian motion. Most of us cannot afford such an investment, so hard choices must be made. Still, without any doubt, there are two fundamental questions that must be considered to a reasonable depth:

- How smooth — or how rough — is a Brownian path?
- How do the paths of Brownian motion relate to those of random walk?

We will find that the wavelet construction of Brownian motion provides a well-honed tool for responding to the *smoothness* part of the first question. The answer to the *roughness* part requires an independent development, but the results are worth the effort. We will find that with probability one, the paths of Brownian motion are not differentiable at any point. This is the geometric reality that forces the stochastic calculus to diverge from the elementary calculus of smooth functions.

Our exploration of the second question has two parts. First, we examine some of the ways that random walk can inform us about Brownian motion. One of the most persistent links is simply analogy, and we explore this link by giving a parallel development for the reflection principle for random walk and Brownian motion. We then give a brief introduction to the more formal bridges between random walk and Brownian motion, especially the Invariance Principle and Donsker's Theorem.

The last part of the chapter provides the most decisive connection. We will see that every unbiased random walk with steps with finite variance can be embedded into Brownian motion in a way that is so explicit that many questions for random walk become corollaries of the corresponding results for Brownian motion. This embedding theory offers one of the most compelling instances in mathematics of the effectiveness with which the continuum can provide insight into the discrete.

5.1. Quantitative Smoothness

One of the most important features of the wavelet sequence $\{\Delta_n\}$ is that it forms a basis for the set of continuous functions on $[0,1]$. Specifically, for any continuous function $f : [0,1] \to \mathbb{R}$ there is a *unique* sequence of constants $\{c_n\}$ such that

$$f(t) = f(0) + \sum_{n=0}^{\infty} c_n \Delta_n(t) \text{ with uniform convergence.} \tag{5.1}$$

Moreover, the coefficients in the expansion (5.1) are determined by a delightfully simple linear operation on f.

To compute the coefficients c_n, we first note that $c_0 = f(1) - f(0)$, and, if we set $g(t) = f(t) - c_0\Delta_0(t) - f(0)$, then we find $c_1 = g(\frac{1}{2})$. For $n \geq 1$ we can continue to determine the coefficients by successive evaluations at dyadic rationals of the residual function at each successive step of the approximation. The net effect can be summarized in a simple formula for the coefficients c_n with $n \geq 1$:

$$c_n = f\left(\frac{k+\frac{1}{2}}{2^j}\right) - \frac{1}{2}\left[f\left(\frac{k}{2^j}\right) + f\left(\frac{k+1}{2^j}\right)\right], \tag{5.2}$$

where as before $n \geq 1$ is written uniquely as $n = 2^j + k$ for $j \geq 0$ and $0 \leq k < 2^j$.

The local nature of the wavelet coefficients (5.2) stands in considerable contrast to the coefficients of a classical Fourier expansion. Each classical Fourier coefficient is given by an integral that depends on the entire domain of f, but as n becomes larger the coefficient defined by equation (5.2) depends on an increasingly small part of the domain. This locality is the main reason wavelet coefficients can provide an efficient measure of the smoothness.

Hölder Continuity

To make our measure of smoothness precise, we first recall that a function $f : [a, b] \to \mathbb{R}$ is said to be *Hölder continuous of order* $0 < \alpha < 1$ if there is a constant c such that

$$|f(s) - f(t)| \leq c|s - t|^\alpha \text{ for all } a < s < t < b.$$

The set of all such functions is denoted by $C^\alpha[a, b]$, and the basic connection between the smoothness of f and the smallness of its wavelet coefficients is given by the following lemma.

LEMMA 5.1. *If the coefficients $\{c_n\}$ of the wavelet expansion (5.1) satisfy the bound $|c_n| \leq 2^{-\alpha j}$ for all n with $n = 2^j + k$, $0 \leq k < 2^j$, and $0 \leq n < \infty$, then $f \in C^\alpha[0, 1]$.*

PROOF. There is no loss of generality if we assume that $c_0 = 0$, and in that case we can write

$$f(s) - f(t) = \sum_{j=0}^{\infty} D_j(s, t),$$

where

$$D_j(s, t) = \sum_{2^j \leq n < 2^{j+1}} c_n\{\Delta_n(s) - \Delta_n(t)\}.$$

The key observation is that we have two bounds on $D_j(s, t)$:

$$|D_j(s, t)| \leq \begin{cases} 2^{-\alpha j} \\ 2^{-\alpha j} \cdot 2^{j+1}|s - t|. \end{cases} \tag{5.3}$$

Both of the bounds in (5.3) lean on the fact that we have $\Delta_n(u) = 0$ for all but at most one value of $n \in [2^j, 2^{j+1})$. The first inequality then calls on the fact that $0 \leq \Delta_n(u) \leq 1$, and the second uses the observation that Δ_n is a piecewise linear function with maximum slope 2^{j+1}, so we have

$$|\Delta_n(s) - \Delta_n(t)| \leq 2^{j+1}|s - t|.$$

The second bound in (5.3) is sharper when s is near t, so to take advantage of this distinction we first make a decomposition that is valid for all k:

$$|f(s) - f(t)| \leq \sum_{0 \leq j \leq k} 2^{-\alpha j} \cdot 2^{j+1}|s-t| + \sum_{j:j>k} 2^{-\alpha j}$$

$$\leq 2|s-t|\frac{2^{(1-\alpha)(k+1)}-1}{2^{1-\alpha}-1} + \frac{2^{-\alpha(k+1)}}{1-2^{-\alpha}}.$$

If we now choose k so that $2^{-k-1} \leq |s-t| \leq 2^{-k}$, we find there is a $c = c(\alpha)$ such that $|f(s) - f(t)| \leq c|s-t|^{\alpha}$, exactly as required to complete the proof of the lemma. □

Back to Brownian Motion

For us, the most important consequence of Lemma 5.1 is that it provides a precise tool by which to gauge the smoothness of the paths of Brownian motion. Here, we first recall that the coefficients in our wavelet series for Brownian motion are given by $\lambda_n Z_n$, where the $\{Z_n\}$ are standard independent Gaussians, and the real sequence $\{\lambda_n\}$ is given by $\lambda_0 = 1$ and $\lambda_n = \frac{1}{2} \cdot 2^{-j/2}$, where n and j are related by $n = 2^j + k$ with $0 \leq k < 2^j$. We further recall that in the course of confirming the continuity of Brownian motion, we also proved that there is a random variable C such that

$$|Z_n| \leq C\sqrt{\log n} \quad \text{for all } n \geq 2$$

and

$$P(C < \infty) = 1.$$

If we apply these bounds here, we find that the coefficient c_n in the wavelet expansion of Brownian motion are bounded by

$$C(\omega)\sqrt{(\log n)}\, 2^{-j/2} \leq C(\omega)\sqrt{(j+1)}\, 2^{-j/2},$$

and for any $\alpha < \frac{1}{2}$ the last term is bounded by $2^{-\alpha j}$ for all sufficiently large n. Finally we can apply Lemma 5.1 to obtain the basic result on the Hölder continuity of Brownian paths.

THEOREM 5.1. *For any $0 \leq \alpha < \frac{1}{2}$, the paths of Brownian motion are in $C^{\alpha}[0, \infty)$ with probability one.*

5.2. Not Too Smooth

The estimates used in the proof of Theorem 5.1 break down when $\alpha = \frac{1}{2}$, but this is no accident — for $\alpha > \frac{1}{2}$ we find a dramatic shift. In fact, we will shortly find that with probability one there is no point in $[0, 1]$ at which a Brownian path is Hölder continuous of order $\alpha > \frac{1}{2}$. This roughness is more than a curiosity. The intrinsic randomness of Brownian motion, its fractal nature, and the required subtlety in the definition of the stochastic integral are all inextricably tied to the fact that the Hölder level $\alpha = \frac{1}{2}$ is a watershed for the smoothness of Brownian motion paths.

THEOREM 5.2. *Let $G(\alpha, c, \epsilon)$ denote the set of all $\omega \in \Omega$ such that for some $s \in [0,1]$ we have*

$$|B_s(\omega) - B_t(\omega)| \le c|s-t|^\alpha \text{ for all } t \text{ with } |s-t| \le \epsilon.$$

If $\alpha > \frac{1}{2}$, then the event $G(\alpha, c, \epsilon)$ has probability zero for all $0 < c < \infty$ and $\epsilon > 0$.

Theorem 5.2 offers a natural balance to Theorem 5.1, but even just taking $\alpha = 1$ provides a result of considerable historical weight — and practical importance. If we let D denote the set of all ω such that there exists some $s \in [0,1]$ for which $f(t) = B_t(\omega)$ is differentiable at s, then we have

$$D \subset \cup_{j=1}^{\infty} \cup_{k=1}^{\infty} C(1, j, 1/k).$$

Since Theorem 5.2 tells us that each of the sets in the union has probability zero, we find that $P(D) = 0$. We record this important result as a corollary.

COROLLARY 5.1. *For all ω except a set of probability zero, the Brownian motion path $B_t(\omega)$ is not differentiable for any $0 \le t \le 1$.*

BACK TO THE PROOF

The proof of Theorem 5.2 is not long, but it has some subtle points. Nevertheless, the main idea is clear; we need to exploit independence by expressing $G(\alpha, c, \epsilon)$ in terms of events that depend explicitly on the increments of the Brownian motion paths. To make this operational, we first break $[0,1]$ into n intervals of the form $[k/n, (k+1)/n]$ with $0 \le k < n$, and we consider the set of all ω for which there is some value s that satisfies the condition:

$$|B_s(\omega) - B_t(\omega)| \le c|s-t|^\alpha \text{ for all } t \text{ with } |s-t| \le \epsilon.$$

The key observation is that $B_t(\omega)$ cannot change very much on any of the intervals that are near s.

To exploit this observation, we choose an integer m that we will regard as a fixed parameter, though we will later find a condition that we will impose on m. In contrast, n will denote an integer that we will eventually let go to infinity. Now, for any $n \ge m$, we define random variables $X_{(n,k)}$ by

$$X_{(n,k)} = \max\left\{|B_{j/n}(\omega) - B_{(j+1)/n}(\omega)| : k \le j < k+m\right\},$$

where $0 \le k \le n-m$. The random variable $X_{(n,k)}$ measures the largest increment of B_t over any individual interval in a *block* $\mathcal{B}$ of m intervals that follow k/n.

Now, if n is sufficiently large that $m/n \le \epsilon$ and if s is contained in one of the intervals of $\mathcal{B}$, then for any interval $[j/n, (j+1)/n] \in \mathcal{B}$, we see that j/n and $(j+1)/n$ are both within $m/n \le \epsilon$ of s, so the triangle inequality and the condition on s give us

$$|B_{j/n} - B_{(j+1)/n}| \le |B_{j/n} - B_s| + |B_s - B_{(j+1)/n}| \le 2c(m/n)^\alpha.$$

The last inequality tells us that if $\omega \in G(\alpha, c, \epsilon)$ then there exists a $0 \le k \le n$ such that $X_{(n,k)} \le 2c(m/n)^\alpha$, or, in other words, we have

$$G(\alpha, c, \epsilon) \subset \left\{\omega : \min_{0 \le k \le n-m} X_{(n,k)} \le 2c(m/n)^\alpha\right\}. \tag{5.4}$$

All of the geometry of our problem is now wrapped up in the inclusion (5.4), and we are left with an easy calculation, although we will still need to take advantage of an opportunity that one might easily miss.

First, we note that all of the $X_{(n,k)}$ have the same distribution, so we see by Boole's inequality and the definition of $X_{(n,k)}$ that

$$P\Big(\min_{0\le k\le n-m} X_{(n,k)} \le 2c(m/n)^\alpha\Big) \le nP\left(X_{(n,1)} \le 2c(m/n)^\alpha\right)$$

$$\le nP\left(|B_{1/n}| \le 2c(m/n)^\alpha\right)^m . \tag{5.5}$$

We then note that $B_{1/n}$ has the same distribution as $n^{-\frac{1}{2}}B_1$, and, since the density of B_1 is bounded by $1/\sqrt{2\pi}$, we then have

$$P(|B_1| \le x) \le \frac{2x}{\sqrt{2\pi}}$$

and

$$nP\big(|B_1| \le 2cm^\alpha n^{\frac{1}{2}-\alpha}\big)^m \le n\big(4cm^\alpha n^{\frac{1}{2}-\alpha}/\sqrt{2\pi}\big)^m \le \left(\frac{4cm^\alpha}{\sqrt{2\pi}}\right)^m n^{1+m(\frac{1}{2}-\alpha)}.$$

Now, we must not forget that we still have some flexibility in our choice of m. In particular, from the very beginning we could have chosen our fixed parameter m so that $m(\alpha-\frac{1}{2}) > 1$. With such a choice, we find that the right-hand side of the last inequality goes to zero as $n \to \infty$, and we see at last that indeed $P(G(\alpha, c, \epsilon)) = 0$.

The Nature of the Argument

This proof will reward careful study by anyone who is interested in the structure of probabilistic arguments. Several rich ideas are embedded in the proof. Still, the pivotal observation was that a single condition could force the occurrence of a large number of independent events. In this case, the existence of a "Hölder s" forced m of the changes $B_{j/n} - B_{(j+1)/n}$ to be small. Such forcing arguments provide a powerful lever whenever they can be found.

The Missing Case, or What About $\alpha = \frac{1}{2}$?

Borderline cases exercise a powerful influence on human curiosity, and many readers will have asked themselves about $\alpha = \frac{1}{2}$. Rest assured, as the following theorem suggests, the behavior at the borderline is understood with great precision.

THEOREM 5.3 (Lévy's Modulus of Continuity). *For the modulus of continuity of Brownian motion on* $[0,1]$,

$$m(\epsilon) = \sup\left\{|B_t - B_s| : 0 < s < t \le 1,\ |s-t| \le \epsilon\right\},$$

we have

$$\limsup_{\epsilon\to 0} \frac{m(\epsilon)}{\sqrt{2\epsilon\log(1/\epsilon)}} = 1 \textit{ with probability one.}$$

This determination of the precise limit supremum of the modulus of continuity tells us more about the smoothness of the Brownian path than we are likely to need in almost any application, so we will not give a proof of Theorem 5.3. Nevertheless, we should note that the tools we have at our disposal would rather quickly show that there is an upper bound on the limit supremum. This fact is even offered as one of the chapter's exercises.

The harder part of Lévy's theorem is to show that 1 is also a *lower bound* on the limit supremum. This fact requires a more refined argument that would definitely test our patience. In any case, Lévy's theorem is not the last word. There are many further refinements, and, even though some of these are of great beauty, they are best left for another time — and another place.

5.3. Two Reflection Principles

Near the end of his long and eventful life, Winston Churchill was asked if he had any regrets. Churchill puffed and pondered for a moment before replying, "Yes, when I look back on the times in Cannes and Monte Carlo when I bet on Red, I would have rather to have bet on Black." In this instance, Churchill may have been more clever than forthcoming, but his quip still delivers a drop of probabilistic wisdom.

Suppose we consider a variation on Churchill's expressed desires and imagine a gambler in a fair game who switches his bet preference only after his winnings have grown to a specified level. In notation, we consider $S_n = X_1 + X_2 + \cdots + X_n$, where the X_i are independent random variables that take the values 1 and -1 with equal probability. We then consider the first time that S_n reaches the value $x > 0$, say $\tau = \min\{n : S_n = x\}$. If S_n represents our gambler's wealth when he bets his natural preferences, then the wealth $\tilde{S}_n$ that he would achieve by switching preferences after reaching level x is given by

$$\tilde{S}_n = \begin{cases} S_n & \text{if } n < \tau \\ S_\tau - (S_n - S_\tau) & \text{if } n \geq \tau. \end{cases} \tag{5.6}$$

The ironic fact — plainly evident to Churchill — is that the wealth process $\tilde{S}_n$ is equivalent to the wealth process S_n; that is, all of the joint distributions of the processes $\{\tilde{S}_n : n \geq 0\}$ and $\{S_n : n \geq 0\}$ are equal. Still, if our aim is other than to make money, there is value in introducing the new process $\tilde{S}_n$.

The first observation is that if $n \geq \tau$ and $S_n > x + y$ for some $y \geq 0$, then we also have $\tilde{S}_n < x - y$. But, since the processes $\{S_n\}$ and $\{\tilde{S}_n\}$ are equivalent, this observation tells us that

$$\begin{aligned} P(\tau \leq n, S_n > x + y) &= P(\tau \leq n, \tilde{S}_n < x - y) \\ &= P(\tau \leq n, S_n < x - y). \end{aligned} \tag{5.7}$$

In our usual notation for the maximal process, $S_n^* = \max\{S_k : 0 \leq k \leq n\}$, we can summarize the last identity by saying that for all $x \geq 0$ and $y \geq 0$ we have

$$P(S_n^* \geq x, S_n > x + y) = P(S_n^* \geq x, S_n < x - y),$$

and since $S_n > x + y$ implies $S_n^* \geq x$, we finally deduce

$$P(S_n > x + y) = P(S_n^* \geq x, S_n < x - y). \tag{5.8}$$

This useful formula is often called the *reflection principle for simple random walk,* although the name might more aptly be used to refer to the assertion that $\{S_n\}$ and $\{\tilde{S}_n\}$ are equivalent processes. The reason for the importance of equation (5.8) is that it gives the joint distribution of (S_n^*, S_n) in terms of our old friend, the distribution of S_n.

USEFUL SPECIALIZATION

One of the ways to get to the essence of equation (5.8) is to consider what happens if we set $y = 0$ to get the simpler relation

$$P(S_n > x) = P(S_n^* \geq x, S_n < x).$$

Because it is trivially true that

$$P(S_n \geq x) = P(S_n^* \geq x, S_n \geq x),$$

we can sum the last two equations to obtain an identity that gives the distribution of S^* in terms of S_n:

$$2P(S_n > x) + P(S_n = x) = P(S_n^* \geq x). \tag{5.9}$$

This simple equation is remarkably informative. To be sure, it yields useful computations for finite values of n. For example, it tells us that the probability that our standard gambler gets ahead by at least \$3 at some time in the first 8 rounds of play is 37/128, or about 0.289.

More surprisingly, the identity can also be used to resolve theoretical questions. For example, if we let $n \to \infty$ in equation (5.9), the central limit theorem tells us that the left-hand side tends to one; so, for the third time, we have proved that

$$P(\tau_x < \infty) = P(S_n^* = x \text{ for some } n) = 1.$$

Similarly, if we replace x by $[\sqrt{n}x]$ in equation (5.9) before we take the limits, we see that the central limit theorem for S_n tells us that for all $x \geq 0$ we have

$$\lim_{n\to\infty} P(S_n^*/\sqrt{n} \geq x) = \lim_{n\to\infty} 2P(S_n/\sqrt{n} \geq x) = 2\{1 - \Phi(x)\}. \tag{5.10}$$

This determination of the asymptotic distribution of the maximum of simple random walk is a pleasing accomplishment, but as soon as it hits the page we are driven to a new question: Does the identity (5.10) hold for all random walks with steps that have mean zero and variance one? We will shortly find that this is indeed the case, but we will first need to bring in Brownian motion as a helpful intermediary.

REFLECTION PRINCIPLES FOR BROWNIAN MOTION

The reflection process $\{\tilde{S}_n\}$ has an obvious analog for Brownian motion, and there is also a natural analog to the equivalence of $\{\tilde{S}_n\}$ and $\{S_n\}$.

PROPOSITION 5.1 (Reflection Principle for Brownian Paths). *If τ is a stopping time with respect to the filtration given by the standard Brownian motion $\{B_t : t \geq 0\}$, then the process $\{\tilde{B}_t : t \geq 0\}$ defined by*

$$\tilde{B}_t = \begin{cases} B_t & \text{if } t < \tau \\ B_\tau - (B_t - B_\tau) & \text{if } t \geq \tau \end{cases} \tag{5.11}$$

is again a standard Brownian motion.

The proof of the reflection principle for random walk can be given by simple discrete mathematics; one just needs to count paths. Unfortunately, there is no analog to path counting for Brownian motion, and, for all of its obviousness, Proposition 5.1 is a bit tricky to prove. If the reflection were to take place at a fixed time t rather than the random time τ, the proof would be no trouble, but there is no escaping the randomness of τ. An honest proof of Proposition 5.1 would require us to develop the so-called *strong Markov property* of Brownian motion, and such a development would be a substantial distraction.

For the moment, we will take Proposition 5.1 as if it were proven. In fact, it yields a treasure trove of useful formulas that we would be foolish to miss. Also, as it turns out, once such formulas are set down, they can almost always be proved by methods that do not call on Proposition 5.1, so there is no real cost to the omission of the proof of the proposition.

JOINT DISTRIBUTION OF B_t AND B_t^*

When we use the reflection principle for Brownian motion, the argument that gave us the refection identity (5.8) for random walk can be repeated almost word-for-word in order to find

$$P(B_t^* \geq x, B_t \leq x - y) = P(B_t > x + y). \tag{5.12}$$

This continuous-time analog of formula (5.8) again owes its importance to the fact that it reduces the joint distribution of (B_t, B_t^*) to the much simpler marginal distribution of B_t.

Now, since $B_t/\sqrt{t}$ is a standard Gaussian, we also see that $P(B_t > x + y)$ is simply $1 - \Phi((x + y)/\sqrt{t})$. If we substitute $u = x - y$ and $v = x$ into the identity (5.12), then for (u, v) in the set $D = \{(u, v) \colon u \in \mathbb{R},\ v \geq \max(0, u)\}$ we have

$$P(B_t < u, B_t^* \geq v) = 1 - \Phi((2v - u)/\sqrt{t}) = \Phi((u - 2v)/\sqrt{t})$$

and since $P(B_t < u, B_t^* < v) = P(B_t < u) - P(B_t < u, B_t^* \geq v)$ we come at last to the elegant formula

$$P(B_t < u, B_t^* < v) = \Phi\left(\frac{u}{\sqrt{t}}\right) - \Phi\left(\frac{u - 2v}{\sqrt{t}}\right) \quad \text{for all } (u, v) \in D. \tag{5.13}$$

For many purposes the distribution function (5.13) is the best way to carry around information about the joint behavior of B_t and B_t^*, but still we should note that two easy differentiations bring us directly to the density of (B_t, B_t^*):

$$f_{(B_t, B_t^*)}(u, v) = \frac{2(2v - u)}{\sqrt{2\pi t^3}} \exp\left(-\frac{(2v - u)^2}{2t}\right) \quad \text{for all } (u, v) \in D. \tag{5.14}$$

In a later chapter, we will see that the density (5.14) plays a key role in the analysis of Brownian motion with drift, but we do not need go so far to find important consequences. For example, we can simply integrate the density (5.14) to find our first derivation of the density of the maximum of Brownian motion.

DENSITY AND DISTRIBUTION OF B_t^*

Naturally, there are more insightful routes to the density of B_t^* than brute force integration, and one of these is to mimic our investigation of the distribution of S_n^* for simple random walk where we exploited the identity (5.8) by specialization to $y = 0$. This time we take $y = 0$ in equation (5.12) to find

$$P(B_t^* \geq x,\ B_t \leq x) = P(B_t > x),$$

and again we have a trivial companion identity:

$$P(B_t^* \geq x,\ B_t > x) = P(B_t > x).$$

Finally, the sum of these two identities gives us the desired representation for the distribution of B_t^*:

$$P(B_t^* \geq x) = 2P(B_t > x) = P(|B_t| \geq x) \text{ for } x \geq 0. \tag{5.15}$$

In other words, the distribution of the maximal process B_t^* is exactly equal to the distribution of the absolute value of Brownian motion $|B_t|$.

This is one of the properties of Brownian motion that deserves to be savored. As processes, $\{B_t^*\}$ and $\{|B_t|\}$ could not be more different. For example, $\{B_t^*\}$ is nondecreasing, and its paths have many long flat spots. The paths of $\{|B_t|\}$ are never monotone. In fact, their local behavior away from zero is identical to that of paths of $\{B_t\}$, so in particular, the paths of $\{|B_t|\}$ are never differentiable even at a single point — much less *flat*. Despite all of these differences, the distributions of B_t^* and $|B_t|$ are equal for each fixed t.

To close our discussion of B_t^* and $|B_t|$, we should compute the density of B_t^*. This is easy, and we will need the result later. Since $B_t/\sqrt{t}$ is a standard Gaussian, we have for all $x \geq 0$ that

$$P(|B_t| > x) = P(|B_t|/\sqrt{t} > x/\sqrt{t}) = 2\{1 - \Phi(x/\sqrt{t})\},$$

and the reflection formula (5.15) then tells us that for $x \geq 0$ we have

$$P(B_t^* \leq x) = 2\Phi(x/\sqrt{t}) - 1.$$

Finally, an easy differentiation provides us with the desired density:

$$f_{B_t^*}(x) = \frac{2}{\sqrt{t}}\phi\left(\frac{x}{\sqrt{t}}\right) = \sqrt{\frac{2}{\pi t}}\, e^{-x^2/2t} \quad \text{for } x \geq 0. \tag{5.16}$$

DENSITY OF THE HITTING TIME τ_a

The distributional results for B_t^* have easy and attractive translations into information for the hitting time of the level $a > 0$:

$$\tau_a = \inf\{t\colon B_t = a\}.$$

Since we have

$$P(B_t^* < a) = P(\tau_a > t),$$

we get

$$P(\tau_a > t) = 2\Phi\left(\frac{a}{\sqrt{t}}\right) - 1$$

and differentiation gives the succinct formula:

$$f_{\tau_a}(t) = \frac{a}{t^{3/2}}\,\phi\left(\frac{a}{\sqrt{t}}\right) \quad \text{for } t \geq 0. \tag{5.17}$$

This formula will answer any question you have about the distribution of τ_a, but the only application that we will make right now is to observe that it gives a very simple upper bound on the tail probability $P(\tau_a \geq t)$. Specifically, since $\phi(x) \leq \frac{1}{2}$ for all $x \in \mathbb{R}$, we see that

$$P(\tau_a \geq t) \leq \int_t^\infty \frac{a}{2s^{3/2}}\, ds = \frac{a}{\sqrt{t}} \quad \text{for all } a \geq 0,\ t > 0. \tag{5.18}$$

One could hardly ask for a simpler result, and we will find in a later chapter that it is also very useful.

5.4. The Invariance Principle and Donsker's Theorem

When one has a handy law for simple random walk such as the reflection principle identity (5.8), or it special case (5.9), one is virtually forced to ask if there is an analogous law for Brownian motion. In fact, the analogous law is almost always valid, and there are several systematic methods for converting identities for random walk into identities for Brownian motion, and vice versa.

One of the oldest and best of these methods is based on the approximation of Brownian motion by the linear interpolation of a general random walk. To describe this approximation, we first take a general sequence $\{X_n\}$ of independent identically distributed random variables with mean zero and variance one, and then we interpolate the random walk $S_n = X_1 + X_2 + \cdots + X_n$ in order to define a piecewise linear stochastic process for all $t \geq 0$ by setting $S_0 = 0$ and taking

$$S(n,t) = S_n + (t-n)X_{n+1} \quad \text{when } n < t \leq n+1. \tag{5.19}$$

This process can now be scaled in space and time in order to obtain a process that approximates Brownian motion on $[0,1]$:

$$B_t^{(n)} = S(n,nt)/\sqrt{n} \quad \text{for } 0 \leq t \leq 1. \tag{5.20}$$

The suggestion that $B_t^{(n)}$ approximates Brownian motion is intuitive enough, but some work is needed to make this idea precise. First, we should note that the central limit theorem and a bit of easy analysis tell us that

$$\lim_{n\to\infty} P(B_t^{(n)} \leq x) = P(B_t \leq x),$$

and with a little more effort we can even prove the convergence of the joint distributions,

$$\begin{aligned}\lim_{n\to\infty} P(B_{t_1}^{(n)} \leq x_1, B_{t_2}^{(n)} \leq x_2, \cdots, B_{t_d}^{(n)} \leq x_d)\\ = P(B_{t_1} \leq x_1, B_{t_2} \leq x_2, \cdots, B_{t_d} \leq x_d).\end{aligned}$$

These results tell us that the finite dimensional distributions of the process $\{B_t^{(n)}\}$ do approximate those of $\{B_t\}$, but there is even more to the approximation. For example, by the finite distributional results and a bit of additional work, we can show

$$P\Big(\max_{0\leq t\leq T} B_t^{(n)} \leq x\Big) \to P\Big(\max_{0\leq t\leq T} B_t \leq x\Big) \tag{5.21}$$

or even show

$$P\Big(\max_{0\leq t\leq T} B_t^{(n)} \leq x,\, B_T^{(n)} \leq y\Big) \to P\Big(\max_{0\leq t\leq T} B_t \leq x,\, B_T \leq y\Big). \tag{5.22}$$

Formulas such as (5.21) and (5.22) can be produced in almost endless varieties, and we will soon need to find some way to consolidate them. Still, before we take up that consolidation, we should recall how even the humble identity (5.21) can lead to a remarkable general principle.

The Original Invariance Principle

Perhaps the most instructive inference one can draw from the limit (5.21) is that it implies a general *invariance principle*. Since the limit one finds in (5.21) does not depend on the underlying random walk, the simple equality (5.21) actually contains a number of interesting identities.

For example, if we take a random walk that is based on a general sequence $\{X_i\}$ with $E(X_i) = 0$ and $E(X_i^2) = 1$, we will find the same limit (5.21) that we get for the simple random walk based on the independent variables $\{Y_i\}$ where $P(Y_i = 1) = P(Y_i = -1) = \frac{1}{2}$. As a corollary of (5.21), we have the identity

$$\lim_{n\to\infty} P\Big(n^{-\frac{1}{2}} \max_{1\le k\le n} \sum_{i=1}^{k} X_i \le x\Big) = \lim_{n\to\infty} P\Big(n^{-\frac{1}{2}} \max_{1\le k\le n} \sum_{i=1}^{k} Y_i \le x\Big), \tag{5.23}$$

and we already know from equation (5.10) that

$$\lim_{n\to\infty} P\Big(n^{-\frac{1}{2}} \max_{1\le k\le n} \sum_{i=1}^{k} Y_i \le x\Big) = 2\Phi(x) - 1 \text{ for all } x \ge 0. \tag{5.24}$$

By the identity (5.23), we have the same limit law for $\{X_i\}$ that we have for $\{Y_i\}$, and this proves our earlier conjecture about the asymptotic distribution of the maximum of a general random walk.

The beauty of equation (5.23) is that it tells us that the limit law (5.24) is *invariant* under the change from the simple $\{Y_i\}$ to the general $\{X_i\}$. In a broad range of circumstances, identities such as (5.23) can be used to show that distributional limit laws are invariant under the change of the underlying distribution. This means among other things that one can use methods of discrete mathematics to prove limit laws for simple random walk and then almost automatically translate these results to limit laws that apply to all random walks with steps X_i with $E(X_i) = 0$ and $\mathrm{Var}(X_i) = 1$.

DONSKER'S THEOREM

As we noted earlier, there are countless variations on the identity (5.21). One can stir in minima, intermediate maxima, or whatever. Fortunately, one does not need to pursuc proofs of each of these individual formulas. There is a single theorem that nicely houses the whole menagerie.

If we consider the space $C[0,1]$ of all continuous functions on $[0,1]$, then $C[0,1]$ is a complete metric space with respect to the norm $||f||_\infty = \sup_{0\le t\le 1} |f(t)|$. To say that $H : C[0,1] \to \mathbb{R}$ is a continuous function on $C[0,1]$ then means nothing more than to say that if $f_n \to f$ uniformly on $[0,1]$, then we have convergence of the real sequence $H(f_n) \to H(f)$. The archetypes for such continuous functions are none other than our friends the maximum function $H(f) = \max_{0\le t\le 1} f(t)$ and the point evaluation function $H(f) = f(1)$. The remarkable fact is that whenever we can work out the distribution of H applied to the Brownian motion paths, then we can work out the corresponding asymptotic distribution for H applied to the linear interpolation of simple random walk, or vice versa.

THEOREM 5.4 (Donsker's Invariance Principle).

For any continuous function $H : C[0,1] \to \mathbb{R}$, the interpolated and scaled random walk $\{B_t^{(n)} : 0 \le t \le 1\}$ satisfies

$$\lim_{n\to\infty} P[H(B_{(\cdot)}^{(n)}) \le x] = P[H(B_{(\cdot)}) \le x]. \tag{5.25}$$

The proof of this result is not difficult, but to do a proper job would take us a bit too far out of our way. Still, this theorem is a landmark in the theory of probability, and it provides one of the basic justifications for calling Brownian motion the most important stochastic process.

5.5. Random Walks Inside Brownian Motion

At moments, the connection between Brownian motion and simple random walk goes beyond analogy or approximation — in some contexts, the two processes have exactly the same structure. For example, if A and B are positive integers and τ is the first hitting time of level A or level $-B$, then the formulas we found for the hitting probability and the expected hitting time are exactly the same for Brownian motion and for simple random walk:

$$(5.26)\quad P(B_\tau = A) = \frac{B}{A+B} = P(S_\tau = A) \quad \text{and} \quad E[\tau] = AB = E[S_\tau^2] = E[B_\tau^2].$$

One of the remarkable consequences of the hitting probability formula (5.26) is that it tells us that there is a simple random walk that lives on every Brownian path. Specifically, if we let $\tau_1 = \inf\{t : B_t \notin (-1,1)\}$ and set

$$(5.27)\qquad \tau_{n+1} = \inf\{t > \tau_n : B_t - B_{\tau_n} \notin (-1,1)\}$$

for $n \geq 1$, then the sequence $B_{\tau_1}, B_{\tau_2}, \ldots, B_{\tau_n}, \ldots$ has the same joint distributions as $S_1, S_2, \ldots, S_n, \ldots$, where as usual $S_n = X_1 + X_2 + \cdots + X_n$ and the X_i are independent random variables with $P(X_n = 1) = P(X_n = -1) = \frac{1}{2}$. Thus, simple random walk can be studied in any detail we choose by looking at Brownian motion at a well chosen set of stopping times.

This is a very pleasing situation. It also prompts the immediate conjecture that any unbiased walk with steps with finite variance can be embedded in an analogous way. This is indeed a theorem that we will soon prove, but first we will take a brief look at an amusing consequence of the embedding provided by the stopping times (5.27).

An Illustrative Limit Theorem

Because S_n has the same distribution as B_{τ_n} and since B_n is exactly Gaussian with mean 0 and variance n, if we show that $B_{\tau_n}/\sqrt{n}$ and $B_n/\sqrt{n}$ do not differ too much we can deduce that $S_n/\sqrt{n}$ is approximately Gaussian. This is not a great deduction; rather, it is a strange path to a result that has been known for two centuries as the DeMoivre–Laplace approximation of the binomial distribution. Still, the path offers more than quaint charm. The proof points to a general plan that has led to many important developments and remains a tool of active use in current research.

To use the embedding, we need some information about the embedding times $\{\tau_n\}$. The main observation is that the increments $\delta_i = \tau_i - \tau_{i-1}$ are independent, identically distributed, and have expectation $E(\delta_i) = 1$. This is all we know about the $\{\tau_n\}$, but it is all we need to establish the closeness of B_{τ_n} and B_n. Now, given any $\delta > 0$ and $\epsilon > 0$, we first note that

$$\begin{aligned}P(|B_{\tau_n} - B_n| \geq \epsilon\sqrt{n}) &= P(|B_{\tau_n} - B_n| \geq \epsilon\sqrt{n} \text{ and } |\tau_n - n| < \delta n)\\ &\quad + P(|B_{\tau_n} - B_n| \geq \epsilon\sqrt{n} \text{ and } |\tau_n - n| \geq \delta n)\\ &\leq P\Big(\sup_{s:|s-n|\leq \delta n} |B_s - B_n| \geq \epsilon\sqrt{n}\Big) + P(|\tau_n - n| \geq \delta n).\end{aligned}$$

Now, the term $P(|\tau_n - n| \geq \delta n)$ is no trouble at all, since the law of large numbers tells us that it goes to zero as $n \to \infty$. Also, the second term is easily addressed by symmetry and Doob's maximal inequality

$$P\Big(\sup_{s:|s-n|\leq \delta n} |B_s - B_n| \geq \epsilon\sqrt{n} \Big) \leq 2P\Big(\sup_{0\leq s\leq \delta n} |B_s| \geq \epsilon\sqrt{n} \Big)$$
$$\leq 2E(B_{\delta n}^2)/(\epsilon^2 n) = 2\delta/\epsilon^2.$$

We see therefore that

$$\lim_{n\to\infty} P(|B_{\tau_n} - B_n| \geq \epsilon\sqrt{n}) \leq 2\delta/\epsilon^2, \tag{5.28}$$

and since δ can be chosen as small as we like in the bound (5.28), we see that

$$\lim_{n\to\infty} P(|B_{\tau_n} - B_n| \geq \epsilon\sqrt{n}) = 0. \tag{5.29}$$

Because $B_n/\sqrt{n}$ has the standard Gaussian distribution for all n, we see now (or, after reviewing the remark attached to Exercise 5.2) that the limit (5.29) gives us our central limit theorem for $S_n = B_{\tau_n}$:

$$\lim_{n\to\infty} P(B_{\tau_n}/\sqrt{n} \leq x) = \Phi(x).$$

Extending the Class of Embedded Variables[1]

If we want to embed a random variable that takes on only two values, we already have a good idea how to proceed. If the values of X are $-a < 0$ and $0 < b$, then the assumption that $E(X) = 0$ forces us to the conclusion that

$$P(X = -a) = b/(a+b) \text{ and } P(X = b) = a/(a+b);$$

so, if we simply let $\tau = \inf\{t : B_t \notin (-a, b)\}$, then our earlier work tells us that we do indeed reach the required conclusions of the theorem.

We could now consider random variables that take on only three values, but we may as well take a bold step and consider a random variable X that takes on an arbitrary finite number of values. To parallel our first exploration, we suppose that there are nonnegative reals $\{a_i\}$ and $\{b_j\}$ such that the distribution of X is given by

$$P(X = -a_i) = \alpha_i \text{ for } i = 1, 2, \ldots m$$

and

$$P(X = b_j) = \beta_j \text{ for } j = 1, 2, \ldots n.$$

This time, the condition that $E(X) = 0$ will not determine the distribution completely, but it will impose an essential relationship:

$$\sum_{i=1}^{m} a_i\alpha_i = \sum_{j=1}^{n} b_j\beta_j. \tag{5.30}$$

Randomized Stopping Times

At this point, there are several ideas one might try, but perhaps the most natural plan is to look for a way to randomize our earlier method. Explicitly, this

[1]Readers in a hurry can skip directly to Theorem 5.5 and the succinct proof outlined in Exercise 5.1. The present discussion offers a slightly unconventional approach which leans on more linear algebra than might be to everyone's taste. The relationship to other approaches is discussed in the bibliographical notes.

suggests that in order to construct τ we should search for a pair of integer-valued random variables I and J, so that if we take

$$\tau = \inf\{t\colon B_t \notin (-a_I, b_J)\},$$

then B_τ will have the same distribution as X. What makes this a reasonable plan is that it places at our disposal an entire probability distribution,

$$P(I = i, J = j) = p_{ij}.$$

To determine the constraints on $\{p_{ij}\}$, we first note that the definition of τ gives us

$$\begin{aligned} P(B_\tau = -a_i) &= \sum_{j=1}^{n} P(B_\tau = -a_i \mid I = i, J = j)\, p_{ij} \\ &= \sum_{j=1}^{n} \frac{b_j}{a_i + b_j}\, p_{ij}, \end{aligned} \tag{5.31}$$

and

$$\begin{aligned} P(B_\tau = b_j) &= \sum_{i=1}^{m} P(B_\tau = b_j \mid I = i, J = j)\, p_{ij} \\ &= \sum_{i=1}^{m} \frac{a_i}{a_i + b_j}\, p_{ij}. \end{aligned} \tag{5.32}$$

Some Linear Algebra

The structure of these equations is made most evident in vector notation. If we consider $\{a_i\}$,$\{\alpha_i\}$,$\{b_j\}$, and $\{\beta_j\}$ as column vectors $\mathbf{a}, \boldsymbol{\alpha}, \mathbf{b}$, and $\boldsymbol{\beta}$; and if we introduce a matrix $Q = \{q_{ij}\}$ by taking

$$q_{ij} = p_{ij}/(a_i + b_j),$$

then the requirement that B_τ has the same distribution as X can be written very neatly as

$$\boldsymbol{\alpha} = Q\mathbf{b} \quad \text{and} \quad \boldsymbol{\beta}^T = \mathbf{a}^T Q. \tag{5.33}$$

Moreover, the mean zero condition (5.30) also has the simple vector form

$$\mathbf{a}^T \boldsymbol{\alpha} = \mathbf{b}^T \boldsymbol{\beta}. \tag{5.34}$$

Solving for Q

We are quite accustomed to solving for $\mathbf{b}$ in equations such as $\boldsymbol{\alpha} = Q\mathbf{b}$, but this time we face the slightly odd task of solving for the matrix Q given $\boldsymbol{\alpha}$ and $\mathbf{b}$. As we would expect, this is even easier. In fact, we can immediately check that

$$Q_v = \frac{\boldsymbol{\alpha}\mathbf{v}^T}{\mathbf{v}^T\mathbf{b}}$$

is a solution of the matrix equation $\boldsymbol{\alpha} = Q\mathbf{b}$ for any $\mathbf{v}$ such that $\mathbf{v}^T\mathbf{b} \neq 0$, and, in the same way,

$$Q_w = \frac{\mathbf{w}\boldsymbol{\beta}^{\mathbf{T}}}{\mathbf{a}^T\mathbf{w}}$$

is a solution of the second equation of (5.33) for any w such that $\mathbf{a}^T\mathbf{w} \neq 0$.

To find one Q that solves both of the equations (5.33), we only need to exploit the freedom we have in choosing v and w. When we equate the formulas for Q_v and Q_w, we get

$$\frac{\boldsymbol{\alpha}\mathbf{v}^T}{\mathbf{v}^T\mathbf{b}} = \frac{\mathbf{w}\boldsymbol{\beta}^T}{\mathbf{a}^T\mathbf{w}}; \tag{5.35}$$

and, since we can make the numerators match if we take $\mathbf{v} = \boldsymbol{\beta}^{\mathbf{T}}$ and $\mathbf{w} = \boldsymbol{\alpha}$, we only need to check that this choice also makes the denominators check. Fortunately, this is an immediate consequence of equation (5.34). The bottom line is that we can satisfy both equations of the pair (5.33) if we take

$$Q = \gamma^{-1}\boldsymbol{\alpha}\boldsymbol{\beta}^{\mathbf{T}}, \tag{5.36}$$

where γ is the common value of $\mathbf{a}^T\boldsymbol{\alpha}$ and $\mathbf{b}^T\boldsymbol{\beta}$.

And Q Really Works

Thus far, we have seen that $Q = (q_{ij})$ solves equations (5.33) and (5.34), but we still must check that $p_{ij} = q_{ij}(a_i + b_j)$ is a bona fide probability distribution. The solution (5.36) tells us $q_{ij} = \gamma^{-1}\alpha_i\beta_j \geq 0$, so we only need to note that

$$\sum_{i=1}^{m}\sum_{j=1}^{n} p_{ij} = \sum_{j=1}^{n}\beta_j\gamma^{-1}\sum_{i=1}^{m}a_i\alpha_i + \sum_{i=1}^{m}\alpha_i\gamma^{-1}\sum_{j=1}^{n}b_j\beta_j = 1,$$

where in the last step we used the definition of γ and the fact that the sum of all the α_i's and β_j's is 1. The bottom line is that our simple choices have been fortuitous choices, and our construction of Q has produced an honest probability distribution $\{p_{ij}\}$.

Checking $E(\tau) = \text{Var}(X)$

To complete our treatment of the discrete case, we only need to calculate the expectation of τ and show that it equals $\text{Var}(X)$. There is a temptation to use the martingale $M_t = B_t^2 - t$ to argue via Doob's stopping time theorem and the dominated convergence theorem that $M_{t\wedge\tau}$ is a bounded martingale and consequently $E(\tau) = E(B_\tau^2) = \text{Var}(X)$. One can make this argument work, but it is not automatic; τ is a randomized stopping time, not just a plain vanilla stopping time. To make the argument honest, we would need to extend Doob's theorem to cover the case of randomized stopping times.

A much quicker way to show that $E(\tau) = \text{Var}(X)$ is to exploit the explicit form of τ to give a direct computational check:

$$\begin{aligned}
E(\tau) &= \sum_{i=1}^{m}\sum_{j=1}^{n} a_i b_j p_{ij} = \sum_{i=1}^{m}\sum_{j=1}^{n} a_i b_j (a_i + b_j)\alpha_i\beta_j/\gamma \\
&= \sum_{i=1}^{m}\sum_{j=1}^{n} a_i^2\alpha_i\beta_j b_j/\gamma + \sum_{i=1}^{m}\sum_{j=1}^{n} a_i\alpha_i\beta_j b_j^2/\gamma \\
&= \sum_{i=1}^{m} a_i^2\alpha_i\sum_{j=1}^{n}\beta_j b_j/\gamma + \sum_{j=1}^{n} b_j^2\beta_j\sum_{i=1}^{m} a_i\alpha_i/\gamma = \text{Var}(X).
\end{aligned}$$

FINAL STEP: CONTINUOUS CASE

There are two natural plans for extending our embedding of discrete random variables to the case of general distributions. The most direct route uses the idea that any random variable can be approximated as closely as we like in distribution by random variables that take on only finitely many values. This plan works, but it is not labor-free; many details must be checked. The second plan is to prove the general case by mimicking the construction that was discovered in the discrete case. The details of the second plan turn out to be both easy and instructive, as one can verify by working Exercise 5.1.

Our discussion of the embedding of a random variable in Brownian motion is almost complete. We only need to summarize the main conclusions in a formal theorem.

THEOREM 5.5 (Skorohod Embedding Theorem). *If X is a random variable with mean zero and variance $\sigma^2 < \infty$, there is a randomized stopping time τ such that*

$$P(B_\tau \leq x) = P(X \leq x) \quad \textit{for all} \quad x \in \mathbb{R}$$

and

$$E(\tau) = \sigma^2.$$

Finally, as we noted in the introduction to the embedding problem, we are much more interested in embedding a whole process than a single random variable. Fortunately, the embedding of random walk just requires the repeated application of the same process that permitted us to embed a single random variable. The most useful form of the embedding construction is given in the following corollary.

COROLLARY 5.2 (Embedding of Random Walk). *For any sequence of independent, identically distributed random variables $\{X_i : 1 \leq i < \infty\}$ such that $E(X_i) = 0$ and $E(X_i^2) = \sigma^2$, there is a sequence $\{\tau_i : 1 \leq i < \infty\}$ of independent, identically distributed, nonnegative random variables with $E(\tau_i) = \sigma^2$ such that the random walk*

$$S_n = X_1 + X_2 + \cdots + X_n$$

and the process defined by

$$\tilde{S}_n = B_{\tau_1 + \tau_2 + \cdots + \tau_n}$$

are equivalent; that is, all of the joint distributions of

$$\{S_n : 1 \leq n < \infty\} \textit{ and } \{\tilde{S}_n : 1 \leq n < \infty\}$$

are equal.

SOME PERSPECTIVE ON METHODS

The proof of the Skorohod Embedding theorem is often given by producing a recipe for τ as a magician might produce a rabbit from a hat — as a surprise and without any indication of how we might find our own rabbits. The path we have taken here is not as short, but it should provide a more reliable guide to the trail one might follow to pursue an attractive idea.

Here we began with the obvious, but striking, fact that one can embed simple random walk into a Brownian motion by use of stopping times. We then looked at the simplest possible extension of this embedding, and we observed that we could also embed any unbiased walk with steps that take on only two values. Creeping along, we then considered walks with steps that could take on an arbitrary finite number of values.

There is something to this process of incremental investigation that is both sensible and satisfying. There is no silver bullet that slays all mathematical problems, and even well-motivated investigations may grind to an unsatisfying halt. Still, when progress is possible, many investigators have found that it comes most quickly when they regularly remind themselves of Pólya's fundamental question of problem solving: "What is the simplest question you can't solve ?"

5.6. Exercises

The first exercise outlines a proof of the general case of the Skorohod embedding theorem. One can think of Section 5.5 as a warmup and an introduction to this problem. The next exercise illustrates the curious fact that the Skorohod theorem provides a logically independent proof of the central limit theorem. The last exercise then offers the opportunity to use the explicit construction of Brownian motion to make the first step toward a proof of Lévy's modulus of continuity theorem.

EXERCISE 5.1 (General Case of Theorem 5.5). The same idea that we discovered for discrete random variables can be applied to general distributions. Specifically, if F is the distribution of X, we let

$$\gamma = -\int_{-\infty}^{0} x\,dF(x) = \int_{0}^{\infty} x\,dF(x)$$

and define the joint distribution of (I, J) by

$$P(I \le s, J \le t) = \gamma^{-1} \int_{-\infty}^{s} \int_{0}^{t} (u - v) dF(u) dF(v) \quad s < 0 \le t.$$

If we rewrite this definition in differential terms

$$P(I \in ds, J \in dt) = \gamma^{-1}(t - s) F\{ds\} F\{dt\} \quad s < 0 \le t, \tag{5.37}$$

we see that it is the continuous parallel to our earlier choice of

$$p_{ij} = \gamma^{-1}(a_i + b_j)\alpha_i \beta_j.$$

Define τ as before as the first exit time of the random interval (I, J), and check both the conclusions of Theorem 5.5.

EXERCISE 5.2 (Central Limit Theorem via Skorohod). The classical central limit theorem says that if $\{X_i\}$ is a sequence of independent identically distributed random variables with $E(X_i) = 0$ and $\text{Var}X_i = 1$, then the sum $S_n = X_1 + X_2 + \cdots + X_n$ satisfies

$$\lim_{n \to \infty} P(S_n/\sqrt{n} \le x) = \frac{1}{2\pi} \int_{-\infty}^{x} e^{-u^2/2}\,du \equiv \Phi(x).$$

Use Skorohod embedding to give a clear, careful, and complete proof of the classical central limit theorem.

Remark: For a complete proof, one needs a clear connection between convergence in probability and convergence in distribution. As a warmup for the exercise, one should first check that if

$$\lim_{n\to\infty} P(|X_n - Y_n| \geq \epsilon) = 0 \quad \text{for all} \quad \epsilon > 0,$$

and $P(Y_n \leq x) = F(x)$, then $\lim_{n\to\infty} P(X_n \leq x) = F(x)$ for all x such that F is continuous at x.

EXERCISE 5.3 (Modulus of Continuity — Version 1.0). Show that the argument that was used to prove Theorem 5.1 may be pressed a bit harder to show that there is a random variable C such that $P(C < \infty) = 1$ and

$$|B_{t+h}(\omega) - B_t(\omega)| \leq C(\omega)\sqrt{h\log(1/h)}$$

for all ω and all $0 \leq t, t+h \leq \frac{1}{2}$.

CHAPTER 6

Itô Integration

The Itô integral carries the notion of a martingale transform from discrete time into continuous time. The construction gives us a systematic method for building new martingales, and it leads to a new calculus for stochastic processes, the consequences of which turn out to be more far reaching than anyone could have possibly expected. The Itô calculus is now well established as one of the most useful tools of probability theory.

The main goal of this chapter is to provide an honest — yet well motivated — construction of the Itô integral,

$$I(f)(\omega) = \int_0^T f(\omega, t)\, dB_t. \tag{6.1}$$

Because the paths of Brownian motion do not have bounded variation, this integral cannot be interpreted as an ordinary Riemann integral. Instead, we must rely on a more subtle limit process.

In a nutshell, the idea is to define the integral on a class of rudimentary functions and to extend the definition to a larger class by a continuity argument. Unfortunately, one small nutshell does not do justice to the full story. If we want an integral that offers real insight into the theory of stochastic processes, we must be prepared to do some fancy footwork.

6.1. Definition of the Itô Integral: First Two Steps

For the integral (6.1) to have any hope of making sense, the integrand must meet some basic measurability and integrability requirements. First, we consider measurability. To begin, we let $\mathcal{B}$ denote the smallest σ-field that contains all of the open subsets of $[0, T]$; that is, we let $\mathcal{B}$ denote the set of *Borel* sets of $[0, T]$. We then take $\{\mathcal{F}_t\}$ to be the standard Brownian filtration, and for each $t \geq 0$ we take $\mathcal{F}_t \times \mathcal{B}$ to be the smallest σ-field that contains all of the product sets $A \times B$ where $A \in \mathcal{F}_t$ and $B \in \mathcal{B}$. Finally, we say $f(\cdot, \cdot)$ is *measurable* if $f(\cdot, \cdot) \in \mathcal{F}_T \times \mathcal{B}$, and we will say that $f(\cdot, \cdot)$ is *adapted* provided that $f(\cdot, t) \in \mathcal{F}_t$ for each $t \in [0, T]$.

In the first stage of our development of the Itô integral, we will focus on integrands from the class $\mathcal{H}^2 = \mathcal{H}^2[0, T]$ that consists of all measurable adapted functions f that satisfy the integrability constraint

$$E\left[\int_0^T f^2(\omega, t)\, dt\right] < \infty. \tag{6.2}$$

Incidentally, one should note that the expectation (6.2) is actually a double integral and that $\mathcal{H}^2$ is a closed linear subspace of $L^2(dP \times dt)$.

To anticipate the definition of the Itô integral, we first consider what we would expect the integral to be in the simplest cases. For example, if we take $f(\omega, t)$ to be the indicator of the interval $(a, b] \subset [0, T]$, then for the integral to be worthy of its name we would insist that

$$I(f)(\omega) = \int_a^b dB_t = B_b - B_a. \tag{6.3}$$

Now, since we fully expect the Itô integral to be linear, our insistence on the identity (6.3) already tells us how we must define I for a relatively large class of integrands. To define this class, we let $\mathcal{H}_0^2$ denote the subset of $\mathcal{H}^2$ that consists of all functions of the form

$$f(\omega, t) = \sum_{i=0}^{n-1} a_i(\omega) 1(t_i < t \le t_{i+1}), \tag{6.4}$$

where $a_i \in \mathcal{F}_{t_i}$, $E(a_i^2) < \infty$, and $0 = t_0 < t_1 < \cdots < t_{n-1} < t_n = T$.

Linearity and equation (6.3) give us no choice about the definition of I on $\mathcal{H}_0^2$. For functions of the form (6.4), we are forced to define I on $\mathcal{H}_0^2$ by

$$I(f)(\omega) = \sum_{i=0}^{n-1} a_i(\omega)\{B_{t_{i+1}} - B_{t_i}\}. \tag{6.5}$$

We would now like to show that we can extend the domain of I from $\mathcal{H}_0^2$ to all of $\mathcal{H}^2$, and to complete this extension we need to know that $I : \mathcal{H}_0^2 \to L^2(dP)$ is a continuous mapping. This is indeed the case, and the keystone that caps the arch is the following fundamental lemma.

Lemma 6.1 (Itô's Isometry on $\mathcal{H}_0^2$). *For $f \in \mathcal{H}_0^2$ we have*

$$||I(f)||_{L^2(dP)} = ||f||_{L^2(dP \times dt)}. \tag{6.6}$$

Proof. The proof could not be simpler; we just compute both norms. To calculate $||f||_{L^2(dP\times dt)}$, we note that for f of the form (6.4) we have

$$f^2(\omega, t) = \sum_{i=0}^{n-1} a_i^2(\omega) 1(t_i < t \le t_{i+1}),$$

so

$$E\left[\int_0^T f^2(\omega, t)\, dt\right] = \sum_{i=0}^{n-1} E(a_i^2)(t_{i+1} - t_i).$$

To calculate $||I(f)||_{L^2(dP)}$, we first multiply out the terms in the definition (6.5) of $I(f)$ and observe that a_i is independent of $B_{t_{i+1}} - B_{t_i}$. The cross terms have expectation zero, so we finally see

$$\begin{aligned} E[I(f)^2] &= \sum_{i=0}^{n-1} E\Big(a_i^2 (B_{t_{i+1}} - B_{t_i})^2\Big) \\ &= \sum_{i=0}^{n-1} E(a_i^2)(t_{i+1} - t_i). \end{aligned}$$

These calculations tell us that the two sides of equation (6.6) are indeed equal, so the proof of the lemma is complete. □

Cauchy Sequences and Approximating Sequences

Itô's isometry establishes the essential fact that I maps $\mathcal{H}_0^2$ continuously into $L^2(dP)$, but an unexpected precision is thrown into the bargain. Because I preserves distances when it maps the metric space $\mathcal{H}_0^2$ into $L^2(dP)$, we get for free that I takes a Cauchy sequence in $\mathcal{H}_0^2$ into a Cauchy sequence in $L^2(dP)$. This handy fact makes light work of the extension step once we know that we can approximate elements of $\mathcal{H}^2$ by elements of $\mathcal{H}_0^2$, as the following lemma guarantees.

Lemma 6.2 ($\mathcal{H}_0^2$ is Dense in $\mathcal{H}^2$). *For any $f \in \mathcal{H}^2$, there exists a sequence $\{f_n\}$ with $f_n \in \mathcal{H}_0^2$ such that*

$$||f - f_n||_{L^2(dP\times dt)} \to 0 \text{ as } n \to \infty.$$

Lemma 6.2 is almost intuitive enough to believe without proof, and, to keep the logical development of the Itô integral as transparent as possible we will not prove the lemma at this point. Rather, in the last section of this chapter, we will prove an approximation theorem that has Lemma 6.2 as an immediate corollary.

Logical Heart of the Second Step

Now, given any $f \in \mathcal{H}^2$, the last lemma tells us there is a sequence $\{f_n\} \subset \mathcal{H}_0^2$ such that f_n converges to f in $L^2(dP \times dt)$. Also, for each n the integrals $I(f_n)$ are well-defined elements of $L^2(dP)$ that are given explicitly by formula (6.5). The natural plan is then to define $I(f)$ as the limit of the sequence $I(f_n)$ in $L^2(dP)$. That is, we take

$$I(f) = \lim_{n\to\infty} I(f_n), \tag{6.7}$$

where the detailed interpretation of equation (6.7) is that the random variable $I(f)$ is an element of $L^2(dP)$ such that $||I(f_n) - I(f)||_{L^2(dP)} \to 0$ as $n \to \infty$.

To be sure that this definition is legitimate, we only need to make a couple of checks. First, we need to show that $||f - f_n||_{L^2(dP\times dt)} \to 0$ implies that $I(f_n)$ converges in $L^2(dP)$. This is quite easy because the convergence of $f - f_n$ to zero in $L^2(dP \times dt)$ tells us that $\{f_n\}$ is a Cauchy sequence in $L^2(dP \times dt)$, and the Itô isometry then tells us that $\{I(f_n)\}$ is also a Cauchy sequence in $L^2(dP)$. But $L^2(dP)$ is a complete metric space, and completeness just means that every Cauchy sequence must converge. In particular, we see that $I(f_n)$ converges to some element of $L^2(dP)$, which we then denote by $I(f)$.

Our second check requires that we show that I is well defined in the sense that if f_n' is another sequence with the property that $||f - f_n'||_{L^2(dP\times dt)} \to 0$, then $I(f_n')$ has the same limit in $L^2(dP)$ as $I(f_n)$. This is also immediate, since the triangle inequality tells us that any such f_n' must satisfy $||f_n - f_n'||_{L^2(dP\times dt)} \to 0$, and the Itô isometry then tells us that $||I(f_n) - I(f_n')||_{L^2(dP)} \to 0$.

Intermezzo: Further Itô Isometries

Now that the Itô integral has been defined on $\mathcal{H}^2$, we should note that Lemma 6.1 has an immediate extension to $\mathcal{H}^2$. The space $\mathcal{H}^2$ is the natural domain of the Itô isometry, but, of course, the $\mathcal{H}^2$ result could not even be stated until we had made good use of Lemma 6.1 to extend I from a mapping from $\mathcal{H}_0^2$ into $L^2(dP)$ to a mapping from $\mathcal{H}^2$ into $L^2(dP)$.

THEOREM 6.1 (Itô's Isometry on $\mathcal{H}^2[0,T]$). *For $f \in \mathcal{H}^2[0,T]$, we have*

$$||I(f)||_{L^2(dP)} = ||f||_{L^2(dP\times dt)}. \tag{6.8}$$

PROOF. All we need to do is to line up the application of Lemma 6.1. First, we choose $f_n \in \mathcal{H}_0^2$ so that $||f_n - f||_{L^2(dP\times dt)} \to 0$ as $n \to 0$. The triangle inequality for the $L^2(dP \times dt)$ norm then tells us that $||f_n||_{L^2(dP\times dt)} \to ||f||_{L^2(dP\times dt)}$. Similarly, since $I(f_n)$ converges to $I(f)$ in $L^2(dP)$, the triangle inequality in $L^2(dP)$ tells us that $||I(f_n)||_{L^2(dP)} \to ||I(f)||_{L^2(dP)}$. But by Lemma 6.1, we have

$$||I(f_n)||_{L^2(dP)} = ||f_n||_{L^2(dP\times dt)},$$

so taking the limit of this identity as $n \to \infty$ completes the proof of the theorem. □

A second variation of the Itô isometry that is often useful is the conditional version that is given by the next proposition.

PROPOSITION 6.1. *For any $b \in \mathcal{H}^2$ and any $0 \le s \le t$, we have*

$$E\left[\left(\int_s^t b(\omega,u)\,dB_u\right)^2 \middle|\, \mathcal{F}_s\right] = E\left(\int_s^t b^2(\omega,u)\,du \,\middle|\, \mathcal{F}_s\right). \tag{6.9}$$

PROOF. We could almost assert that this identity is obvious from the unconditional version, but we might soon wonder if it is *really* obvious or not. In any case, an honest proof is quick enough if we just note that equation (6.9) is equivalent to saying that for all $A \in \mathcal{F}_s$ we have

$$E\left[1_A\left(\int_s^t b(\omega,u)\,dB_u\right)^2\right] = E\left(1_A\int_s^t b^2(\omega,u)\,du\right). \tag{6.10}$$

Now, there is no reason to doubt obviousness of this equation; it follows immediately from the unconditional Itô isometry applied to the modified integrand

$$\tilde{b}(\omega,u) = \begin{cases} 0 & u \in [0,s] \\ 1_A b(\omega,u) & u \in (s,t]. \end{cases} \tag{6.11}$$

□

One of the reasons the last proposition is useful is that it implies that for any $b \in \mathcal{H}^2$ the process defined by

$$M_t = \left(\int_0^t b(\omega,u)\,dB_u\right)^2 - \int_0^t b^2(\omega,u)\,du$$

is a martingale. This gives us a large and useful class of martingales that generalize our trusted friend $B_t^2 - t$.

6.2. Third Step: Itô's Integral as a Process

The construction of the map $I\colon \mathcal{H}^2 \mapsto L^2(dP)$ moves us a long way forward, but to obtain a theory with the power to help us represent stochastic processes we need a map that takes a process to a *process* — not to a random variable. Fortunately, this richer view is within easy reach, and we will see that the Itô integral even provides us with a continuous martingale. We only need to find a proper way to view a whole continuum of Itô integrals at a single glance.

The natural idea is to look for a systematic way to introduce the time variable. For this purpose, we use the truncation function $m_t \in \mathcal{H}^2[0,T]$ defined by the simple formula

$$m_t(\omega, s) = \begin{cases} 1 & \text{if } s \in [0,t] \\ 0 & \text{otherwise.} \end{cases}$$

Now, for $f \in \mathcal{H}^2[0,T]$ the product $m_t f$ is also in $\mathcal{H}^2[0,T]$ for all $t \in [0,T]$, so $I(m_t f)$ is a well-defined element of $L^2(dP)$. One then might reasonably suppose that a good candidate for the process version of Itô's integral could be given by

$$X_t'(\omega) = I(m_t f)(\omega). \tag{6.12}$$

This is almost correct, but, if we take a hard look, we can also see that it is almost meaningless.

What's the Problem?

The problem is that for each $0 \le t \le T$ the integral $I(m_t f)$ is only defined as an element of $L^2(dP)$, so the value of $I(m_t f)$ can be specified arbitrarily on any set $A_t \in \mathcal{F}_t$ with $P(A_t) = 0$. In other words, the definition of $I(m_t f)$ is ambiguous on the null set A_t.

If we only had to consider a countable number of such A_t, this ambiguity would not be a problem because the union of a countable number of sets of probability zero is again a set of measure zero. Unfortunately, $[0,T]$ is an uncountable set and the union of the A_t over all t in $[0,T]$ might well be all of Ω. The bottom line is that if we are too naive in the way we build a process out of the integrals $I(m_t f)(\omega)$, then our construction could be ambiguous for all $\omega \in \Omega$. This would certainly be unacceptable.

What's the Solution?

Naturally, there is a way out of this quagmire; and, in fact, we can construct a *continuous martingale* X_t such that for all $t \in [0,T]$ we have

$$P(X_t = I(m_t f)) = 1. \tag{6.13}$$

The process $\{X_t\colon t \in [0,T]\}$ then gives us exactly what we want from a process version of the Itô integral. Thus, after all is said and done, the idea behind the "process" $X_t' = I(m_t f)(\omega)$ was not so terribly foolish; it only called for a more delicate touch.

THEOREM 6.2 (Itô Integrals as Martingales). *For any $f \in \mathcal{H}^2[0,T]$, there is a process $\{X_t\colon t \in [0,T]\}$ that is a continuous martingale with respect to the standard Brownian filtration $\mathcal{F}_t$ such that the event*

$$\{\omega\colon X_t(\omega) = I(m_t f)(\omega)\} \tag{6.14}$$

has probability one for each $t \in [0,T]$.

PROOF. We begin by taking a hint from our construction of the Itô integral on $[0,T]$, and we recall that the $\mathcal{H}_0^2$ approximation lemma gives us a sequence of functions $f_n \in \mathcal{H}_0^2[0,T]$ such that $||f - f_n||_{L^2(dP\times dt)} \to 0$. For $t \in [0,T]$, we also have $m_t f_n \in \mathcal{H}_0^2[0,T]$, and, since the Itô integral I is defined explicitly for elements of $\mathcal{H}_0^2[0,T]$, we can define a new process $X_t^{(n)}$ for all $n \ge 1$, $\omega \in \Omega$, and $0 \le t \le T$ by taking

$$X_t^{(n)}(\omega) = I(m_t f_n)(\omega).$$

We can even give an *explicit formula* for $X_t^{(n)}(\omega)$ because the definition (6.5) of $I(\cdot)$ on $\mathcal{H}_0^2$ tells us that for $t_k < t \le t_{k+1}$ we have

$$X_t^{(n)} = a_k(\omega)(B_t - B_{t_k}) + \sum_{i=0}^{k-1} a_i(\omega)(B_{t_{i+1}} - B_{t_i}). \tag{6.15}$$

Now, each $X_t^{(n)}$ is a continuous $\mathcal{F}_t$-adapted martingale, so for any $n \ge m$ we can simply apply Doob's L^2 maximal inequality to the continuous submartingale $M_t = |X_t^{(n)} - X_t^{(m)}|$ to find

$$\begin{aligned} P\Big(\sup_{0\le t\le T} |X_t^{(n)} - X_t^{(m)}| \ge \epsilon\Big) &\le \frac{1}{\epsilon^2} E(|X_T^{(n)} - X_T^{(m)}|^2) \\ &\le \frac{1}{\epsilon^2} ||f_n - f_m||^2_{L^2(dP\times dt)}, \end{aligned} \tag{6.16}$$

where in the second inequality we used Itô's isometry.

Because f_n converges to f in $L^2(dP \times dt)$, we can choose an increasing subsequence n_k such that

$$\max_{n\ge n_k} ||f_n - f_{n_k}||^2_{L^2(dP\times dt)} \le 2^{-3k};$$

so, in particular, taking $\epsilon = 2^{-k}$ in the bound (6.16) provides us with

$$P\Big(\sup_{0\le t\le T} |X_t^{(n_{k+1})} - X_t^{(n_k)}| \ge 2^{-k}\Big) \le 2^{-k}.$$

By the Borel–Cantelli lemma, we then have a set Ω_0 of probability one together with a random variable C such that $C(\omega) < \infty$ for all $\omega \in \Omega_0$ and

$$\sup_{0\le t\le T} |X_t^{(n_{k+1})}(\omega) - X_t^{(n_k)}(\omega)| \le 2^{-k} \qquad \text{for all } k \ge C(\omega). \tag{6.17}$$

Since 2^{-k} is summable, the bound (6.17) tells us that for all $\omega \in \Omega_0$ the sequence $\{X_t^{n_k}(\omega)\}$ is a Cauchy sequence in the uniform norm on $C[0,T]$. Thus, for each $\omega \in \Omega_0$, there is a continuous function $t \mapsto X_t(\omega)$ such that

$$X_t^{(n_k)}(\omega) \to X_t(\omega) \text{ uniformly on } [0,T].$$

Since the $\mathcal{F}_t$ martingales $\{X_t^{(n_k)}\}$ also converge in $L^2(dP)$ to $\{X_t\}$, the martingale identity for $\{X_t\}$ follows from the corresponding identity for the processes $\{X_t^{(n_k)}\}$. The bottom line is that $\{X_t\}$ is a continuous $\mathcal{F}_t$ martingale, just as we claimed.

To prove the last part of the theorem, we note that $m_t f_{n_k} \to m_t f$ in $L^2(dP \times dt)$, and therefore the Itô isometry tells us that $I(m_t f_{n_k}) \to I(m_t f)$ in $L^2(dP)$. Since we already have that

$$X_t^{(n_k)} = I(m_t f_{n_k}) \to X_t \text{ in } L^2(dP),$$

the uniqueness of $L^2(dP)$ limits tells us that $||X_t - I(m_t f)||_{L^2(dP)} = 0$ for each $t \in [0,T]$. This is precisely what we needed to deduce that the event (6.14) has probability one, so the proof is complete. □

6.3. The Integral Sign: Benefits and Costs

In his research notes for October 29, 1675, Leibniz wrote the memorable line:

UTILE ERIT SCRIBIT $\int$ PRO OMNIA.

Prior to this note, Leibniz had used the word OMNIA to capture the notion that he subsequently denoted with his integral sign. His note suggests that it would be useful to write $\int$ for OMNIA, and few people have been so right for so long.

Until now, we usually avoided the integral sign when writing stochastic integrals. This discipline helped to underscore the logical development of the Itô integral and to keep conflicting intuitions at a respectful distance. Now, we are close enough to the end of the development to be free to use the notation that Leibniz found so useful and that our own training makes so natural.

If $f \in \mathcal{H}^2[0,T]$ and if $\{X_t \colon 0 \le t \le T\}$ is a continuous martingale such that $P(X_t = I(m_t f)) = 1$ for all $0 \le t \le T$, then we will write

$$X_t(\omega) = \int_0^t f(\omega, s)\, dB_s \text{ for all } 0 \le t \le T. \tag{6.18}$$

We have gone to considerable lengths to define the left-hand side of equation (6.18), and the symbol on the right-hand side is nothing more than an evocative shorthand for the process defined on the left-hand side. Still, notation does make a difference, as one sees in the crisp restatement of the Itô isometry:

$$f \in \mathcal{H}^2 \Rightarrow E\left[\left(\int_0^t f(\omega, s)\, dB_s\right)^2\right] = E\left[\int_0^t f^2(\omega, s)\, ds\right] \quad \text{for all } t \in [0,T].$$

6.4. An Explicit Calculation

Perhaps the most natural way to confirm one's mastery of the construction of the Itô integral in $\mathcal{H}^2$ is to work out a concrete example. The easiest non-trivial example is given by taking $f(\omega, s) = B_s$, and we will find that the abstract definition of $I(m_t f)$ produces the engaging formula

$$X_t = \int_0^t B_s\, dB_s = \frac{1}{2}B_t^2 - \frac{1}{2}t. \tag{6.19}$$

This formula differs from the one we would expect from the usual calculus because of the presence of the extra term $-t/2$. We will find that such terms are characteristic of the Itô integral, and they turn out to have important probabilistic interpretations.

Before we dig into the construction of the integral, we should build some intuition by calculating the mean and variance of the two sides of equation (6.19). The martingale property of the Itô integral tells us $E(X_t) = 0$, and since $E(B_t^2) = t$ the right-hand side also has expectation zero — so far, so good. Next, we consider the variances. By the Itô isometry, we have

$$\operatorname{Var}(X_t) = E\left[\int_0^t B_s^2\, ds\right] = \frac{1}{2}t^2,$$

where we evaluated the last integral by interchanging integration and expectation and by using $E(B_s^2) = s$. Finally, when we expand the right-hand side of equation (6.19) and use $E(B_t^4) = 3t^2$ we again find a variance of $t^2/2$; so, as far as mean and variance are concerned, formula (6.19) is feasible.

Our variance check also confirms that $f(\omega, s) = B_s(\omega)$ is in $\mathcal{H}^2$, so we are also on solid ground when we start to work through the general construction. Our first step in the process is to set $t_i = iT/n$ for $0 \le i \le n$ and to define

$$f_n(\omega, t) = \sum_{i=0}^{n-1} B_{t_i} 1(t_i < t \le t_{i+1});$$

so, f_n is in $\mathcal{H}_0^2$, and f_n gives us a natural candidate for an $\mathcal{H}_0$ approximation to f. To confirm that $f_n(\omega, t) \to f(\omega, t) = B_t(\omega)$ in $L^2(dP \times dt)$, we have a straightforward computation:

$$\begin{aligned} ||f - f_n||^2_{L^2(dP\times dt)} &= E\left[\int_0^T \sum_{i=0}^{n-1} (B_t - B_{t_i})^2 1(t_i < t \le t_{i+1})\, dt\right] \\ &= \sum_{i=0}^{n-1} \int_{t_i}^{t_{i+1}} (t - t_i)\, dt \\ &= \frac{1}{2}\sum_{i=0}^{n-1} (t_{i+1} - t_i)^2 = \frac{1}{2}\frac{T^2}{n}. \end{aligned} \tag{6.20}$$

The computation confirms that $||f - f_n||_{L^2(dP\times dt)} \to 0$, and, as a consequence, we also have for the time-truncated functions that $||m_t(f - f_n)||_{L^2(dP\times dt)} \to 0$. Therefore, if we let $k_n(t) = k(t) = \max\{i\colon t_{i+1} \le t\}$, then by the definition of the Itô integral for a function in $\mathcal{H}^2$ we have

$$\int_0^t B_s\, dB_s = I(m_t f) = \lim_{n\to\infty} I(m_t f_n) =$$

$$\lim_{n\to\infty} \left\{ \sum_{i \le k(t)} B_{t_i}(B_{t_{i+1}} - B_{t_i}) + B_{t_{k(t)+1}}(B_t - B_{t_{k(t)+1}}) \right\}. \tag{6.21}$$

Also, to be perfectly explicit about the interpretation of the last equation, we note that the equality signs simply mean "equal for all ω in a set of probability one" and the limits are taken in the sense of $L^2(dP)$.

To clean up the last limit, we first observe that the last summand satisfies

$$0 \le E[B^2_{t_{k(t)+1}}(B_t - B_{t_{k(t)+1}})^2] = t_{k(t)+1}(t - t_{k(t)+1}) \le tT/n,$$

so the little scrap that is given as the last summand of equation (6.21) is asymptotically negligible. The more amusing observation is that we can rewrite the basic summands in a way that sets up telescoping:

$$B_{t_i}(B_{t_{i+1}} - B_{t_i}) = \frac{1}{2}(B^2_{t_{i+1}} - B^2_{t_i}) - \frac{1}{2}(B_{t_{i+1}} - B_{t_i})^2.$$

When we apply this identity in equation (6.21), we find

$$\lim_{n\to\infty} I(m_t f_n) = \lim_{n\to\infty} \frac{1}{2} B^2_{t_{k(t)+1}} - \frac{1}{2} \lim_{n\to\infty} \sum_{i\le k(t)} (B_{t_{i+1}} - B_{t_i})^2,$$

and, when we note that $B^2_{t_{k(t)+1}} \to B_t^2$ as $n \to \infty$, we see that equation (6.21) reduces to the pleasing representation

$$\int_0^t B_s\, dB_s = \frac{1}{2}B_t^2 - \frac{1}{2}\lim_{n\to\infty} \sum_{i\le k(t)} (B_{t_{i+1}} - B_{t_i})^2.$$

To complete the derivation of our integration formula (6.19), we only need to show that the last limit is equal to t. If we set

$$Y_n = \sum_{i \le k(t)} (B_{t_{i+1}} - B_{t_i})^2, \tag{6.22}$$

then we have $E(Y_n) = k(t)T/n$ and consequently $|E(Y_n) - t| \le 2T/n$. Further, since $E[(B_t - B_s)^4] = 3(t-s)^2$ we see that $\mathrm{Var}((B_t - B_s)^2)$ is just $2(t-s)^2$, so the simple bound $k(t) \le nt/T$ and the independence of the summands in equation (6.22) tell us

$$\mathrm{Var}(Y_n) = 2k(t)(T/n)^2 \le 2tT/n.$$

Finally, the last estimate implies that $Y_n \to t$ in $L^2(dP)$, exactly as required to complete the derivation of formula (6.19).

6.5. Pathwise Interpretation of Itô Integrals

One of the interesting psychological consequences of the use of Leibniz's integral sign is the increased inclination to treat stochastic integrals as ordinary integrals. In particular, the integral notation can easily seduce us into thinking about the integral as if it were defined on a path-by-path basis. Sometimes, pathwise reasoning is valid, and at other times it is bogus. Care is the watchword, and the most reliable tool for resolving disputes is to go back to the definition.

The next theorem provides a useful illustration of the risks and rewards that come from the integral sign. The result is one that we might easily guess, and, if our pathwise imagination is overactive, we might even think the result is obvious. Nevertheless, when we look at what is required in an honest proof, we see that our imagination did not do justice to the theorem's real meaning.

THEOREM 6.3 (Pathwise Interpretation of the Integral). *If $f \in \mathcal{H}^2$ is bounded and if ν is a stopping time such that $f(\omega, s) = 0$ for almost all $\omega \in \{\omega : s \le \nu\}$, then*

$$X_t(\omega) = \int_0^t f(\omega, s)\, dB_s = 0$$

for almost all $\omega \in \{\omega : t \le \nu\}$.

AN INTUITIVE, BUT BOGUS, ARGUMENT

If τ is *any* random variable with values in $[0, T]$, then it makes perfect sense to consider the random variable defined by $X_{\tau \wedge t}$, and it is even fine to write

$$X_{\tau \wedge t} = \int_0^{\tau \wedge t} f(\omega, s)\, dB_s. \tag{6.23}$$

Also, since we often use indicator functions to express the limits in ordinary integrals, we may then be quite tempted to rewrite (6.23) as

$$X_{\tau \wedge t} = \int_0^t f(\omega, s) 1(s \le \tau)\, dB_s. \tag{6.24}$$

Now, if $f(\omega, s) = 0$ almost everywhere on the set $\{s \le \tau\}$, we see that the integrand in (6.24) is identically zero, and we can conclude that $X_{\tau \wedge t}$ is also equal to zero.

This argument seems to prove Theorem 6.3, and it may even seem to provide a more general result. We never assumed τ to be a stopping time. Unfortunately, Theorem 6.3 is not true in such generality, and the argument contains two errors.

The easiest error to notice is that if τ is not a stopping time, then $f(\omega, s)1(s \leq \tau)$ need not be adapted, so the integral of equation (6.24) would not even be defined.

Moreover, even if we take τ to be a stopping time, there is still a gap in our reasoning despite the fact that equation (6.24) is well defined. The transition from equation (6.23) to equation (6.24) may look like a simple substitution, but, on careful review of the definition of the Itô integral, one will find that the justification of the substitution is a bit subtle.

An Honest Proof of Theorem 6.3

An honest proof is not difficult; one mainly needs to attend to the difference between the way the Itô integral is defined for integrands in $\mathcal{H}_0^2$ and the way it is defined in general. First, we recall that for any $f \in \mathcal{H}_0^2$ we may write f as a sum

$$f(\omega, s) = \sum_{i=0}^{n} a_i(\omega) 1(t_i < s \leq t_{i+1}), \tag{6.25}$$

where $a_i(\omega) \in \mathcal{F}_{t_i}$ for all $i = 0, 1, 2, ..., n$. From this representation we see that if $f(\omega, s) = 0$ for all (ω, s) such that $s \leq \nu(\omega)$, then we also have $a_i(\omega) = 0$ for all i such that $t_i < \nu(\omega)$. Also, the Itô integral of f up to t is given by the explicit formula

$$\int_0^t f(\omega, s)\, dB_s = I(m_t f)(\omega) = \sum_{i=1}^{m-1} a_i(\omega)(B_{t_{i+1}} - B_{t_i}) + a_m(\omega)(B_t - B_{t_m}),$$

where m is determined by the condition $t \in (t_m, t_{m+1}]$. Now, if $f(\omega, s) = 0$ for all $s \leq \nu(\omega)$, we see from our observation about the $\{a_i\}$ that $I(m_t f)(\omega) = 0$ for all (ω, t) satisfying $t \leq \nu(\omega)$. In other words, we have found a straightforward verification of Theorem 6.3 for $f \in \mathcal{H}_0^2$.

To begin the passage $\mathcal{H}_0^2$ to $\mathcal{H}^2$, we first recall that for any $f \in \mathcal{H}^2[0, T]$ there is a sequence $\{f_n\}$ of elements of $\mathcal{H}_0^2$ such that $f_n \to f$ in $L^2(dP \times dt)$. Moreover, since we assume that $|f|$ is bounded by B, we may even take f_n with the representation

$$f_n(\omega, s) = \sum_{i=0}^{2^n - 1} a_i(\omega) 1(t_i < s \leq t_{i+1})$$

where $|a_i(\omega)| \leq B$ and $t_i = iT/2^n$ for $0 \leq i \leq 2^n$. The only bad news is that our hypothesis that $f(\omega, s) = 0$ for all (ω, s) such that $s \leq \nu(\omega)$, does not guarantee that our approximations will share this property. Thus, to exploit our hypothesis, we need to modify the $\{f_n\}$ a bit. If we set

$$\hat{f}_n(\omega, s) = \sum_{i=0}^{2^n - 1} a_i(\omega) 1(t_i < s \leq t_{i+1}) 1(\nu \leq t_i), \tag{6.26}$$

then the $\mathcal{F}_{t_i}$ measurability of the indicator function $1(\nu \leq t_i)$ tells us that $\hat{f}_n$ is again an element of $\mathcal{H}_0^2$. The passage from $\mathcal{H}_0^2$ to $\mathcal{H}^2$ will follow quickly once we check that $\hat{f}_n$ the three basic properties:

- $\hat{f}_n(\omega, s) \to f(\omega, s) = f(\omega, s)1(\nu \leq t)$ in $L^2_{L^2(dP \times dt)}$,
- $\hat{f}_n(\omega, s) = 0$ for all (ω, s) such that $s \leq \nu$, and
- $I(m_t \hat{f}_n)(\omega) = 0$ for all ω such that $t \leq \nu$.

Since f_n converges to f in $L^2(dP \times dt)$ and since $f(\omega, s) = f(\omega, s)1(\nu < s)$, we see that $f_n(\omega, s)1(\nu < s)$ converges to $f(\omega, s)$ in $L^2(dP \times dt)$. Thus, to prove the first itemized property, we just need to show that the difference

$$\Delta_n(\omega, s) = f_n(\omega, s)1(\nu \le s) - \hat{f}_n(\omega, s).$$

satisfies $||\Delta_n||_{L^2(dP\times dt)} \to 0$. To check this fact, we first note that by the definition of $\hat{f}_n$ and f_n, the difference $\Delta_n(\omega, s)$ is zero except possibly on one interval $(t_i, t_{i+1}]$ with $\nu(\omega) \in (t_i, t_{i+1}]$. Thus, we have

$$\int_0^T |\Delta_n(\omega, s)|^2 \, ds \le 2^{-n} \max_i |a_i(\omega)|^2 \le 2^{-n} B^2,$$

and this bound is more than we need to show $||\Delta_n||_{L^2(dP\times dt)} \to 0$. Finally, the second itemized property of $\hat{f}_n$ is immediate from its definition, and the third property follows from the second property together with the fact that $\hat{f}_n \in \mathcal{H}_0^2$.

The pieces are now all in place to complete the proof Theorem 6.3. First, we note that for each fixed t the construction of the Itô integral for $f \in \mathcal{H}^2$ tells us

$$X_t = \int_0^t f(\omega, s) \, dB_s = I(m_t f)(\omega) = \lim_{n\to\infty} I(m_t \hat{f}_n)(\omega) \tag{6.27}$$

for all ω except possibly for a set of probability zero. We have already seen that $I(m_t \hat{f}_n)(\omega) = 0$ for all (ω, t) such that $t \le \nu$, so the preceding identity tells us $X_t = 0$ for all ω such that $t \le \nu$, except for a set E_t of probability zero. If we apply this fact for all rational t and use the fact that $X_t(\omega)$ is a continuous process, we see that except for a set of probability zero, we have we have $X_t = 0$ for all $t \le \nu$. Thus, the proof of Theorem 6.3 is complete.

With a bit more work, we can extend Theorem 6.3 to arbitrary elements of $\mathcal{H}^2$ and rephrase Theorem 6.3 in terms of the persistence under integration of an identity. This retooling of Theorem 6.3 is quite useful in the next two chapters.

PERSISTENCE OF IDENTITY

THEOREM 6.4 (Persistence of Identity). *If f and g are elements of $\mathcal{H}^2$ and if ν is a stopping time such that $f(\omega, s) = g(\omega, s)$ for almost all $\omega \in \{\omega \colon s \le \nu\}$, then the integrals*

$$X_t(\omega) = \int_0^t f(\omega, s) \, dB_s \text{ and } Y_t(\omega) = \int_0^t g(\omega, s) \, dB_s$$

are equal for almost all $\omega \in \{\omega \colon t \le \nu\}$.

PROOF. If we take $f_n = f1(|f| \le n)$ and $g_n = g1(|g| \le n)$, then the dominated convergence theorem tells us that

$$f_n \to f \text{ and } g_n \to g \text{ in } L^2(dP \times dt), \tag{6.28}$$

so by the Itô isometry we also have

$$\int_0^t f_n(\omega, s) \, dB_s \to X_t \text{ and } \int_0^t g_n(\omega, s) \, dB_s \to Y_t \text{ in } L^2(dP). \tag{6.29}$$

Now, by Theorem 6.3, we already know that

$$\int_0^t f_n(\omega, s) \, dB_s = \int_0^t g_n(\omega, s) \, dB_s \text{ almost everywhere on } \{\omega \colon t \le \nu\},$$

so (6.29) finally tells us that $X_t = Y_t$ almost everywhere on $\{\omega \colon t \le \nu\}$. □

6.6. Approximation in $\mathcal{H}^2$

Our definition of the Itô integral on $\mathcal{H}^2$ used the fact that $\mathcal{H}_0^2$ is a dense subset of $\mathcal{H}^2$, and the main goal of this section is to prove an explicit approximation theorem that provides the desired density result as an immediate corollary. The proof of this approximation theorem is interesting for several reasons, not the least of which is that it illustrates the important fact that one can sometimes approximate a "nonmartingale" with a martingale. Still, this proof does require a detour from our main path, and the reader who is pressed for time should know that the techniques of this section will not be used in subsequent chapters.

Approximation Operator

Given any integer $n \geq 1$, the approximation operator A_n is defined to be the mapping from $\mathcal{H}^2$ into $\mathcal{H}_0^2$ that is given by taking

$$A_n(f) = \sum_{i=1}^{2^n-1} \left\{ \frac{1}{t_i - t_{i-1}} \int_{t_{i-1}}^{t_i} f(\omega, u)\, du \right\} 1(t_i < t \leq t_{i+1}), \tag{6.30}$$

where $t_i = iT/2^n$ for $0 \leq i \leq 2^n$. Our interest in this operator is explained by the following theorem and its immediate corollary — $\mathcal{H}_0^2$ is a dense subset of $\mathcal{H}^2$.

Theorem 6.5 (Approximation Theorem). *The approximation operator A_n defines a bounded linear mapping from $\mathcal{H}^2$ into $\mathcal{H}_0^2$ such that*

$$||A_n(f)||_\infty \leq ||f||_\infty, \tag{6.31}$$

$$||A_n(f)||_{L^2(dP\times dt)} \leq ||f||_{L^2(dP\times dt)}, \tag{6.32}$$

and

$$\lim_{n\to\infty} ||A_n(f) - f||_{L^2(dP\times dt)} = 0 \quad \textit{for all } f \in \mathcal{H}^2. \tag{6.33}$$

First Checks

The operator A_n is clearly linear. To verify that $A_n(f) \in \mathcal{H}_0^2$, we first consider measurability. First, note that for all $0 < i < 2^n$, the coefficients

$$a_i(\omega) = \frac{1}{t_i - t_{i-1}} \int_{t_{i-1}}^{t_i} f(\omega, u)\, du \tag{6.34}$$

are in $\mathcal{F}_{t_i}$. For $t \in [0, T2^{-n}]$ we have $A_n(f)(t) = 0$, while for $1 \leq i < 2^n$ and $t_i < t \leq t_{i+1}$ we have $A_n(f)(t) = a_i(\omega) \in \mathcal{F}_{t_i} \subset \mathcal{F}_t$. Therefore, for all $0 \leq t \leq T$, we have $A_n(f)(t) \in \mathcal{F}_t$. Next, to show that $A_n(f) \in \mathcal{H}_0^2$, we only need to prove that $E[(A_n(f))^2] < \infty$, and this follows immediately if we show $E(a_i^2) < \infty$. By Jensen's inequality, we have

$$a_i^2(\omega) \leq \frac{1}{t_i - t_{i-1}} \int_{t_{i-1}}^{t_i} f(\omega, u)^2\, du, \tag{6.35}$$

and since the right-hand side is bounded by $(t_i - t_{i-1})^{-1}||f||^2_{L^2(dP\times dt)} < \infty$, we find $E(a_i^2) < \infty$. This bound tells us that $A_n(f) \in \mathcal{H}_0^2$, exactly as required.

The proof of the L^∞ bound (6.31) is trivial, and the proof of the L^2 bound (6.32) is not much harder. When we square $A_n(f)$ and integrate over $[0, T]$, we find that inequality (6.35) gives us

$$A_n^2(f) = \sum_{i=1}^{2^n-1} 1(t_i < t \le t_{i+1}) a_i^2(\omega),$$

so, when we take expectations and apply the bound (6.35), we find

$$E\left[\int_0^T A_n^2(f)\,dt\right] = E\left[\sum_{i=1}^{2^n-1}(t_{i+1}-t_i)a_i^2(\omega)\right] \le E\left[\int_0^T f(\omega,t)^2\,dt\right],$$

precisely the required L^2 bound (6.32).

A Martingale Cousin

We now come to the crux of Theorem 6.5, the proof of the limit (6.33). Here, it is fruitful to introduce a new operator that looks a lot like A_n but which has an interpretation as a martingale. This new operator B_n is defined for $f \in \mathcal{H}^2$ by taking $t_i = iT/2^n$ for $0 \le i \le 2^n$ and setting

$$B_n(f) = \sum_{i=1}^{2^n}\left\{\frac{1}{t_i - t_{i-1}}\int_{t_{i-1}}^{t_i} f(\omega,u)\,du\right\} 1(t_{i-1} < t \le t_i). \tag{6.36}$$

The operator B_n is closely related to A_n, although one should note that B_n no longer maps $\mathcal{H}^2$ into $\mathcal{H}_0^2$; the coefficients have been shifted in such a way that the coefficient of the indicator of the interval $(t_{i-1}, t_i]$ is no longer measurable with respect to $\mathcal{F}_{t_{i-1}}$. This means that $B_n(f)$ no longer has the measurability that is required of a function in $\mathcal{H}_0^2$. In compensation, we have for all $f \in L^2(dP \times dt)$ and all $\omega \in \Omega$ that the process defined by

$$\{M_n(\cdot) = B_n(f)(\omega,\cdot) : n \ge 0\}$$

is a martingale, or at least it is a martingale once we spell out the filtration and the probability space where the martingale property holds.

In fact, our new martingale is conceptually identical to the one introduced in Exercise 4.7, but in this case we have some extra baggage since $B_n(\omega, t)$ is a function of two variables. Our task is thus to show that when we fix ω we obtain a martingale in the t variable that exactly replicates the martingale of Exercise 4.7. This is a little tedious to write out, but it is easy to think through.

To punch out the formalities, we first fix $\omega \in \Omega$ and then define a new probability space $(\Omega', \mathcal{F}', Q)$ by taking the base space $\Omega' = \omega \times [0,T]$ and a new σ-field $\mathcal{F}' = \{(\omega, A) : A \in \mathcal{B}\}$. We then define a probability measure Q on $\mathcal{F}'$ by taking $Q(\omega, A) = P(\Omega \times A)/T$ for all $A \in \mathcal{B}$. Finally, the filtration for our martingale is taken to be $\{\mathcal{G}_n\}$, where $\mathcal{G}_n$ is the smallest σ-field contained in $\mathcal{F}'$ such that all of the functions of the form

$$\sum_{i=1}^{2^n} c_i 1(t_{i-1} < t \le t_i) \text{ where } c_i \in \mathbb{R} \tag{6.37}$$

are $\mathcal{G}_n$-measurable.

The benefit of this choice of $\mathcal{G}_n$ is that it gives us a representation of $B_n(f)(\omega,\cdot)$ as a conditional expectation

$$E_Q(f(\omega,t) \mid \mathcal{G}_n) = B_n(f)(\omega,t), \tag{6.38}$$

so the martingale property for M_n comes to us as a consequence of the tower property of conditional expectations. We are now ready to prove a lemma that tells us $B_n(f)$ is a good approximation to f provide that $f \in \mathcal{H}^2 \subset L^2(dP \times dt)$ is bounded.

LEMMA 6.3. *If $f \in \mathcal{H}^2$ and $|f(\omega,t)| \le B$ for all (ω,t), then*

$$||B_n(f) - f||_{L^2(dP\times dt)} \to 0 \text{ as } n \to \infty. \tag{6.39}$$

PROOF. If we fix ω, our interpretation of $B_n(f)(\omega,\cdot)$ as a martingale lets us apply the L^2-bounded martingale convergence theorem to conclude that $B_n(f)(\omega,t)$ converges for all $t \in [0,T]$ except possibly a set of measure zero. We denote this limit by $B_\infty(f)(\omega,t)$ and note by the bound on f this limit is bounded by B. We can then apply the dominated convergence theorem to find

$$\lim_{n\to\infty} \int_A B_n(f)(\omega,t)\,dt = \int_A B_\infty(f)(\omega,t)\,dt \text{ for all } A \in \mathcal{B}. \tag{6.40}$$

We also know directly from the definition of the sequence $\{B_n\}$ that

$$\int_A B_m(f)(\omega,t)\,dt = \int_A f(\omega,t)\,dt \tag{6.41}$$

for all $m \ge n$ and all A such that $(\omega, A) \in \mathcal{G}_n$. From equations (6.40) and (6.41), we can conclude (e.g., by Exercise 6.5) that for all $\omega \in \Omega$ we have

$$B_\infty(f)(\omega,t) = f(\omega,t) \tag{6.42}$$

for all $t \in [0,T]$ except a set of measure zero. By a second application of the DCT, we therefore have for all $\omega \in \Omega$ that

$$\lim_{n\to\infty} \int_0^T |B_n(f)(\omega,t) - f(\omega,t)|^2\,dt = 0. \tag{6.43}$$

Finally, we can take expectations in equation (6.43) and apply the DCT one last time to complete the proof of the limit (6.39) and the lemma. □

THE MARTINGALE MEETS THE APPROXIMATION

The key connection between A_n and B_n is given by the following lemma.

LEMMA 6.4. *For any $f \in \mathcal{H}^2$ and any fixed integer m,*

$$\lim_{n\to\infty} ||(A_n(B_m(f)) - B_m(f)||_{L^2(dP\times dt)} = 0.$$

PROOF. Two basic observations yield the lemma. First, we note that by direct consideration of the definition of B_m we have that $B_m(f)(\omega,t)$ is constant on each of the intervals $(i2^{-m}T, (i+1)2^{-m}T]$ for $i = 0, 1, \ldots, 2^m - 1$, so from the definition of A_n we find for $n \ge m$ and all ω that

$$A_n(B_m(f))(\omega,t) = B_m(f)(\omega,t) \text{ for all } t \notin \cup_{1\le i<2^m}[i2^{-m}T,\, i2^{-m}T + 2^{-n}T].$$

In view of this identity, we have

$$\lim_{n\to\infty} A_n(B_m(f))(\omega,t) = B_m(f)(\omega,t)$$

for all (ω,t) such that $t \ne i2^{-m}T$.

Second, we note that the converging terms are dominated. In particular, for all $n \ge m$, we have

$$|A_n(B_m(f))(\omega,t)| \le |B_m(f)(\omega,t)| + |B_m(f)(\omega,t-2^{-n}T)|,$$

and $|B_m(f)(\omega,t)| \in L^2(dP \times dt)$, so for all $n \geq 1$ the difference $|A_n(B_m(f))(\omega,t) - B_m(f)(\omega,t)|^2$ is bounded by a fixed integrable function. The DCT then completes the proof of the lemma.

PROOF OF THE APPROXIMATION THEOREM: LAST STEP

We only need prove the limit result (6.33). First, we consider a bounded $f \in \mathcal{H}^2$. By the triangle inequality and the L^2 bound (6.32), we have

$$\begin{aligned}||A_n(f) - f||_{L^2(dP\times dt)} &\leq ||A_n(f - B_m(f))||_{L^2(dP\times dt)} + ||A_n(B_m(f)) - f||_{L^2(dP\times dt)}\\ &\leq ||(f - B_m(f))||_{L^2(dP\times dt)} + ||A_n(B_m(f)) - f||_{L^2(dP\times dt)},\end{aligned}$$

and by Lemma 6.4 we have $||A_n(B_m(f)) - B_m(f)||_{L^2(dP\times dt)} \to 0$, so for all m we have

$$\limsup_{n\to\infty} ||A_n(f) - f||_{L^2(dP\times dt)} \leq 2||(f - B_m(f))||_{L^2(dP\times dt)}.$$

Since m is arbitrary, this inequality and Lemma 6.3 tell us that

$$\lim_{n\to\infty} ||A_n(f) - f||_{L^2(dP\times dt)} = 0$$

for all bounded $f \in \mathcal{H}^2$.

Now, at last, we consider a possibly unbounded $f \in \mathcal{H}^2$. For any $\epsilon > 0$, we have a bounded $f_0 \in \mathcal{H}^2$ such that $||f - f_0||_{L^2(dP\times dt)} \leq \epsilon$, and now we can check that the contraction property of A_n will complete the proof. Specifically, by the triangle inequality and the L^2 bound (6.32), we have

$$\begin{aligned}||A_nf - f||_{L^2(dP\times dt)} &= ||A_nf_0 - f_0 + A(f - f_0) - (f - f_0)||_{L^2(dP\times dt)}\\ &\leq ||A_nf_0 - f_0||_{L^2(dP\times dt)} + ||A(f - f_0)||_{L^2(dP\times dt)} + ||(f - f_0)||_{L^2(dP\times dt)}\\ &\leq ||A_nf_0 - f_0||_{L^2(dP\times dt)} + 2||(f - f_0)||_{L^2(dP\times dt)}.\end{aligned}$$

If we now let $n \to \infty$, the boundedness of f_0 and the first half of our argument tell us $||A_nf_0 - f_0||_{L^2(dP\times dt)}$ goes to zero. Finally, the arbitrariness of ϵ completes the proof of the fundamental limit result (6.33) and the proof of Theorem 6.5. □

6.7. Exercises

The first three exercises offer "must do" practice with the Itô isometry and the distinction between ordinary and stochastic integrals. The last two exercises are more technical. In particular, Exercise 6.5 sketches the proof of an intuitive result that we needed in our proof of the approximation theorem, and Exercise 6.4 suggests an alternative way to find a dense subset of $\mathcal{H}^2$ in $\mathcal{H}^2_0$.

EXERCISE 6.1. Use the Itô isometry to calculate the variances of

$$\int_0^t |B_s|^{\frac{1}{2}}\, dB_s \text{ and } \int_0^t (B_s + s)^2\, dB_s.$$

EXERCISE 6.2. The integrals

$$I_1 = \int_0^t B_s\, ds \text{ and } I_2 = \int_0^t B_s^2\, ds$$

are *not* stochastic integrals, although they are random variables. For each ω the integrands are nice continuous functions of s and the ds integration is just the traditional calculus integration. Find the mean and variance of the random variables I_1 and I_2.

EXERCISE 6.3. For any fixed t the random variable B_t has the same distribution as $X_t = \sqrt{t}Z$, where $Z \sim N(0,1)$, but as processes B_t and X_t could not be more different. For example, the paths of X_t are differentiable for all $t > 0$ with probability one, but the paths of B_t are not differentiable for any t with probability one. For another difference (and similarity), show that for a bounded continuous f the processes

$$U_t = \int_0^t f(B_s)\,ds \text{ and } V_t = \int_0^t f(X_s)\,ds$$

will have the same expectations but will not in general have the same variance.

EXERCISE 6.4. Let C denote the set of all measurable, adapted functions $f(t,\omega)$ that are continuous in the t variable. Show that if we assume C is dense in $\mathcal{H}^2$, then we can construct $\{f_n\} \subset \mathcal{H}_0^2$ such that $||f - f_n||_{L^2(dP\times dt)} \to 0$ without appealing to the approximation theorem.

EXERCISE 6.5 (A Uniqueness Theorem). Suppose that f and g are integrable functions that are $\mathcal{F}$ measurable. Suppose that we have a sequence of σ-fields $\{\mathcal{F}_n\}$ such that $\mathcal{F}_n \subset \mathcal{F}_{n+1}$ and for which

$$\int_A f(x)\,dx = \int_A g(x)\,dx \text{ for all } A \in \mathcal{F}_n. \tag{6.44}$$

Show that if $\mathcal{F}$ is the smallest σ-field that contains $\mathcal{F}_n$ for all n, then

$$f(x) = g(x) \text{ for all } x \text{ except a set of measure zero.}$$

The reader who wants a hint might first note that $\{x : f > g\} \in \mathcal{F}$. Next, it may be useful to show that equation (6.44) holds for all $A \in \mathcal{F}$. Here one may want to consider the set $\mathcal{G}$ of all $A \in \mathcal{F}$ for which we have equation (6.44) and then show that $\mathcal{G}$ is a σ-field.

CHAPTER 7

Localization and Itô's Integral

If $f\colon \mathbb{R} \to \mathbb{R}$ is a continuous function, then any truly convenient theory of stochastic integration should have no trouble with the definition of the integral

$$\int_0^T f(B_t)\,dB_t. \tag{7.1}$$

The Itô integral will ultimately meet this test, but so far it has only been defined for integrands that satisfy the integrability constraint:

$$E\left[\int_0^T f^2(B_t)\,dt\right] < \infty. \tag{7.2}$$

Unfortunately, this inequality can fail even for a perfectly pleasant continuous function such as $f(x) = \exp(x^4)$. The bottom line is that if we want stochastic integration to work in the most natural way, then we must find some way to circumvent the integrability constraint.

Luckily, there is a way out of the trap — even a nice, general way. Frequently one needs to relax an integrability constraint, and it is often the case that the method of *localization* will do the trick. Here, we will see that an appropriate use of localization will permit us to extend the Itô integral to a class of integrands that easily contains all of the continuous functions of Brownian motion; so, as we hoped, the Itô integral (7.1) can be defined without imposing any explicit integrability conditions.

After completing this final step of the definition of the Itô integral, we examine two special cases of importance. In the first of these, we see that the Itô integral of a continuous function of Brownian motion can be written as the limit of Riemann sums. In the second case, we see that the Itô integral of a deterministic function is always a Gaussian process. This last fact lets us build an example that shows why $\mathcal{L}^2_{\text{LOC}}$ is a natural home for Itô integration, and it also gives us a large class of martingales that can be viewed as Brownian motions with a timescale that speeds up and slows down. Finally, we will collect some simple tools that help us relate local martingales to honest ones.

7.1. Itô's Integral on $\mathcal{L}^2_{\text{LOC}}$

We begin by considering the class $\mathcal{L}^2_{\text{LOC}} = \mathcal{L}^2_{\text{LOC}}[0,T]$ of all of the adapted, measurable functions $f\colon \Omega \times [0,T] \mapsto \mathbb{R}$ such that

$$P\left(\int_0^T f^2(\omega,t)\,dt < \infty\right) = 1. \tag{7.3}$$

This class of functions certainly contains $\mathcal{H}^2$. Also, for any continuous $g : \mathbb{R} \to \mathbb{R}$, we have $f(\omega,t) = g(B_t) \in \mathcal{L}^2_{\text{LOC}}$ because the continuity of Brownian motion implies that for each ω the mapping $t \mapsto g(B_t(\omega))$ yields a bounded function on $[0,T]$. Still, the real wisdom behind $\mathcal{L}^2_{\text{LOC}}$ is that it relates nicely to stopping times. To make this relationship explicit, we first require a definition.

DEFINITION 7.1 (Localizing Sequence for $\mathcal{H}^2$).
An increasing sequence of stopping times is called an $\mathcal{H}^2[0,T]$ *localizing sequence for* f *provided that*

$$f_n(\omega,t) = f(\omega,t)1(t \le \nu_n) \in \mathcal{H}^2[0,T] \text{ for all } n \tag{7.4}$$

and that

$$P\left(\bigcup_{n=1}^{\infty}\{\omega\colon \nu_n = T\}\right) = 1. \tag{7.5}$$

One reason that the space $\mathcal{L}^2_{\text{LOC}}$ turns out to be a natural domain for the Itô integral is that any $f \in \mathcal{L}^2_{\text{LOC}}$ has a localizing sequence, as we see in the following proposition.

PROPOSITION 7.1 (Localization in $\mathcal{L}^2_{\text{LOC}}$). *For any* $f \in \mathcal{L}^2_{\text{LOC}}[0,T]$, *the sequence defined by*

$$\tau_n = \inf\left\{s\colon \int_0^s f^2(\omega,t)\,dt \ge n \text{ or } s \ge T\right\} \tag{7.6}$$

is an $\mathcal{H}^2[0,T]$ *localizing sequence for* f.

PROOF. We always have equality of the sets

$$\bigcup_{n=1}^{\infty}\{\omega\colon \tau_n = T\} = \left\{\omega\colon \int_0^T f^2(\omega,t)\,dt < \infty\right\},$$

and for $f \in \mathcal{L}^2_{\text{LOC}}$ the second set has probability one. Furthermore, by the construction of τ_n, we have for $f_n(\omega,t) = f(\omega,t)1(t \le \tau_n)$ that

$$\|f_n\|^2_{L^2(dP\times dt)} \le n,$$

so we certainly have $f_n \in \mathcal{H}^2$ for all n. Consequently, $\{\tau_n\}$ is a localizing sequence as claimed. □

THE $\mathcal{L}^2_{\text{LOC}}$ EXTENSION IN A NUTSHELL

Now, for the construction, we first take any $f \in \mathcal{L}^2_{\text{LOC}}$ and let $\{\nu_n\}$ be any localizing sequence for f. Next, for each n, take $\{X_{t,n}\}$ to be the unique continuous martingale on $[0,T]$ that is a version of the Itô integral of $I(m_t g)$ where $g(\omega,s) = f(\omega,s)1(s \le \nu_n(\omega))$. Finally, we define the Itô integral for $f \in \mathcal{L}^2_{\text{LOC}}[0,T]$ to be the process given by the limit of the processes $\{X_{t,n}\}$ as $n \to \infty$. More precisely, we will show that there is a unique continuous process $\{X_t\colon 0 \le t \le T\}$ such that

$$P\left(X_t = \lim_{n\to\infty} X_{t,n}\right) = 1 \text{ for all } t \in [0,T], \tag{7.7}$$

and we then take the process $\{X_t\}$ to be our Itô integral of $f \in \mathcal{L}^2_{\text{LOC}}$, or, in symbols, we define the Itô integral of f by setting

$$\int_0^t f(\omega,s)\,dB_s \stackrel{\text{def}}{=} X_t(\omega) \text{ for } t \in [0,T]. \tag{7.8}$$

Naturally, some work is required to justify this definition, and the first order of business is to prove that the processes $X_{t,n}$ converge. The next proposition tells us that this convergence actually takes place in the nicest possible way. For any $t \in [0,T]$ and for almost all ω, the sequence $\{X_{t,n}: 1 \le n < \infty\}$ is ultimately constant.

FIRST CHECKS: CONSISTENCY, CONVERGENCE, CONTINUITY

PROPOSITION 7.2 (Sequential Consistency). *For any $f(\omega, s) \in \mathcal{L}^2_{\mathrm{LOC}}[0,T]$ and any localizing sequence $\{\nu_n\}$, if $\{X_{t,n}\}$ is the continuous martingale version of the Itô integrals $I(\,m_t\{f(\omega,s)1(s \le \nu_n)\})$, then for all $t \in [0,T]$ and $n \ge m$ we have*

$$X_{t,n} = X_{t,m} \text{ for almost all } \omega \in \{\omega: t \le \nu_m\}. \tag{7.9}$$

PROOF. Since $\nu_m \le \nu_n$, the two $\mathcal{H}^2$ functions

$$f_m(\omega,t) = f(\omega,t)1(t \le \nu_m) \text{ and } f_n(\omega,t) = f(\omega,t)1(t \le \nu_n)$$

are equal on the set $\{\omega: t \le \nu_m\}$, so (7.9) is immediate by the persistence of identity guaranteed by Theorem 6.4. □

With sequential consistency in hand, the strong convergence of $\{X_{t,n}\}$ to a continuous process is almost immediate. Nevertheless, this convergence is at the heart of our definition of the Itô integral for $f \in \mathcal{L}^2_{\mathrm{LOC}}$, so we will lay out the details.

PROPOSITION 7.3. *There is a continuous process $\{X_t: 0 \le t \le T\}$ such that*

$$P\Big(X_t = \lim_{n\to\infty} X_{t,n}\Big) = 1 \text{ for all } t \in [0,T]. \tag{7.10}$$

PROOF. If we define a random index N by taking

$$N = \min\{n: \nu_n = T\},$$

then by the definition of a localizing sequence we have that $P(N < \infty) = 1$. We then let Ω_0 denote the set of probability one for which all of the functions $t \mapsto X_{t,n}(\omega)$ are continuous.

Finally, for any $\omega \in \Omega_1 \equiv \{N < \infty\} \cap \Omega_0$, we define $X_t(\omega)$ by taking

$$X_t(\omega) = X_{t,N}(\omega).$$

The map defined by $t \mapsto X_{t,N}(\omega)$ is continuous for all $\omega \in \Omega_1$, so $\{X_t\}$ is a continuous process. Moreover, by Proposition 7.2, we have

$$P\Big(\lim_{n\to\infty} X_{t,n} = X_{t,N}\Big) = 1 \text{ for all } t \in [0,T],$$

so the proof of the proposition is complete with $X_t \equiv X_{N,t}$. □

NEXT CHECK: INDEPENDENCE OF THE LOCALIZATION

The Itô integral is now defined unambiguously for all $f \in \mathcal{L}^2_{\mathrm{LOC}}[0,T]$, but, as it sits, the defining limit (7.8) appears to depend on the choice that we make for the localizing sequence $\{\tau_n\}$. The next proposition tells us that in fact there is no such dependence because with probability one every localizing sequence leads to the same limiting value.

PROPOSITION 7.4 (Localizer Independence). *If ν_n and $\hat{\nu}_n$ are both localizing sequences for an $f \in \mathcal{L}^2_{\text{LOC}}[0,T]$, then the corresponding continuous martingale versions $X_{t,n}$ and $\hat{X}_{t,n}$ of the Itô integrals $I(\,m_t\{f(\omega,s)1(s \le \nu_n)\}\,)$ and $I(\,m_t\{f(\omega,s)1(s \le \hat{\nu}_n)\}\,)$ satisfy*

$$\lim_{n\to\infty} X_{t,n} = \lim_{n\to\infty} \hat{X}_{t,n} \tag{7.11}$$

with probability one for each $t \in [0,T]$.

PROOF. If we let $\tau_n = \min(\nu_n, \hat{\nu}_n)$, then by Theorem 6.4 on persistence of identity we have for all $n \ge m$ that

$$\hat{X}_{t,n} = X_{t,n} \text{ almost everywhere on } \{t \le \tau_m\}. \tag{7.12}$$

Also, by Proposition 7.3 we know that $\hat{X}_{t,n}$ and $X_{t,n}$ both converge with probability one, so equation (7.12) tells us that

$$\lim_{n\to\infty} \hat{X}_{t,n} = \lim_{n\to\infty} X_{t,n} \text{ almost everywhere on } \{t \le \tau_m\}. \tag{7.13}$$

Finally, we note that the set

$$\bigcup_{m=1}^{\infty} \{\tau_m = T\} = \bigcup_{m=1}^{\infty} \{\nu_m = T \text{ and } \hat{\nu}_m = T\}$$

has probability one because both ν_m and $\hat{\nu}_m$ are localizing sequences, so we have equation (7.11) just as needed. □

LAST CHECK: PERSISTENCE OF IDENTITY IN $\mathcal{L}^2_{\text{LOC}}$

Next, we check that the Itô integral in $\mathcal{L}^2_{\text{LOC}}$ also has the persistence of identity property that we established in Theorem 6.4 for integrals of functions in $\mathcal{H}^2$. This is a completely straightforward proposition, but it gives nice practice with localization. It also puts the persistence of identity property into the form that makes it ready for immediate application.

PROPOSITION 7.5 (Persistence of Identity in $\mathcal{L}^2_{\text{LOC}}$). *If f and g are elements of $\mathcal{L}^2_{\text{LOC}}$ and if ν is a stopping time such that $f(\omega,s) = g(\omega,s)$ for all $0 \le s \le \nu$, then the integrals*

$$X_t(\omega) = \int_0^t f(\omega,s)\,dB_s \text{ and } Y_t(\omega) = \int_0^t g(\omega,s)\,dB_s$$

are equal for almost all $\omega \in \{\omega : t \le \nu\}$.

PROOF. If we take

$$\tau_n = \inf\{s : |f(\omega,s)| \ge n,\ |g(\omega,s)| \ge n, \text{ or } s \ge T\},$$

then $f(\omega,s)1(s \le \tau_n)$ and $g(\omega,s)1(s \le \tau_n)$ are bounded elements of $\mathcal{H}^2$, so by Theorem 6.4 on the persistence of identity in $\mathcal{H}^2$ we know that the two integrals

$$X_{t,n} = \int_0^t f(\omega,s)1(s \le \tau_n)\,dB_s \text{ and } Y_{t,n} = \int_0^t g(\omega,s)1(s \le \tau_n)\,dB_s$$

have the property that for all $n \ge m$ one has

$$X_{t,n} = Y_{t,n} \text{ for almost all } \omega \in \{\omega : t \le \nu\} \cap \{\omega : t \le \tau_m\}. \tag{7.14}$$

Now, if we let $n \to \infty$ in equation (7.14), we see by Proposition 7.3 that the limit processes $\{X_t\}$ and $\{Y_t\}$ exist and satisfy

$$X_t = Y_t \text{ for almost all } \omega \in \{\omega\colon t \le \nu\} \cap \{\omega\colon t \le \tau_m\}. \tag{7.15}$$

Because τ_n is a localizing sequence, we see that

$$\{\omega\colon t \le \nu\} = \bigcup_{m=1}^{\infty} \{\omega\colon t \le \nu\} \cap \{\omega\colon t \le \tau_m\},$$

so equation (7.15) tells us

$$X_t(\omega) = Y_t(\omega) \text{ for almost all } \omega \in \{\omega\colon t \le \nu\},$$

just as needed. □

PERSPECTIVE ON THE $\mathcal{L}^2_{\text{LOC}}$ EXTENSION

We now have defined the Itô integral on its natural domain, and we can happily consider the Itô integral of any continuous function of Brownian motion. The price of this extension has been one additional layer of abstraction, but the bargain is a sound one. Anytime we feel the need, we can introduce an appropriate localizing sequence that will bring our work back to the familiar ground of $\mathcal{H}^2$. In fact, we will often find it useful to take a localizing sequence that permits us to do our work in a nice subset of $\mathcal{H}^2$, such as the subset of bounded elements.

7.2. An Intuitive Representation

The definition of the Itô integral for a general element of $\mathcal{L}^2_{\text{LOC}}$ required a sustained and somewhat abstract development, but there are cases of importance where a quicker and more concrete representation is possible. In particular, for continuous functions of Brownian motion the Itô integral also has a natural interpretation as the limit of a Riemann sum. This fact adds to our intuition about the Itô integral, and the proof of the theorem also reveals how we can make life easy on ourselves by choosing an appropriate localizing sequence.

THEOREM 7.1 (Riemann Representation). *For any continuous $f\colon \mathbb{R} \to \mathbb{R}$, if we take the partition of $[0,T]$ given by $t_i = iT/n$ for $0 \le i \le n$, then we have*

$$\lim_{n\to\infty} \sum_{i=1}^{n} f(B_{t_{i-1}})(B_{t_i} - B_{t_{i-1}}) = \int_0^T f(B_s)\,dB_s, \tag{7.16}$$

where the limit is understood in the sense of convergence in probability.

PROOF. The first step of the proof is to introduce a localization that permits us to focus on the representation theorem for a continuous function with compact support. We begin by settling $\tau_M = \min\{\, t\colon |B_t| \ge M \text{ or } t \ge T \,\}$ and by noting that τ_M is a localizing sequence for $f(B_s(\omega)) \in \mathcal{L}^2_{\text{LOC}}[0,T]$. We also note that for any $M > 0$ there is a continuous function f_M that has compact support and such that $f_M(x) = f(x)$ for all $|x| \le M$. The benefit of introducing f_M is that $f_M(B_t) \in \mathcal{H}^2[0,T]$, and we can calculate the Itô integral of $f_M(B_t)$ by using an explicit approximating sequence in $\mathcal{H}^2_0$. Specifically, we will use the sequence defined for $s \in [0,T]$ by

$$\phi_n(\omega, s) = \sum_{i=1}^{n} f_M(B_{t_{i-1}})\, 1(t_{i-1} < s \le t_i).$$

To check that $\{\phi_n\}$ is indeed an approximating sequence for $f_M(B_t)$, we first note that

$$\begin{aligned}
&E\left[\int_0^T (\phi_n(\omega,s) - f_M(B_s))^2\,ds\right] \\
&=E\left[\int_0^T \sum_{i=1}^n (f_M(B_{t_{i-1}}) - f_M(B_s))^2 1(t_{i-1} < s \leq t_i)\,ds\right] \\
&\leq \frac{T}{n}\sum_{i=1}^n E\left[\sup_{\{s:t_{i-1}<s\leq t_i\}} (f_M(B_{t_{i-1}}) - f_M(B_s))^2\right].
\end{aligned}$$

Next, if we let

$$\mu(h) = \sup\{|f_M(x) - f_M(y)| : |x-y| \leq h\},$$

then the fact that f_M is a continuous function with compact support tells us that $\mu(h) \leq B$ for a constant B and that $\mu(h) \to 0$ as $h \to 0$. If we then set

$$M_i = \sup_{\{s:t_{i-1}<s\leq t_i\}} |B_{t_{i-1}} - B_s|,$$

we also have the basic bound

$$E\left[\sup_{\{s:t_{i-1}<s\leq t_i\}} (f_M(B_{t_{i-1}}) - f_M(B_s))^2\right] \leq E(\mu^2(M_i)). \tag{7.17}$$

The random variables defined by $\mu^2(M_i)$ are all bounded by B^2, and the uniform continuity of $B_t(\omega)$ on the compact interval $[0,T]$ gives us

$$\sup_{1\leq i\leq n} \mu^2(M_i) \to 0 \text{ as } n \to \infty \text{ for all } \omega. \tag{7.18}$$

Finally, the dominated convergence theorem tells us that the expectations in inequality (7.17) converge to zero as $n \to \infty$, so ϕ_n is indeed an approximating sequence for f_M.

By the Itô isometry, we now know that $I(\phi_n) \to I(f_M)$ in $L^2(dP)$, but we also have the explicit formula

$$I(\phi_n) = \sum_{i=1}^n f_M(B_{t_{i-1}})(B_{t_i} - B_{t_{i-1}}),$$

so Itô's isometry tells us that we have a Riemann representation for f_M:

$$\int_0^t f_M(B_s)\,dB_s = \lim_{n\to\infty} \sum_{i=1}^n f_M(B_{t_{i-1}})(B_{t_i} - B_{t_{i-1}}), \tag{7.19}$$

where the convergence takes place in $L^2(dP)$.

We are now ready to take up the second step of the proof, which is to exploit the representation (7.19) and our localization to establish the convergence in probability of the Riemann representation (7.16) associated with our original f. The basic connection between f and f_M is that for all $\omega \in \{\omega\colon \tau_M = T\}$ we have the equality $f(B_{t_i}) = f_M(B_{t_i})$ for all $0 \leq i \leq n$, and, as a consequence of the persistence of identity property of the Itô integral, we also have

$$\int_0^t f(B_s)\,dB_s = \int_0^t f_M(B_s)\,dB_s \text{ for almost all } \omega \in \{\omega : \tau_M = T\}. \tag{7.20}$$

All we need to do now is show for all $\epsilon > 0$ that the probability of the event

$$A_n(\epsilon) = \left\{ \omega : \left| \sum_{i=1}^{n} f(B_{t_{i-1}})(B_{t_i} - B_{t_{i-1}}) - \int_0^T f(B_s)\, dB_s \right| \geq \epsilon \right\}$$

goes to zero as n goes to infinity. We begin by breaking the event $A_n(\epsilon)$ into two parts based on the size of τ_M:

$$P(A_n(\epsilon)) \leq P(\tau_M < T) + P(A_n(\epsilon) \cap \{\tau_M = T\}).$$

Because $P(\tau_M < T) \to 0$, we only need to worry about the second term. On the set $\{\omega : \tau_M = T\}$ we can replace $f(B_t)$ by $f_M(B_t)$, so we can apply Chebyshev's inequality to get an upper bound for $P(A_n(\epsilon) \cap \{\tau_M = T\})$ that is given by

$$P\left(\left\{ \omega : \left| \sum_{i=1}^{n} f_M(B_{t_{i-1}})(B_{t_i} - B_{t_{i-1}}) - \int_0^T f_M(B_s)\, dB_s \right| \geq \epsilon \right\} \cap \{\tau_M = T\} \right)$$

$$\leq \frac{1}{\epsilon^2} \left\| \int_0^T f_M(B_s)\, dB_s - \sum_{i=1}^{n} f_M(B_{t_{i-1}})(B_{t_i} - B_{t_{i-1}}) \right\|^2_{L^2(dP)}$$

Finally, because of the Riemann limit result (7.19) for f_M (7.19), we see that the last term goes to zero. The proof of the theorem is therefore complete. □

GAUSSIAN CONNECTION

The next proposition gives another Riemann representation and shows that Itô integrals will give us Gaussian processes if the integrand is nonrandom. We will see later that some of the Gaussian processes that are most important in applications have representations in the form of such integrals.

PROPOSITION 7.6 (Gaussian Integrals). *If $f \in C[0,T]$, then the process defined by*

$$X_t = \int_0^t f(s)\, dB_s \quad t \in [0,T] \tag{7.21}$$

is a mean zero Gaussian process with independent increments and with covariance function

$$\text{Cov}(X_s, X_t) = \int_0^{s \wedge t} f^2(u)\, du. \tag{7.22}$$

Moreover, if we take the partition of $[0,T]$ given by $t_i = iT/n$ for $0 \leq i \leq n$ and choose t_i^* to satisfy $t_{i-1} \leq t_i^* \leq t_i$ for all $1 \leq i \leq n$, then we have

$$\lim_{n\to\infty} \sum_{i=1}^{n} f(t_i^*)(B_{t_i} - B_{t_{i-1}}) = \int_0^T f(s)\, dB_s,$$

where the limit is understood in the sense of convergence in probability.

PROOF. The validity of the Riemann representation is easier than that given in Theorem 7.1, so the proof can be safely omitted. The fact that X_t has independent increments follows from the Riemann representation, as does the variance formula

$$\text{Var}(X_t) = \int_0^t f^2(u)\, du. \tag{7.23}$$

Finally, the covariance formula follows from this representation of the variance and the independence of the increments. □

TIME CHANGE TO BROWNIAN MOTION: SIMPLEST CASE

The last proposition has an immediate corollary that is often useful. It reminds us that in many cases of importance an integral like (7.21) is nothing more than a Brownian motion with a clock that speeds up and slows down according to a deterministic schedule. For the purpose of calculation, we often want to relate such a process more directly to Brownian motion, and the next corollary tells us how.

COROLLARY 7.1. *Suppose that a continuous function $f : [0, \infty) \mapsto \mathbb{R}$ satisfies $f(s) > 0$ for all $s > 0$ and that*

$$\int_0^t f^2(s)\,ds \to \infty \qquad \textit{as } t \to \infty.$$

If we let

$$\tau_t = \inf\left\{u\colon \int_0^u f^2(s)\,ds \geq t\right\} \quad \textit{and} \quad Y_t = \int_0^{\tau_t} f(s)\,dB_s,$$

then the process $\{Y_t\colon 0 \leq t < \infty\}$ is a standard Brownian motion.

PROOF. We have already checked the harder parts. Since τ_t is deterministic, we automatically see that Y_t is a continuous Gaussian process with mean zero. Finally, by the covariance formula (7.22) together with the definition of τ_t, we can easily check that

$$\operatorname{Cov}(Y_s, Y_t) = \min(s, t),$$

and nothing more is required to identify $\{Y_t\}$ as a standard Brownian motion. □

The last corollary has a very important generalization to integrals with nondeterministic integrands. We will develop this generalization in a later chapter.

7.3. Why Just $\mathcal{L}^2_{\mathrm{LOC}}$?

The function space $\mathcal{L}^2_{\mathrm{LOC}}[0, T]$ provides a comfortable home for the Itô integral, but it is only human to ask if we might not be able to find something a bit bigger. In particular, suppose we take a $p \in [1, 2)$ and we define $\mathcal{L}^p_{\mathrm{LOC}}[0, T]$ to be the space of all measurable adapted functions f such that

$$P\left(\int_0^T |f(\omega, s)|^p\,ds < \infty\right) = 1.$$

By Jensen's inequality, we see that $\mathcal{L}^2_{\mathrm{LOC}} \subset \mathcal{L}^p_{\mathrm{LOC}}$ for $1 \leq p < 2$, and one might guess that we could define a useful continuous Itô integral on $\mathcal{L}^p_{\mathrm{LOC}}$. This turns out to be impossible, and Gaussian representation of Proposition 7.6 will help us understand the nature of the difficulty.

Specifically, we take $f(\omega, s) = |1 - s|^{-\frac{1}{2}}$ and note that $f \in \mathcal{L}^p_{\mathrm{LOC}}[0, T]$ for all T and all $1 \leq p < 2$. We also have $f \in \mathcal{L}^2_{\mathrm{LOC}}[0, T]$ if we take $T < 1$, and it is this second property that will help us see that we cannot define a continuous Itô integral for f on the interval $[0, T]$ when $T \geq 1$ even though $f \in \mathcal{L}^p_{\mathrm{LOC}}[0, T]$ for such T when $1 \leq p < 2$.

To see why this is true, we first consider the process given by

$$X_t = \int_0^t |1-s|^{-\frac{1}{2}}\, dB_s \text{ where } t \in [0,1).$$

If we could define an Itô integral X_t' for our f on $[0,T]$ for some $T \geq 1$, then we would need to insist that $X_t = X_t'$ for all $t \in [0,1)$, but, as we will soon see, this modest requirement means that we cannot define $\{X_t'\}$ as a continuous process on $[0,T]$ if $T \geq 1$.

The problem is that X_t goes completely berserk as $t \to 1$. By Proposition 7.6 we know that X_t is Gaussian with mean 0 and variance

$$\text{Var}(X_t) = \int_0^t |1-s|^{-1}\, ds = \log(\frac{1}{1-t}),$$

so if we take $t_n = 1 - 2^{-n}$ for $n = 0, 1, \ldots$, then for all $n \geq 1$ the random variables

$$Z_n = X_{t_n} - X_{t_{n-1}}$$

are independent Gaussians with mean zero and variance $\sigma^2 = \log 2$. When we sum these differences, we then see

$$\int_0^{t_n} |1-s|^{-\frac{1}{2}}\, dB_s = Z_1 + Z_2 + \cdots + Z_n;$$

that is, at time t_n the process X_{t_n} is equal to the nth term of a random walk with Gaussian summands with mean zero and variance one, so as a consequence we have

$$\limsup_{n\to\infty} \int_0^{t_n} |1-s|^{-\frac{1}{2}}\, dB_s = \infty \text{ and } \liminf_{n\to\infty} \int_0^{t_n} |1-s|^{-\frac{1}{2}}\, dB_s = -\infty.$$

The bottom line is that the stochastic process X_t cannot be extended to a continuous function on any interval containing $[0,1]$, despite the fact that the integrand $f(\omega, s) = (1-s)^{-\frac{1}{2}}$ is in $\mathcal{L}^p_{\text{LOC}}[0,T]$ for all $1 \leq p < 2$ and all T.

The lesson to be drawn from this example is that if one hopes to extend the Itô integral to spaces that are larger than $\mathcal{L}^2_{\text{LOC}}$, then at least we know that $\mathcal{L}^p_{\text{LOC}}$ is not the way to go. Fortunately, $\mathcal{L}^2_{\text{LOC}}$ is already roomy enough for all practical purposes.

7.4. Local Martingales and Honest Ones

One of the glorious properties of the Itô integral of a function in $\mathcal{H}^2$ is that it is a martingale. For functions in $\mathcal{L}^2_{\text{LOC}}$ we are no longer quite as lucky, but the $\mathcal{L}^2_{\text{LOC}}$ integrals still come close to being martingales in a sense that is easy to make precise.

DEFINITION 7.2 (Local Martingale). *If a process $\{M_t\}$ is adapted to the filtration $\{\mathcal{F}_t\}$ for all $0 \leq t < \infty$, then $\{M_t\colon 0 \leq t < \infty\}$ is called a* local martingale *provided that there is a nondecreasing sequence $\{\tau_k\}$ of stopping times with the property that $\tau_k \to \infty$ with probability one as $k \to \infty$ and such that for each k the process defined by*

$$M_t^{(k)} = M_{t\wedge\tau_k} - M_0 \quad \text{for } t \in [0,\infty)$$

is a martingale with respect to the filtration $\{\mathcal{F}_t : 0 \leq t < \infty\}$.

One small nuance of this definition is that it does not require M_0 to be integrable. Although this extra freedom will not be exploited here, the definition is given in this form to be consistent with tradition.

Local Almost Anything

Needless to say, one can use the same construction to define a local submartingale or a *local almost anything.* For example, a local L^2-bounded submartingale M_t is a process for which there exist nondecreasing stopping times $\tau_k \to \infty$ such that the processes defined by

$$M_t^{(k)} = M_{t\wedge\tau_k} - M_0, \quad 0 \le t < \infty,$$

are submartingales with

$$\sup_t E\left((M_t^{(k)})^2\right) \le B_k < \infty \text{ for all } k.$$

Connection to Stochastic Integrals

The main reason for isolating the concept of a local martingale is that the Itô integral of an $\mathcal{L}^2_{\text{LOC}}$ function is always a local martingale. We make this notion precise in the following proposition.

Proposition 7.7 (Itô Integrals on $\mathcal{L}^2_{\text{LOC}}$ are Local Martingales).
For any function $f \in \mathcal{L}^2_{\text{LOC}}[0,T]$, there is a continuous local martingale X_t such that

$$P\left(X_t(\omega) = \int_0^t f(\omega, s)\, dB_s\right) = 1.$$

Moreover, one can take the required localizing sequence to be

$$\tau_n(\omega) = \inf\{t : \int_0^t f^2(\omega, s)\, ds \ge n \text{ or } t \ge T\}.$$

The proposition hardly requires proof; one only needs to note that the function $f_k(\omega, s) = f(\omega, s)1(s \le \tau_k)$ is an element of $\mathcal{H}^2[0,T]$. Still, this proposition is very important, and it will be applied constantly in the rest of the text. In fact, the whole construction of the Itô integral was guided by the goal of extending the notion of a martingale transform into continuous time, and — however humble — Proposition 7.7 is a milestone.

Working with Local Martingales

Results for martingales can often be extended to local martingales; the possibility typically hinges on whether one can make the localizing sequence "go away" at the right moment. The traditional dance is easily illustrated by one of our favorite problems — the calculation of the ruin probability.

Proposition 7.8. *If X_t, $0 \le t < \infty$, is a continuous local martingale with $X_0 = 0$ and if the stopping time $\tau = \inf\{t\colon X_t = A \text{ or } X_t = -B\}$ satisfies $P(\tau < \infty) = 1$, then $E(X_\tau) = 0$ and as a consequence*

$$P(X_\tau = A) = \frac{B}{A+B}.$$

PROOF. If we then let τ_k denote a localizing sequence for $\{X_t\}$, then by the definition of a local martingale we have that

$$Y_t^{(k)} \equiv X_{t\wedge\tau_k}$$

is a martingale for each k; so, by the Doob stopping time theorem applied to the martingale $Y_t^{(k)}$, we have that $Y_{t\wedge\tau}^{(k)}$ is also a martingale.

Next, as $t \to \infty$ we see from $P(\tau < \infty) = 1$ that $Y_{t\wedge\tau}^{(k)}$ converges with probability one to $Y_\tau^{(k)}$, and since $\sup_t |Y_{t\wedge\tau}^{(k)}| \le \max(A, B)$ the dominated convergence theorem tells us that

$$E[Y_\tau^{(k)}] = \lim_{t\to\infty} E[Y_{t\wedge\tau}^{(k)}] = 0. \tag{7.24}$$

Finally, by the definition of the τ_k we have $\tau_k \to \infty$ with probability one, so the continuity of $\{X_t\}$ tells us $Y_\tau^{(k)} \to X_\tau$ as $k \to \infty$. Since $\sup_k |Y_\tau^{(k)}| \le \max(A, B)$, this convergence is also dominated, so we can let $k \to \infty$ in equation (7.24) to deduce that $E(X_\tau) = 0$, as claimed. The second conclusion of the proposition then follows by a computation that should by now be quite familiar. □

PATTERNS OF ATTACK

Bare-handed appeal to the definition often serves us well enough when working with local martingales, but it still pays to know a few general results — or patterns of attack. The first problem we consider is the justification of a local martingale version of Doob's stopping time theorem.

PROPOSITION 7.9. *If X_t is a local martingale and τ is a stopping time, then $Y_t = X_{t\wedge\tau}$ is also a local martingale.*

PROOF. We first observe that there is no loss of generality if we assume $X_0 = 0$. Next, by the hypothesis, there is an increasing sequence of stopping times τ_k with $\tau_k \to \infty$ a.s. such that $X_{t\wedge\tau_k}$ is a martingale for each k. Now,

$$Y_{t\wedge\tau_k} = X_{(t\wedge\tau)\wedge\tau_k} = X_{(t\wedge\tau_k)\wedge\tau},$$

and, since $\{X_{t\wedge\tau_k} : 0 \le t < \infty\}$ is an honest martingale, Doob's stopping time theorem tells us that $\{X_{(t\wedge\tau_k)\wedge\tau} : 0 \le t < \infty\}$ is also a martingale. Consequently, $\{Y_{t\wedge\tau_k} : 0 \le t < \infty\}$ is a martingale for each k, and since τ_k is a nondecreasing sequence of stopping with $\tau_k \to \infty$ a.s., we see that $Y_t = X_{t\wedge\tau}$ meets the definition of a local martingale. □

The last proposition required little more than shuffling definitions like dominos, but the next one has some honest grip.

PROPOSITION 7.10. *If X_t is a continuous local martingale and B is a constant such that $|X_t| \le B$ for all $t \ge 0$, then X_t is a martingale.*

PROOF. Again, we start by assuming that $X_0 = 0$ and by taking a nondecreasing sequence of stopping times such that $X_{t\wedge\tau_k}$ is a martingale for each k and for which the sequence $\{\tau_k\}$ increases monotonically to infinity. Next, we consider $s \le t$ and note by the martingale property of $X_{t\wedge\tau_k}$ that

$$E(X_{t\wedge\tau_k} \mid \mathcal{F}_s) = X_{s\wedge\tau_k}. \tag{7.25}$$

Now, since $\tau_k \to \infty$, we have $X_{s\wedge\tau_k} \to X_s$ and $X_{t\wedge\tau_k} \to X_t$. Because we also know that $|X_{t\wedge\tau_k}| \le B < \infty$ for all k, we can take limits on both sides of equation

(7.25) and apply the dominated convergence theorem. This gives us the martingale identity for X_t and completes the proof. □

The preceding proposition is frequently useful, and it also illustrates a basic pattern. We often show that a local martingale is an honest martingale by arguing that we can justify the martingale identity by making an appropriate interchange of limits and expectations. In the last proposition, we saw how the dominated convergence theorem could be used to justify the interchange, and the next proposition shows how one can sometimes use more subtle arguments based on Fatou's lemma.

PROPOSITION 7.11. *Any nonnegative local martingale* $\{X_t\colon 0 \le t \le T\}$ *with* $E(|X_0|) < \infty$ *is also a supermartingale, and if*

$$E(X_T) = E(X_0),$$

then $\{X_t\colon 0 \le t \le T\}$ *is in fact a martingale.*

PROOF. If τ_n is a localizing sequence, then by the local martingale property we have

$$X_{s\wedge\tau_n} = E(X_{t\wedge\tau_n} \mid \mathcal{F}_s) \text{ for all } 0 \le s \le t \le T.$$

If we apply Fatou's lemma as $n \to \infty$, we find

$$X_s \ge E(X_t \mid \mathcal{F}_s) \text{ for all } 0 \le s \le t \le T, \tag{7.26}$$

and this tells us that $\{X_t\}$ is a supermartingale.

Now, if we take expectations in the last inequality we find $E(X_s) \ge E(X_t)$ for all $0 \le s \le t \le T$, so in particular we have

$$E(X_0) \ge E(X_s) \ge E(X_t) \ge E(X_T) \text{ for all } 0 \le s \le t \le T.$$

The hypothesis $E(X_0) = E(X_T)$ then tells us that we must have equality up and down the line. Finally, when we look back at the inequality (7.26), we see that if strict inequality held on a set of positive probability we would have $E(X_s) > E(X_t)$, so, in fact, it must be the case that equality holds with probability one. □

This last proposition will turn out to be very useful for us in later chapters. In particular, the simple "$E(X_T) = E(X_0)$" criterion will provide an essential toehold in our discussion of the famous Novikov condition, which provides a criterion for a stochastic integral to be a martingale.

BUT DON'T JUMP THE GUN

Finally, the last two propositions may seem to suggest that any "sufficiently integrable" local martingale is an honest martingale, but, sadly, this is not the case. One of the exercises of a later chapter will provide an example showing that a local martingale can have moments of all order and still fail to be a martingale. Nonetheless, such examples should not make us paranoid. Much of the time we can do the work we need to do with simple arguments such as those used in Propositions 7.10 and 7.11.

7.5. Alternative Fields and Changes of Time

One of our most trusted tools is the fact that for any martingale $\{X_t, \mathcal{F}_t\}$ and any $\{\mathcal{F}_t\}$ stopping time τ, the stopped process $\{X_{\tau\wedge t}\}$ is also an $\{\mathcal{F}_t\}$ martingale. As it turns out, $\{X_{\tau\wedge t}\}$ is still a martingale with respect to a *smaller* filtration that has some technical and conceptual advantages. In particular, this new filtration will play an essential role in our investigation of the representation of a continuous martingale as a time change of Brownian motion. As usual, a definition starts the ball rolling.

DEFINITION 7.3. *If τ is a stopping time adapted to the filtration $\{\mathcal{F}_t\}$, we define the stopping time σ-field $\mathcal{F}_\tau$ by setting*

$$\mathcal{F}_\tau = \{A \in \cup_{t=0}^{\infty}\mathcal{F}_t : A \cap \{\tau \le t\} \in \mathcal{F}_t \text{ for all } t \ge 0\}. \tag{7.27}$$

The fact that $\mathcal{F}_\tau$ is a σ-field follows directly from the filtration properties of $\{\mathcal{F}_t\}$, and we naturally think of the $\mathcal{F}_\tau$ as a way to represent all of the information up to time τ. Also, as one would expect, the σ-field $\mathcal{F}_\tau$ has many properties that parallel those of the fixed time σ-fields like $\mathcal{F}_t$; in particular, the definition of $\mathcal{F}_\tau$ implies that $\{X_\tau \le y\} \in \mathcal{F}_\tau$ for all y, so X_τ is $\mathcal{F}_\tau$-measurable. The next three results offer variations on the theme that $\mathcal{F}_\tau$ behaves as nicely as $\mathcal{F}_t$ for all practical purposes.

LEMMA 7.1 (Stopping Times and Alternative Fields).
If $\{X_t, \mathcal{F}_t\}$ is a bounded continuous martingale and τ is an $\{\mathcal{F}_t\}$ stopping time, then for any $A \in \mathcal{F}_\tau$ and any $s \le t$, we have

$$E(X_\tau 1_A 1_{\{\tau<s\}}) = E(X_s 1_A 1_{\{\tau<s\}}) = E(X_t 1_A 1_{\{\tau<s\}}). \tag{7.28}$$

Moreover, if X_∞ denotes the limit of X_t as $t \to \infty$, then

$$E(X_\infty | \mathcal{F}_\tau) = X_\tau. \tag{7.29}$$

PROOF. To begin, we let $S(n) = \{s + k/2^n : k \in \mathbb{Z}\} \cap [0, \infty)$, so $\{X_t : t \in S(n)\}$ is a discrete-time martingale. For any $A \in \mathcal{F}_\tau$, the event $A \cap \{\tau < s\}$ is $\mathcal{F}_s$-measurable, so if $\tau_n = \min\{r \in S(n) : r \ge \tau\}$, then we can easily check that $A \cap \{\tau_n = r\} \in \mathcal{F}_r$. Now, by the martingale property of $\{X_t\}$, we can calculate

$$\begin{aligned} E(X_{\tau_n} 1_A 1_{\{\tau_n<s\}}) &= \sum_{r\in S(n), r<s} E(X_r 1_{\{\tau_n=r\}} 1_A) \\ &= \sum_{r\in S(n), r<s} E(X_s 1_{\{\tau_n=r\}} 1_A) \\ &= E(X_s 1_{\{\tau_n<s\}} 1_A) = E(X_t 1_{\{\tau_n<s\}} 1_A). \end{aligned}$$

When $n \to \infty$, we have $1_{\{\tau_n<s\}}(\omega) \to 1_{\{\tau<s\}}(\omega)$ and $X_{\tau_n}(\omega) \to X_\tau(\omega)$ for all ω, so the dominated convergence theorem (DCT) tells us

$$E(X_\tau 1_A 1_{\{\tau<s\}}) = E(X_s 1_A 1_{\{\tau<s\}}) = E(X_t 1_A 1_{\{\tau<s\}}). \tag{7.30}$$

This equation completes the proof of the first assertion (7.28) of the lemma, so when we let $t \to \infty$ and apply the DCT again, we see

$$E(X_\tau 1_A 1_{\{\tau<s\}}) = E(X_\infty 1_A 1_{\{\tau<s\}}).$$

Finally, if we let $s \to \infty$ and invoke the DCT a third time, we find

$$E(X_\tau 1_A) = E(X_\infty 1_A),$$

and, since X_τ is $\mathcal{F}_\tau$-measurable, the last identity is equivalent to equation (7.29) simply by the definition of conditional expectation. □

The preceding lemma has an immediate corollary that shows the stopping time σ-fields interact with martingales in a way that perfectly parallels the interaction of the fixed time σ-fields. The resulting identity is a real workhorse that packs a multitude of concrete facts into a single abstract line.

COROLLARY 7.2 (The Stopping Time Identity). *Suppose that $\{X_t, \mathcal{F}_t\}$ is a bounded continuous martingale. If ν and τ are $\{\mathcal{F}_t\}$ stopping times, then*

$$\nu \leq \tau \quad \Rightarrow \quad X_\nu = E(X_\tau \mid \mathcal{F}_\nu). \tag{7.31}$$

PROOF. To see why this identity holds, we first use the representation (7.29) for X_ν and then use the tower property of conditional expectation to get

$$X_\nu = E(X_\infty|\mathcal{F}_\nu) = E(\, E(X_\infty|\mathcal{F}_\tau)\,|\mathcal{F}_\nu). \tag{7.32}$$

If we apply (7.29) again to replace the last occurrence of $E(X_\infty|\mathcal{F}_\tau)$ with X_τ, we see that equation (7.32) completes the proof of the required implication (7.31). □

The next result follows quickly from the stopping time identity (7.31), but we state it as a proposition because of its importance. Set out by itself, it may look lame, but it has some consequences with real bite, as we will see when we compute the the density of time required for Brownian motion to hit a sloping line.

PROPOSITION 7.12 (Martingale on Alternative Fields). *For any bounded continuous martingale $\{X_t, \mathcal{F}_t\}$ and any $\{\mathcal{F}_t\}$ stopping time τ, the stopped process $\{X_{\tau\wedge t}, \mathcal{F}_{\tau\wedge t}\}$ is also a martingale.*

PROOF. We already know that $X_{\tau\wedge t}$ is $\mathcal{F}_{\tau\wedge t}$-measurable, and the integrability of $X_{\tau\wedge t}$ follows from the boundedness of $\{X_t\}$. Thus, we only need to check the martingale identity

$$E(X_{\tau\wedge t}|\mathcal{F}_{\tau\wedge s}) = X_{\tau\wedge s} \text{ for } 0 \leq s \leq t, \tag{7.33}$$

and if we take $\nu = \tau \wedge s$ and $\sigma = \tau \wedge t$, then we see that the identity (7.33) is a restatement of equation (7.31). □

TIME CHANGE OF A LOCAL MARTINGALE

If $\{\tau_t : 0 \leq t < \infty\}$ a right-continuous nondecreasing process such that each τ_t is a stopping time with respect to the filtration $\{\mathcal{F}_t\}$, then $\{\tau_t\}$ is said to be an $\{\mathcal{F}_t\}$ *time change*. The next proposition tells us that in favorable circumstances one can replace t by τ_t in a local martingale and still preserve the local martingale property. For the moment, this result may seem somewhat bland, but later we will see that it has important consequences.

PROPOSITION 7.13. *Suppose that $\{M_t, \mathcal{F}_t\}$ is a continuous local martingale and suppose that $\{\tau_t\}$ is an $\{\mathcal{F}_t\}$ time change. If $M_t(\omega)$ is constant on the interval $[\tau_{u-}(\omega), \tau_u(\omega)]$ for all $u \geq 0$ and $\omega \in \Omega$, then the process $\{M_{\tau_t}, \mathcal{F}_{\tau_t}\}$ is again a continuous local martingale.*

PROOF. Since $\{M_t, \mathcal{F}_t\}$ is a local martingale, there is a sequence of $\{\mathcal{F}_t\}$ stopping times $\{\sigma_n\}$ with $\sigma_n \to \infty$ for which $\{M_{t\wedge\sigma_n}, \mathcal{F}_t\}$ is a bounded continuous martingale for each n. Also, if we apply the stopping time identity (7.31) to this martingale and the times $\tau_s \leq \tau_t$, we have

$$E(M_{\tau_t\wedge\sigma_n}|\mathcal{F}_{\tau_s}) = M_{\tau_s\wedge\sigma_n}. \tag{7.34}$$

The idea now is to show that the "inverse sequence" $\sigma_n^* = \inf\{t : t \geq \sigma_n\}$ is a localizer that can be used to confirm that if we set $Y_t = M_{\tau_t}$ then $\{Y_t, \mathcal{F}_{\tau_t}\}$ is a local martingale. After checking (as in Exercise 7.4) that $\sigma_n^* \to \infty$ and that σ_n^* is an $\{\mathcal{F}_{\tau_t}\}$ stopping time for which $Y_{t\wedge\sigma_n^*}$ is bounded, one sees that the whole issue boils down to showing $E(Y_{t\wedge\sigma_n^*}|\mathcal{F}_{\tau_s}) = Y_{s\wedge\sigma_n^*}$. Moreover, this identity will follow at once from (7.34) if we show $Y_{t\wedge\sigma_n^*} = M_{\tau_t\wedge\sigma_n}$.

When $t < \sigma_n^*$ we also have $\tau_t < \sigma_n$, so we immediately find that we have $Y_{t\wedge\sigma_n^*} = Y_t = M_{\tau_t} = M_{\tau_t\wedge\sigma_n}$. On the other hand, when $t \geq \sigma_n^*$ the monotonicity of $\{\tau_t\}$ and the definition of σ_n^* give us $\tau_t \geq \tau_{\sigma_n^*} \geq \sigma_n$. Also, by the definition of σ_n^* we have $\sigma_n \in [\tau_{\sigma_n^*-}, \tau_{\sigma_n^*}]$, so the constancy hypothesis tells us $M_{\sigma_n} = M_{\tau_{\sigma_n^*}}$ and consequently

$$t \geq \sigma_n^* \quad \Rightarrow \quad Y_{t\wedge\sigma_n^*} = Y_{\sigma_n^*} = M_{\tau_{\sigma_n^*}} = M_{\sigma_n} = M_{\tau_t\wedge\sigma_n}.$$

Thus, we always have $Y_{t\wedge\sigma_n^*} = M_{\tau_t\wedge\sigma_n}$, and the proof is complete. □

7.6. Exercises

A brief chapter deserves a brief collection of exercises. The first gives basic practice in the application of the simplest time change to Brownian motion, and it also isolates a large class of stochastic integrals that permit precise determination of their distribution. The next two problems then offer practice with the notion of localization, and finally Exercise 7.4 puts a spotlight on Proposition 7.13 by suggesting a very simple "counterexample."

EXERCISE 7.1. Find a function τ_t such that the processes

$$X_t = \int_0^t e^s \, dB_s \text{ and } Y_t = B_{\tau_t}$$

are equivalent processes on the infinite interval $0 \leq t < \infty$. Check your solution by making independent computations of $E(X_t^2)$ and $E(Y_t^2)$. Finally, use the time-change representation to calculate $E(X_t^4)$ and $P(X_t \geq 1)$. Would these calculations be as easy without the representation?

EXERCISE 7.2. Show that if X_t is any continuous martingale and ϕ is any convex function, then $Y_t = \phi(X_t)$ is always a local submartingale. Give an example that shows Y_t need not be an honest submartingale.

EXERCISE 7.3. Show that if X_t is a continuous local submartingale such that

$$E\Big(\sup_{0\leq s\leq T} |X_s|\Big) < \infty, \tag{7.35}$$

then $\{X_t : 0 \leq t \leq T\}$ is an honest submartingale. Show how this result implies our earlier result that a bounded local martingale is a martingale. In Exercise 8.5 of the next chapter, we will find that the condition (7.35) cannot be replaced with the uniform integrability of $\{X_t : 0 \leq t \leq T\}$.

EXERCISE 7.4. (a) First, complete the checks suggested in Proposition 7.13 by showing σ_n^* is an $\mathcal{F}_{\tau_t}$-stopping time and $\sigma_n^* \to \infty$. (b) Next, consider Brownian motion $\{B_t\}$ and the filtration $\bar{\mathcal{F}} = \sigma\{B_s : s \leq t\}$. If $\tau_t = \inf\{s : B_s \geq t\}$, then $\{\tau_t\}$ is an $\bar{\mathcal{F}}$ time change, but $Y_t = B_{\tau_t} = t$ is certainly *not* a local martingale. Explain how this example is compatible with Proposition 7.13.

CHAPTER 8

Itô's Formula

When we compute the familiar integrals of Newton and Leibniz, we almost invariably call on the fundamental theorem of calculus — only a few integrals can be done comfortably by direct appeal to the definition. The situation with the Itô integral is parallel, and the Itô calculus would be stopped dead in its tracks if we could not find an appropriate analog to the traditional fundamental theorem of calculus. One of the great charms of Itô integration is that the required analog comes with an unexpected twist — and several probabilistic interpretations.

THEOREM 8.1 (Itô's Formula — Simplest Case). *If* $f: \mathbb{R} \mapsto \mathbb{R}$ *has a continuous second derivative, then*

$$f(B_t) = f(0) + \int_0^t f'(B_s)\, dB_s + \frac{1}{2}\int_0^t f''(B_s)\, ds. \tag{8.1}$$

The most striking feature of this formula is the presence of the second integral, without which we would just have a formal transcription of the usual fundamental theorem of calculus. This difference is no mere detail; it turns out to be rich in probabilistic meaning because the two integrals capture entirely different features of the process $f(B_t)$.

The first integral of equation (8.1) has mean zero, so the second integral is forced to capture all of the information about the drift of $f(B_t)$. We will see later that the first integral holds up its end of the bargain by capturing all of the essential information about the local variability of $f(B_t)$. Thus, in a way that may take a while to master, Itô's formula has a second interpretation as a decomposition of $f(B_t)$ into components that are representative of noise and signal.

One baseline observation that we should not omit is that both the integrals on the right-hand side make sense simply because of the continuity of f' and f''. We do not need to check any integrability conditions on $f'(B_t)$ or $f''(B_t)$, and the parsimonious phrasing of Itô's formula is our first dividend on the investment we made to extend the Itô integral from $\mathcal{H}^2$ to $\mathcal{L}^2_{\text{LOC}}$.

Finally, to avoid any confusion, we should note that formula (8.1) just begins a long chain of results that share the name *Itô's formula*, or *Itô's lemma*. We start with (8.1) because it is the simplest element of the chain. It also captures the essence of all of the subsequent formulas.

8.1. Analysis and Synthesis

A large part of human knowledge has been brought to light by the complementary processes of *analysis* and *synthesis* — or of breaking apart and putting together. The proof of Itô's formula follows this paradigm to the letter. We first break $f(B_t) - f(0)$ into a sum of small pieces of the form $f(B_{t_i}) - f(B_{t_{i-1}})$, and these pieces are further decomposed with the help of Taylor's formula. These steps complete the analysis process. Next, we interpret the limit of the sums of small terms as integrals, and when we put these pieces back together we complete the synthesis step. Rather remarkably, we arrive at a formula for $f(B_t) - f(0)$ that is both simple and powerful.

The implementation of the plan is not difficult, although, if one is honest at each step, some patience is needed. To go as directly as possible to the heart of the problem, we first assume that f has compact support. After we deal with this case, we will show how a localization argument can be used to obtain the full result. We start by setting $t_i = it/n$ for $0 \leq i \leq n$, and then we note that the telescoping of the summed differences gives us the representation:

$$f(B_t) - f(0) = \sum_{i=1}^{n} \{f(B_{t_i}) - f(B_{t_{i-1}})\}, \tag{8.2}$$

a formula that is itself a discrete variant of the fundamental theorem of calculus.

Because of random variation, the differences $B_{t_i} - B_{t_{i-1}}$ may not all be small, but enough of them should be small enough, for enough of the time, for a two-term Taylor approximation of $f(B_{t_i}) - f(B_{t_{i-1}})$ to be effective. To make this plan concrete, we use Taylor's formula in the remainder form which says that if f has a continuous second derivative, then for all real x and y we have

$$f(y) - f(x) = (y-x)f'(x) + \frac{1}{2}(y-x)^2 f''(x) + r(x,y), \tag{8.3}$$

where the remainder term $r(x,y)$ is given by

$$r(x,y) = \int_x^y (y-u)(f''(u) - f''(x))\, du.$$

From this formula for $r(x,y)$ and the continuity of f'' we can easily check that $|r(x,y)| \leq (y-x)^2 h(x,y)$ where h is uniformly continuous, bounded, and $h(x,x) = 0$ for all x. These modest properties are all that we will ever need to know about $r(x,y)$ and $h(x,y)$.

Synthesis of Three Terms

The telescoping sum (8.2) can then be rewritten as a sum of three terms, A_n, B_n, and C_n, where the first two terms are given by

$$A_n = \sum_{i=1}^{n} f'(B_{t_{i-1}})(B_{t_i} - B_{t_{i-1}}) \text{ and } B_n = \frac{1}{2}\sum_{i=1}^{n} f''(B_{t_{i-1}})(B_{t_i} - B_{t_{i-1}})^2,$$

and the third term C_n satisfies

$$|C_n| \leq \sum_{i=1}^{n} (B_{t_i} - B_{t_{i-1}})^2 h(B_{t_{i-1}}, B_{t_i}). \tag{8.4}$$

Because f' is continuous, we know from the Riemann representation given in the preceding chapter that

$$A_n \xrightarrow{p} \int_0^t f'(B_s)\, dB_s.$$

The proof of the lemma therefore boils down to the analysis of B_n and C_n.

When we write B_n as a centered sum

$$\frac{1}{2}\sum_{i=1}^{n} f''(B_{t_{i-1}})(t_i - t_{i-1}) + \frac{1}{2}\sum_{i=1}^{n} f''(B_{t_{i-1}})\left\{(B_{t_i} - B_{t_{i-1}})^2 - (t_i - t_{i-1})\right\},$$

we find by the continuity of $f''(B_s(\omega))$ as a function of s that the first summand converges as an ordinary integral for all ω:

$$\lim_{n\to\infty}\sum_{i=1}^{n} f''(B_{t_{i-1}})(t_i - t_{i-1}) = \int_0^t f''(B_s)\,ds.$$

If we denote the second summand of B_n by $\widetilde{B}_n$, we then find by the orthogonality of the summands that

$$\begin{aligned} E(\widetilde{B}_n^2) &= \frac{1}{4}\sum_{i=1}^{n} E\left[f''(B_{t_{i-1}})^2\left\{(B_{t_i} - B_{t_{i-1}})^2 - (t_i - t_{i-1})\right\}^2\right] \\ &\le \frac{1}{4}\|f''\|_\infty^2 \sum_{i=1}^{n} E[\{(B_{t_i} - B_{t_{i-1}})^2 - (t_i - t_{i-1})\}^2] = \frac{t^2}{2n}\|f''\|_\infty^2, \end{aligned}$$

where in the last step we used the fact that $B_{t_i} - B_{t_{i-1}} \sim N(0, t/n)$ and consequently $\mathrm{Var}((B_{t_i} - B_{t_{i-1}})^2) = 2t^2/n^2$. By Markov's inequality, the last bound on $E(\widetilde{B}_n^2)$ is more than we need to show $\widetilde{B}_n \xrightarrow{p} 0$, so all that is left to prove the theorem is the estimation of the remainder term C_n.

ESTIMATION OF THE REMAINDER TERM

There are more subtle ways to argue that $C_n \xrightarrow{p} 0$, but there is also a certain charm to the Neanderthal approach that begins by applying the Cauchy inequality to the summands in the bound (8.4) on C_n to find

$$E(|C_n|) \le \sum_{i=1}^{n} E[(B_{t_i} - B_{t_{i-1}})^4]^{\frac{1}{2}} E[h^2(B_{t_{i-1}}, B_{t_i})]^{\frac{1}{2}}. \tag{8.5}$$

The first factor in the sum is easily calculated since $B_{t_i} - B_{t_{i-1}} \sim N(0, t/n)$ gives us

$$E[(B_{t_i} - B_{t_{i-1}})^4] = 3t^2/n^2. \tag{8.6}$$

To estimate the second factor, we first note by the uniform continuity of h and the fact that $h(x,x) = 0$ for all x, we have for each $\epsilon > 0$ a $\delta = \delta(\epsilon)$ such that $|h(x,y)| \le \epsilon$ for all x, y with $|x - y| \le \delta$, so we also have

$$\begin{aligned} E[h^2(B_{t_{i-1}}, B_{t_i})] &\le \epsilon^2 + \|h\|_\infty^2 P(|B_{t_i} - B_{t_{i-1}}| \ge \delta) \\ &\le \epsilon^2 + \|h\|_\infty^2 \delta^{-2} E(|B_{t_i} - B_{t_{i-1}}|^2) \\ &= \epsilon^2 + \|h\|_\infty^2 \delta^{-2} t/n. \end{aligned} \tag{8.7}$$

When we apply the bounds given by equations (8.6) and (8.7) to the sum in inequality (8.5), we find

$$E(|C_n|) \le n(3t^2/n^2)^{\frac{1}{2}}(\epsilon^2 + \|h\|_\infty^2 \delta^{-2} t/n)^{\frac{1}{2}}$$

and consequently

$$\limsup_{n\to\infty} E(|C_n|) \le 3^{\frac{1}{2}} t\epsilon.$$

By the arbitrariness of ϵ, we finally see $E(|C_n|) \to 0$ as $n \to \infty$, so Markov's inequality tells us that we also have $C_n \xrightarrow{p} 0$.

Covering All of the Bases

The proof of the lemma is now essentially complete for f with compact support, but there are some bookkeeping issues that deserve attention if we are to be absolutely clear. First, we have seen that for any given $t \in \mathbb{R}^+$ the sums A_n and B_n converge in probability to the two integral terms of Itô's formula (8.1), and we have seen that C_n converges in probability to zero. Now, if we we fix $t \in \mathbb{R}^+$, we can choose a subsequence n_j such that A_{n_j}, B_{n_j}, and C_{n_j} all converge with probability one, so in fact we see that Itô's formula (8.1) holds with probability one for each fixed $t \in \mathbb{R}^+$. Finally, if we then apply this fact for each rational and if we also observe that both sides of Itô's formula are continuous, then we see that there is a set Ω_0 with $P(\Omega_0) = 1$ such that for each $\omega \in \Omega_0$ we have Itô's formula for all $t \in \mathbb{R}^+$.

Finishing Touch: The Localization Argument

To obtain the general result, we first recall that for any $f \in C^2(\mathbb{R})$ there is an $f_M \in C^2$ with compact support such that $f(x) = f_M(x)$ for all $|x| \le M$. We have already proved the Itô formula for C^2 functions with compact support, so we have the identity

$$f_M(B_t) - f_M(0) = \int_0^t f_M'(B_s)\,dB_s + \frac{1}{2}\int_0^t f_M''(B_s)\,ds. \tag{8.8}$$

Now, if we take $\tau_M = \min\{t \colon |B_t| \ge M\}$, then for all $\omega \in \{s \le \tau_M\}$ we have $f'(B_s) = f_M'(B_s)$, and by Proposition 7.5 on the persistence of identity in $\mathcal{L}^2_{\mathrm{LOC}}$ we also have

$$\int_0^t f'(B_s)\,dB_s = \int_0^t f_M'(B_s)\,dB_s \quad \text{for } \omega \in \{t \le \tau_M\}.$$

Also, it is elementary that for $\omega \in \{t \le \tau_M\}$ we have

$$f(B_t) = f_M(B_t) \text{ and } \int_0^t f''(B_s)\,ds = \int_0^t f_M''(B_s)\,ds.$$

When we apply these identities in equation (8.8), we find that for all $\omega \in \{t \le \tau_M\}$ that we have

$$f(B_t) - f(0) = \int_0^t f'(B_s)\,dB_s + \frac{1}{2}\int_0^t f''(B_s)\,ds. \tag{8.9}$$

Finally, we note that $\tau_M \to \infty$ with probability one, so equation (8.9) also holds with probability one, and the proof of Itô's formula is complete. □

8.2. First Consequences and Enhancements

Without Itô's formula our only way to calculate or interpret an Itô integral is by working our way back to its definition. This tedious reduction can now be replaced by a process that is almost always simpler and more concrete. For example, if we take an $F \in C^2(\mathbb{R})$ for which $F' = f$ and $F(0) = 0$, then Itô's formula can be written as a formula for the integral of f:

$$\int_0^t f(B_s)\,dB_s = F(B_t) - \frac{1}{2}\int_0^t f'(B_s)\,ds. \tag{8.10}$$

The most compelling feature of the identity (8.10) is that the right-hand side can be evaluated for each sample path $\{B_s(\omega)\colon 0 \le s \le t\}$, so equation (8.10) gives us a new way to understand the Itô integral on the left-hand side; we now have an interpretation of this Itô integral on a pathwise, or ω-by-ω, basis. We already knew that the Itô integral could be viewed as a process, but the general definition of the integral was a global one without any immediate connection to the individual sample paths. In contrast, the right-hand side of equation (8.10) assigns a value directly to each ω. This view can be particularly useful when we think about the simulation of processes defined by Itô integrals, since otherwise all we can do is pick some Riemann approximation and just hope that for the application at hand approximation does a suitable job of reflecting the behavior of the true integral.

Now, to get down to the calculation of interesting integrals, the obvious first step is to specialize equation (8.10). When $f(B_s) \equiv 1$, both sides simply reduce to B_t, which is rather boring, but, when we take $f(B_s) = B_s$, we find the engaging formula

$$\int_0^t B_s\,dB_s = \frac{1}{2}B_t^2 - \frac{1}{2}t. \tag{8.11}$$

Naturally, we recognize this integral as an old friend from Chapter 6, where it was derived at the cost of considerable labor. We should also notice that B_s is an element of $\mathcal{H}^2$, so its Itô integral is a martingale, and the representation (8.11) therefore tells us that $B_t^2 - t$ is a martingale, a fact that we obtained in Chapter 4 by explicit calculation.

Beyond Space to Space and Time

The first two of the three classic Brownian motion martingales have popped out of Itô's formula so quickly that one is almost forced to inquire about the third member of the classic trio:

$$M_t = \exp(\alpha B_t - \alpha^2 t/2).$$

Now, an interesting problem presents itself. The process M_t is no longer a function of B_t alone; it is a joint function of t and B_t. Nevertheless, we can easily accommodate such functions if we develop Itô's formula a little further, and it turns out that this further development is especially fruitful. Before stating the required extension, we first recall that if a function $(t, x) \mapsto f(t, x) \in \mathbb{R}$ has m continuous derivatives in t and n continuous derivatives in x, then we write $f \in C^{m,n}(\mathbb{R}^+ \times \mathbb{R})$.

THEOREM 8.2 (Itô's Formula with Space and Time Variables). *For any function $f \in C^{1,2}(\mathbb{R}^+ \times \mathbb{R})$, we have the representation*

$$f(t, B_t) = f(0,0) + \int_0^t \frac{\partial f}{\partial x}(s, B_s)\, dB_s + \int_0^t \frac{\partial f}{\partial t}(s, B_s)\, ds + \frac{1}{2}\int_0^t \frac{\partial^2 f}{\partial x^2}(s, B_s)\, ds.$$

PROOF. The proof follows the same pattern as the proof of the simplest Itô formula. In particular, by the same localization argument that we used in the proof of Theorem 8.1, we can assume without loss of generality that f has compact support in $\mathbb{R}^+ \times \mathbb{R}$. We begin the real work by writing $f(t, B_t) - f(0,0)$ as a telescoping sum and by writing the summands in terms of Taylor's formula with remainder. In this case, the appropriate expansion of $f(t, y)$ is given by a Taylor series about the point (s, x), so we can rewrite $f(t, y)$ as

$$f(s,x) + (t-s)\frac{\partial f}{\partial t}(s,x) + (y-x)\frac{\partial f}{\partial x}(s,x) + \frac{1}{2}(y-x)^2\frac{\partial^2 f}{\partial x^2}(s,x) + r(s,t,x,y),$$

where the remainder term now satisfies the bound

$$|r(s,t,x,y)| \leq (y-x)^2 h(x,y,s,t) + (t-s)k(x,y,s,t),$$

where h and k are bounded, uniformly continuous functions that equal zero when $x = y$ and $s = t$.

After the Taylor expansion is applied, the terms are collected to provide a sum of four finite sums. The first three have limits that give the three integrals that appear in Itô's formula. The fourth sum contains the remainder terms, and by following the pattern of Theorem 8.1, one can show that this sum converges to zero in expectation. The final steps of the proof are then identical to those used to complete the proof of Theorem 8.1. □

MARTINGALES AND CALCULUS

Perhaps the central benefit of Itô's formula is that it leads to many powerful connections between the theory of martingales and the well-oiled machinery of differential equations. The next proposition gives one the simplest, but handiest, of these connections.

PROPOSITION 8.1 (Martingale PDE Condition). *If $f \in C^{1,2}(\mathbb{R}^+ \times \mathbb{R})$ and*

$$\frac{\partial f}{\partial t} = -\frac{1}{2}\frac{\partial^2 f}{\partial x^2}, \tag{8.12}$$

then $X_t = f(t, B_t)$ is a local martingale. Moreover, if

$$E\left[\int_0^T \left\{\frac{\partial f}{\partial x}\right\}^2 (t, B_t)\, dt\right] < \infty \tag{8.13}$$

then X_t is a martingale on $0 \leq t \leq T$.

PROOF. The proposition just puts our earlier work into a neat package. By Theorem 8.2, any $f \in C^{1,2}(\mathbb{R}^+ \times \mathbb{R})$ that satisfies the partial differential equation (8.12) gives us an integral representation

$$f(t, B_t) = f(0,0) + \int_0^t \frac{\partial f}{\partial x}(s, B_s)\, dB_s.$$

Also, since $f \in C^{1,2}(\mathbb{R}^+ \times \mathbb{R})$, we see that f_x is continuous, so the integrand is an element of $\mathcal{L}^2_{\mathrm{LOC}}$. As a result, the Itô integral gives us a well-defined local martingale. Finally, if condition (8.13) holds, the integrand is an element of $\mathcal{H}^2$, so the integral is an honest martingale. □

FIRST EXAMPLES

To see how the last proposition can be applied, we first consider one of our favorite processes:

$$M_t = \exp(\alpha B_t - \alpha^2 t/2). \tag{8.14}$$

From our earlier work, we already know that M_t is a martingale, but the PDE approach reveals this fact in a different light. To begin, we note that we can write $M_t = f(B_t, t)$ where $f(x,t) = \exp(\alpha x - \alpha^2 t/2)$. By direct calculation, we have

$$\frac{\partial f}{\partial t} = -\frac{1}{2}\alpha^2 f \text{ and } \frac{\partial^2 f}{\partial x^2} = \alpha^2 f,$$

so the martingale PDE condition (8.12) is satisfied, and we see that M_t is a local martingale. Moreover, the $\mathcal{H}^2$ condition (8.13) is also immediate, so in fact we have an honest martingale, as we expected.

The same method will also verify that the processes defined by $M_t = B_t^2 - t$ and $M_t = B_t$ are martingales. One only has to note that $f(t,x) = x^2 - t$ and $f(t,x) = x$ satisfy the PDE condition (8.12), and in both cases we have $f(t, B_t) \in \mathcal{H}^2$. In fact, we will find in Exercise 8.4 that these two martingales are members of a countable family of martingales that can be obtained from the Taylor expansion of the moment generating function martingale (8.14).

NEW GROUND: $\mathrm{Cov}(\tau, B_\tau)$

The last examples show how easily we can check if $M_t = f(t, B_t)$ is a local martingale, but the PDE method only starts to shine when we need to invent some new martingale in order to solve a concrete problem. For example, suppose we want to compute $\mathrm{Cov}(\tau, B_\tau)$, where as usual we have $\tau = \inf\{t\colon B_t = A \text{ or } B_t = -B\}$ for $A > 0$ and $B > 0$. We already know that $E(B_\tau) = 0$, so to determine the covariance we only need to compute $E(\tau B_\tau)$. The natural idea is to find some martingale M_t of the form $f(t, B_t)$ where $f(t,x) = tx + g(t,x)$ and where we know how to compute $E[g(\tau, B_\tau)]$. This would be the case if $g(t,x)$ contained only powers of x, say $g(t,x) = ax^3 + bx^2 + cx + d$. The PDE condition for a function of the form $f(t,x) = tx + ax^3 + bx^2 + cx + d$ only requires that $a = -1/3$ and $b = 0$. The choice of c and d is unconstrained, so we just set $c = d = 0$ to find that taking $f(t,x) = tx - x^3/3$ makes $M_t = f(t, B_t)$ a local martingale. We can then argue as before to establish that $M_{t\wedge\tau}$ is in fact a bounded martingale and that $E(M_\tau) = 0$. When it is expanded, the last equation says

$$\mathrm{Cov}(\tau, B_\tau) = E(\tau B_\tau) = \frac{1}{3}E(B_\tau^3),$$

and, when we use our old formula $P(B_\tau = A) = B/(A+B)$ to help compute $E(B_\tau^3)$, we quickly find

$$\mathrm{Cov}(\tau, B_\tau) = \frac{1}{3}AB(A - B). \tag{8.15}$$

The covariance formula (8.15) is another marvel of simplicity. It also rewards looking back. First, we see that the covariance is zero if and only if $A = B$. This is quite sensible from the point of view of symmetry. It is more amusing that we get a positive covariance precisely for $A > B$. This tells us that in order to have hit the more distant boundary first, we are likely to have taken more time for our trip. This is also perfectly intuitive.

Brownian Motion with Drift: The Ruin Problem

The PDE method will help us find martingales to fit almost any purpose. One of our favorite purposes is the solution of ruin problems, and the PDE method is the tool of choice when facing a new ruin problem for a process that can be related to Brownian motion.

If B_t denotes standard Brownian motion, then the process X_t defined by

$$X_t = \mu t + \sigma B_t$$

is called Brownian motion with a constant drift rate μ and an instantaneous variance σ^2. This process is one of the most useful in applied probability for the simple reason that X_t gives a first approximation to many other processes with relatively constant drift and relatively constant variability.

The ruin problem for X_t is the calculation of $P(X_\tau = A)$, where as usual

$$\tau = \inf\{t\colon X_t = A \text{ or } X_t = -B\} \text{ with } A > 0 \text{ and } B > 0.$$

The martingale approach to the calculation of $P(X_\tau = A)$ invites us to find a function $h(\cdot)$ on the interval $[-B, A]$ such that for all $t \geq 0$ the process $M_t = h(X_t)$ is a bounded martingale. If we can also arrange matters so that $h(A) = 1$ and $h(-B) = 0$, then the stopping time identity tells us

$$E(M_0) = E(M_\tau) \text{ and } P(X_\tau = A) = h(0),$$

since $E(M_0) = h(0)$ and $E(M_\tau) = P(X_\tau = A)$. The bottom line is that Itô's formula tells us how to reduce the ruin problem to a boundary value problem for a differential equation. This is a very general, and very useful, reduction.

In our immediate case, we want to find a martingale of the form $h(\mu t + \sigma B_t)$, so the function $f(t, x) = h(\mu t + \sigma x)$ needs to satisfy the PDE condition (8.12). Since we have $f_t(t, x) = \mu h'(\mu t + \sigma x)$ and $f_{xx}(t, x) = \sigma^2 h''(\mu t + \sigma x)$, the PDE becomes an ODE for h:

$$h''(x) = -(2\mu/\sigma^2)h'(x) \text{ where } h(A) = 1 \text{ and } h(-B) = 0.$$

This is one of the easiest of ODEs, and we quickly find

$$h(x) = \frac{\exp(-2\mu x/\sigma^2) - \exp(2\mu B/\sigma^2)}{\exp(-2\mu A/\sigma^2) - \exp(2\mu B/\sigma^2)}.$$

When we evaluate this formula at $x = 0$, we find a formula for the ruin probability for Brownian motion with drift that we can summarize as a proposition.

Proposition 8.2 (Ruin Probability for Brownian Motion with Drift).
If $X_t = \mu t + \sigma B_t$ and $\tau = \inf\{t\colon X_t = A \text{ or } X_t = -B\}$ where $A > 0$ and $B > 0$, then we have

$$P(X_\tau = A) = \frac{\exp(-2\mu B/\sigma^2) - 1}{\exp(-2\mu(A+B)/\sigma^2) - 1}. \tag{8.16}$$

Looking Back — and Down the Street

Equation (8.16) is one that deserves space in active memory. It offers considerable insight into many stochastic processes and real-world situations. In the simplest instance, if we confront any stochastic process with constant drift and variability, we almost always gain some insight by comparing that process to a Brownian motion with corresponding drift and variability.

Perhaps the best way to build some intuition about such comparisons is to revisit the ruin problem for biased random walk. For the biased random walk $S_n = X_1 + X_2 + \cdots + X_n$ the mean of each step is $p - q$ and the variance is $1 - (p-q)^2$. When we consider the Brownian motion with drift $\mu = p - q$ and instantaneous variance $\sigma^2 = 1 - (p-q)^2$, we again have a process with a unit time step that is comparable to that of biased random walk in terms of mean and variance. While we have not established any a priori guarantee that the ruin probabilities of the two processes will have anything in common, we can see from Table 8.1 that they turn out to be remarkably close.

Table 8.1. Biased RW versus Brownian motion with Drift: Chance of Hitting Level 100 Before Hitting Level −100

p	$-\mu$	σ^2	Biased RW	BM with Drift
0.495	0.01	0.9999	0.119196	0.119182
0.490	0.02	0.9996	0.0179768	0.017958
0.480	0.04	0.9984	0.000333921	0.000331018
0.470	0.06	0.9964	6.05614×10^{-6}	5.88348×10^{-6}

For a long time, we have known a simple formula for the ruin probabilities for biased random walk, so we did not really need to use the parameter-matching approximation. In fact, our two formulas tell us that if we take μ and σ so that

$$q/p = \exp(-2\mu/\sigma^2),$$

then the ruin probabilities for Brownian motion with drift exactly match those for for the biased walk. As a consequence, any differences in Table 8.1 can be attributed to the difference between q/p and $\exp(-2(p-q)/\{1-(p-q)^2\})$. Obviously, there is no good reason to use equation (8.16) and mean-variance matching in order to estimate the ruin probabilities for biased random walk, although for problems that are even a bit more complicated one almost always gains useful insight from equation (8.16).

To be sure, estimates based on equation (8.16) and mean-variance matching are never perfect, but in a real-world problem where the ratio μ/σ^2 is small, the errors due to modeling defects or inadequate parameter estimation are almost always larger than the error that comes from using (8.16) as an approximation to the ruin probabilities. If you must make a quick bet on a ruin probability, the simple formula (8.16) is often a wise place to start.

Exponential Distribution of the Supremum

The ruin probability formula (8.16) has a further consequence that is simply too nice not to take a moment to appreciate. If we consider $A = \lambda$ and $B \to \infty$, then (8.16) rather delightfully tells us that for $X_t = -\mu t + \sigma B_t$ with $\mu > 0$ we have

$$P\Big(\sup_{0 \le t < \infty} X_t \ge \lambda\Big) = \exp(-2\mu\lambda/\sigma^2),$$

or, in other words, the random variable $Y = \sup_{0\le t<\infty} X_t$ is simply an exponential random variable with mean $\sigma^2/2\mu$. Every streetwise probabilist should keep this fact in active memory. It is perfect for back-of-the-envelope calculations.

Shorthand Notation

If $f \in C^{1,2}(\mathbb{R}^+ \times \mathbb{R})$, then Itô's Formula tells us that the process $X_t = f(t, B_t)$ can be written as

$$X_t = X_0 + \int_0^t \frac{\partial f}{\partial x}(s, B_s)\, dB_s + \int_0^t \frac{\partial f}{\partial t}(s, B_s)\, ds + \frac{1}{2}\int_0^t \frac{\partial^2 f}{\partial x^2}(s, B_s)\, ds, \tag{8.17}$$

and because the three integrals use up so much of the page, we will usually prefer to write equation (8.17) in the shorthand

$$dX_t = \frac{\partial f}{\partial x}(t, B_t)\, dB_t + \frac{\partial f}{\partial t}(t, B_t)\, dt + \frac{1}{2}\frac{\partial^2 f}{\partial x^2}(t, B_t)\, dt. \tag{8.18}$$

This natural notation is used universally, and it helps in ways that go beyond efficient transcription. Just as in ordinary calculus, our intuition is often well served by thinking about dX_t as an analog of $X_{t+\epsilon} - X_t$ for some small positive ϵ. Nevertheless, we must keep in mind that *we have not given any definition of dX_t except as shorthand.* Any equations that we write with dX_t absolutely *must* carry the promise that all the terms have a transcription back into the fully defined integrals used in the long form (8.17) of Itô's formula. Still, this promise should not make us paranoid; many useful calculations can be performed with complete honesty while using this shorthand.

8.3. Vector Extension and Harmonic Functions

Standard Brownian motion in $\mathbb{R}^d$ is defined to be the vector-valued process given by

$$\vec{B}_t = (B_t^1, B_t^2, \ldots, B_t^d),$$

where the one-dimensional component processes $\{B_t^k : 0 \le t < \infty\}$ are independent standard Brownian motions. When $d = 2$, we speak of planar Brownian motion, the mathematical model for the movement of those grains of pollen that Robert Brown first observed under his microscope in the months of June, July, and August 1827.

For Brownian motion in $\mathbb{R}^d$, the appropriate extension of Itô's formula takes advantage of the compact notation of vector analysis. If $f : \mathbb{R}^+ \times \mathbb{R}^d \mapsto \mathbb{R}$, we view f

as a function of a time variable $t \in \mathbb{R}^+$ and a space variable $x \in \mathbb{R}^d$, and we will find that the Itô formula for $f(t, \vec{B}_t)$ is neatly expressed in terms of the space-variable gradient

$$\nabla f = \left(\frac{\partial f}{\partial x_1}, \frac{\partial f}{\partial x_2}, \dots, \frac{\partial f}{\partial x_d} \right)$$

and the corresponding space-variable Laplacian

$$\Delta f = \nabla \cdot \nabla f = \frac{\partial^2 f}{\partial x_1^2} + \frac{\partial^2 f}{\partial x_2^2} + \cdots + \frac{\partial^2 f}{\partial x_d^2}.$$

THEOREM 8.3 (Itô's Formula —Vector Version). *If $f \in C^{1,2}(\mathbb{R}^+ \times \mathbb{R}^d)$ and $\vec{B}_t$ is standard Brownian motion in $\mathbb{R}^d$, then*

$$df(t, \vec{B}_t) = f_t(t, \vec{B}_t)\, dt + \nabla f(t, \vec{B}_t) \cdot d\vec{B}_t + \frac{1}{2} \Delta f(t, \vec{B}_t)\, dt.$$

The proof of this version of Itô's formula can be safely omitted. If we follow the familiar pattern of telescope expansion and Taylor approximation, the only novelty that we encounter is the need to use the independence of the component Brownian motions when we estimate the cross terms that appear in our Taylor expansion. Now, once we have the new Itô formula, we naturally have a criterion for $f(t, \vec{B}_t)$ to be a local martingale.

PROPOSITION 8.3 (Martingale PDE Condition for $\mathbb{R}^d$). *If $f \in C^{1,2}(\mathbb{R}^+ \times \mathbb{R}^d)$ and $\vec{B}_t$ is standard Brownian motion in $\mathbb{R}^d$, then $M_t = f(t, \vec{B}_t)$ is a local martingale provided that*

$$f_t(t, \vec{x}) = -\frac{1}{2} \Delta f(t, \vec{x}).$$

CONNECTION TO HARMONIC FUNCTIONS

If we specialize the conclusion of Proposition 8.3 to functions that depend only on $\vec{x}$, we see that $M_t = f(\vec{B}_t)$ is a local martingale provided that

$$\Delta f = 0. \tag{8.19}$$

Functions that satisfy the equation (8.19) are called *harmonic* functions, and they somehow manage to show up in almost every branch of science and engineering. They are also of great importance in pure mathematics, especially because they have a close connection to the theory of analytic functions. We cannot pursue these remarkable connections to any depth, but there are a few elementary facts that everyone must know.

One particularly handy fact is that the real and imaginary parts of an analytic function are harmonic functions; in particular, the real and imaginary parts of a complex polynomial are harmonic functions. To appreciate one concrete example, just consider

$$z^3 = (x + iy)^3 = (x^3 - 3xy^2) + i(3yx^2 - y^3).$$

We can quickly check that $u(x, y) = x^3 - 3xy^2$ and $v(x, y) = 3yx^2 - y^3$ are both harmonic. Moreover, these polynomials are homogeneous of order three, $u(\lambda x, \lambda y) = \lambda^3 u(x, y)$, and an analogous calculation will provide harmonic polynomials that are homogeneous of any order.

Even a little exploration of examples such as these will show that there is rare magic in the connections among martingales, harmonic functions, and complex variables. Our first application of this magic will be to develop an understanding of the recurrence — or transience — of Brownian motion in $\mathbb{R}^d$. We will use harmonic functions that were familiar to Newton and made famous by Laplace.

Recurrence in $\mathbb{R}^2$

If we calculate the Laplacian of the function defined on $\mathbb{R}^2 - \{0\}$ by

$$f(\vec{x}) = \log|\vec{x}| = \log\sqrt{x_1^2 + x_2^2},$$

then we will find that $\Delta f(\vec{x}) = 0$; so, for any $0 < r < R$, we see that f is a harmonic function on the annulus $A = \{\vec{x}\colon r \le |\vec{x}| \le R\}$. We can then rescale f to get a harmonic function

$$h(\vec{x}) = \frac{\log R - \log|\vec{x}|}{\log R - \log r}$$

with values on the boundary of A given by

$$h(\vec{x}) = 1 \text{ for } |\vec{x}| = r \text{ and } h(\vec{x}) = 0 \text{ for } |\vec{x}| = R.$$

Next, we take $\vec{x} \in A$ and let

$$\tau_r = \inf\{\, t\colon |\vec{B}_t| = r\,\} \text{ and } \tau_R = \inf\{\, t\colon |\vec{B}_t| = R\,\}.$$

To keep our notation tidy, we first let $P_{\vec{x}}(\cdot) = P(\cdot \mid B_0 = \vec{x})$. We then note that for $\tau = \tau_r \wedge \tau_R$ we have $P_{\vec{x}}(\tau < \infty) = 1$ and $E_{\vec{x}}[h(B_\tau)] = P_{\vec{x}}(\tau_r < \tau_R)$. The familiar stopping time argument then tells us that we have $E_{\vec{x}}[h(B_\tau)] = h(\vec{x})$, or in detail,

$$P_{\vec{x}}(\tau_r < \tau_R) = \frac{\log R - \log|\vec{x}|}{\log R - \log r} \text{ where } r < |\vec{x}| < R. \tag{8.20}$$

This is a marvelous formula. One of its immediate implications is that if we start a Brownian motion at any $\vec{x} \in \mathbb{R}^2$, then in a finite amount of time the process will hit a ball of nonzero radius located anywhere in $\mathbb{R}^2$. Specifically, from our understanding of Brownian motion in $\mathbb{R}$, we know that $\tau_R \to \infty$ as $R \to \infty$, so, by the dominated convergence theorem applied to the hitting probability formula (8.20), we have

$$P_{\vec{x}}(\tau_r < \infty) = 1 \text{ for all } |\vec{x}| \ge r > 0. \tag{8.21}$$

This formula expresses the *recurrence property* of Brownian motion in $\mathbb{R}^2$. We have seen before that we have an analogous property for Brownian motion in $\mathbb{R}$, but shortly we will show that when $d \ge 3$ Brownian motion in $\mathbb{R}^d$ is no longer recurrent.

Transience in $\mathbb{R}^d$ for $d \ge 3$

We can make a prefectly analogous calculation in $\mathbb{R}^d$ by starting with the function

$$f(\vec{x}) = |\vec{x}|^{2-d}$$

that is harmonic on the annulus $A = \{\vec{x}\colon r \le |\vec{x}| \le R\} \subset \mathbb{R}^d$. In this case, we find that the hitting probability formula is given by the simple formula

$$P_{\vec{x}}(\tau_r < \tau_R) = \frac{R^{2-d} - |\vec{x}|^{2-d}}{R^{2-d} - r^{2-d}},$$

but now when we let $R \to \infty$ we find

$$P_{\vec{x}}(\tau_r < \infty) = \left(\frac{r}{|\vec{x}|}\right)^{d-2}.$$

This formula tells us that if we start a Brownian motion at a point outside of a sphere in $\mathbb{R}^d$, then for $d \ge 3$ there is positive probability the process will never hit the sphere. Pólya described the relationship between the recurrence property of Brownian motion in $\mathbb{R}^2$ and the transience in $\mathbb{R}^3$ by saying "a drunken man will always find his way home, but a drunken bird may not."

8.4. Functions of Processes

If a process X_t is given to us as a stochastic integral that we may write in shorthand as

$$dX_t = a(\omega, t)\,dt + b(\omega, t)\,dB_t, \tag{8.22}$$

then it is natural to *define* the dX_t integral of $f(t, \omega)$ by setting

$$\int_0^t f(\omega, s)\,dX_s = \int_0^t f(\omega, s)a(\omega, s)\,ds + \int_0^t f(\omega, s)b(\omega, s)\,dB_s, \tag{8.23}$$

provided that $f(\omega, t)$ is an integrand for which the last two integrals make sense; that is, we are happy to write equation (8.22) or equation (8.23) whenever $f(\omega, t)$ is adapted and it has the integrability conditions:

- $f(\omega, s)\,a(\omega, s) \in L^1(dt)$ for all ω in a set of probability one and
- $f(\omega, s)b(\omega, s) \in \mathcal{L}^2_{\mathrm{LOC}}$.

Now, we come to a very natural question: If the process X_t can be written as a stochastic integral of the form (8.23) and $g(t, y)$ is a smooth function, can we then write the process $Y_t = g(t, X_t)$ as a dX_t integral?

Itô's formula tells us that this is the case if X_t is Brownian motion, and we will soon obtain a positive answer to the general question. We begin our approach to this problem by first considering two instructive examples. These examples will lead us to some computational rules that will become our daily companions.

Chain Rules and the Box Calculus

If X_t is Brownian motion with general drift and variance, then we have our choice whether to write X_t as a stochastic integral or as a function of Brownian motion:

$$dX_t = \mu\,dt + \sigma\,dB_t,\ X_0 = 0 \quad \text{or } X_t = \mu t + \sigma B_t.$$

A similar situation prevails when we consider Itô's formula for a function of t and X_t. Specifically, if we have $Y_t = f(t, X_t)$, then we can also write $Y_t = g(t, B_t)$ where we take g to be defined by $g(t,x) = f(t, \mu t + \sigma x)$. When we apply Itô's formula to the representation $Y_t = g(t, B_t)$, we find

$$dY_t = g_t(t, B_t)\,dt + g_x(t, B_t)\,dB_t + \frac{1}{2} g_{xx}(t, B_t)\,dt,$$

and the chain rule gives us $g_t(t,x) = f_t(t, \mu t + \sigma x) + f_x(t, \mu t + \sigma x)\mu$, $g_x(t,x) = f_x(t, \mu t + \sigma x)\sigma$, and $g_{xx}(t,x) = f_{xx}(t, \mu t + \sigma x)\sigma^2$, so in terms of f we have

$$\begin{aligned} dY_t &= \{f_t(t, X_t) + \mu f_x(t, X_t)\}\,dt + \sigma f_x(t, X_t)\,dB_t + \frac{1}{2}\sigma^2 f_{xx}(t, X_t)\,dt \\ &= f_t(t, X_t)\,dt + f_x(t, X_t)\,dX_t + \frac{1}{2}\sigma^2 f_{xx}(t, X_t)\,dt. \end{aligned} \tag{8.24}$$

The last expression could be called the Itô formula for functions of Brownian motion with drift, and it might be worth remembering — except for the fact that there is a simple recipe that will let us write equation (8.24) without adding to our mental burdens.

The rule depends on a formalism that is usually called the *box calculus*, though the term *box algebra* would be more precise. This is an algebra for the set $\mathcal{A}$ of linear combinations of the formal symbols dt and dB_t. In this algebra, adapted functions are regarded as scalar values and addition is just the usual algebraic addition. Products are then computed by the traditional rules of associativity and transitivity together with a multiplication table for the special symbols dt and dB_t.

TABLE 8.2. BOX ALGEBRA MULTIPLICATION TABLE

$\cdot$	dt	dB_t
dt	0	0
dB_t	0	dt

If we use the *centered dot* "$\cdot$" to denote multiplication in the box algebra, then we can work out the product

$$(a\,dt + b\,dB_t) \cdot (\alpha\,dt + \beta\,dB_t)$$

as a sum that nicely simplifies,

$$a\alpha\,dt \cdot dt + a\beta\,dt \cdot dB_t + b\alpha\,dB_t \cdot dt + b\beta\,dB_t \cdot dB_t = b\beta\,dt.$$

Now, let us reconsider the formula (8.24), which we called the Itô formula for Brownian motion with drift. We proved this formula by an honest application of the usual Itô Formula for Brownian motion, and it gives a simple example of computation with the box calculus.

When we use the shorthand that defines $dX_t \cdot dX_t$ as

$$dX_t \cdot dX_t = \sigma^2\,dt$$

in accordance with the rules of the Box Calculus, we see that equation (8.24) can be written as

$$df(t, X_t) = f_t(t, X_t)\,dt + f_x(t, X_t)\,dX_t + \frac{1}{2} f_{xx}(t, X_t)\,dX_t \cdot dX_t. \tag{8.25}$$

This general formula for functions of $X_t = \mu t + \sigma B_t$ is easier to remember than the orignal version (8.24), but the really remarkable aspect of the formula is that it applies to processes that are much more general than $X_t = \mu t + \sigma B_t$. We will shortly find that it applies to essentially all of the processes that we have seen.

Nevertheless, before we look to the general case, it is worthwhile to examine a second example where the chain rule can be used to justify the box calculus version of Itô's formula (8.25).

Box Calculus and Functions of Geometric Brownian Motion

The process defined by $X_t = \exp(\alpha t + \sigma B_t)$ is known as *geometric Brownian motion*, and we will find it to be one of the most useful stochastic processes. Since X_t is a function of Brownian motion and time, a standard application of the space-time version of Itô's formula given by Theorem 8.2 tells us that X_t satisfies

$$dX_t = \left(\alpha + \frac{1}{2}\sigma^2\right) X_t\, dt + \sigma X_t\, dB_t. \tag{8.26}$$

Now, we consider a process defined by taking $Y_t = f(t, X_t)$. As before, this process can be written as a function of Brownian motion $Y_t = g(t, B_t)$. This time we have $g(t,x) = f(t, \exp(\alpha t + \sigma x))$, and, if we are patient, we can again calculate dY_t by Theorem 8.2. First, by the direct application of the chain rule, we have

$$g_t(t,x) = f_t(t, \exp(\alpha t + \sigma x)) + f_x(t, \exp(\alpha t + \sigma x)) \exp(\alpha t + \sigma x)\alpha$$

and

$$g_x(t,x) = f_x(t, \exp(\alpha t + \sigma x)) \exp(\alpha t + \sigma x)\sigma.$$

Finally, $g_{xx}(t,x)$ is the ugliest worm

$$\sigma^2 \exp(\alpha t + \sigma x)\{f_x(t, \exp(\alpha t + \sigma x)) + f_{xx}(t, \exp(\alpha t + \sigma x)) \exp(\alpha t + \sigma x)\}.$$

By these calculations and an application of Itô's space-time formula, we find that the process $Y_t = g(t, B_t) = f(t, X_t)$ satisfies

$$\begin{aligned} dY_t =& g_t(t, B_t)\, dt + g_x(t, B_t)\, dB_t + \frac{1}{2} g_{xx}(t, B_t)\, dt \\ =& \{f_t(t, X_t) + \alpha X_t f_x(t, X_t)\}\, dt + \sigma X_t f_x(t, X_t)\, dB_t \\ &+ \frac{1}{2}\{\sigma^2 X_t f_x(t, X_t) + \sigma^2 X_t^2 f_{xx}(t, X_t)\}\, dt. \end{aligned}$$

When we collect terms and recall the formula (8.26) for dX_t, we finally find a formula that could be called Itô's Formula for functions of geometric Brownian motion:

$$dY_t = f_t(t, X_t)\, dt + f_x(t, X_t)\, dX_t + \frac{1}{2} f_{xx}(t, X_t)\sigma^2 X_t^2\, dt. \tag{8.27}$$

Now, our task is to see how the identity (8.27) may be rewritten in the language of the box calculus. From the formula (8.26) for dX_t and the Box Calculus rules, we see that $dX_t \cdot dX_t$ is shorthand for $\sigma^2 X_t^2\, dt$, so the definition of the box product tells us that the formula (8.27) also may be written as

$$df(t, X_t) = f_t(t, X_t)\, dt + f_x(t, X_t)\, dX_t + \frac{1}{2} f_{xx}(t, X_t)\, dX_t \cdot dX_t. \tag{8.28}$$

This is *exactly* the same as the formula (8.25) that we found for functions of Brownian motion with drift. Moreover, this coincidence is not an accident. By working Exercise 8.6, one can confirm that the formula (8.28) is valid whenever X_t is a smooth function of time and Brownian motion. In the next section, we will find that the formula (8.28) is actually valid for a much larger class of processes.

8.5. The General Itô Formula

The chain rule method for the derivation of formulas such as (8.24) and (8.28) is limited in its range of applicability, but the same identities hold for essentially all of the processes that we can write as stochastic integrals. To make this notion precise, we need a formal definition of what we will call *standard processes.*

DEFINITION 8.1. *We say that a process* $\{X_t\colon 0 \le t \le T\}$ *is* standard *provided that* $\{X_t\}$ *has the integral representation*

$$X_t = x_0 + \int_0^t a(\omega, s)\,ds + \int_0^t b(\omega, s)\,dB_s \text{ for } 0 \le t \le T,$$

and where $a(\cdot,\cdot)$ *and* $b(\cdot,\cdot)$ *are adapted, measurable processes that satisfy the integrability conditions*

$$P\left(\int_0^T |a(\omega, s)|\,ds < \infty\right) = 1 \text{ and } P\left(\int_0^T |b(\omega, s)|^2\,ds < \infty\right) = 1.$$

The standard processes provide a natural home for the theory of Itô integration, and one of the clearest expressions of this fact is that Itô's formula continues to apply. Any smooth function of a standard process is again a standard process with an explicit representation as a stochastic integral.

THEOREM 8.4 (Itô's Formula for Standard Processes).
If $f \in C^{1,2}(\mathbb{R}^+ \times \mathbb{R})$ *and* $\{X_t\colon 0 \le t \le T\}$ *is a standard process with the integral representation*

$$X_t = \int_0^t a(\omega, s)\,ds + \int_0^t b(\omega, s)\,dB_s, \quad 0 \le t \le T,$$

then we have

$$f(t, X_t) = f(0,0) + \int_0^t \frac{\partial f}{\partial t}(s, X_s)\,ds + \int_0^t \frac{\partial f}{\partial x}(s, X_s)\,dX_s + \frac{1}{2}\int_0^t \frac{\partial^2 f}{\partial x^2}(s, X_s)\,b^2(\omega, s)\,ds.$$

When we look at Theorem 8.4 in the language of the box calculus, it tells us that for the process $Y_t = f(t, X_t)$ we have

$$dY_t = f_t(t, X_t)\,dt + f_x(t, X_t)\,dX_t + \frac{1}{2} f_{xx}(t, X_t)\,dX_t \cdot dX_t. \tag{8.29}$$

This is exactly the formula that we found by ad hoc means for Brownian motion with drift and for geometric Brownian motion. The thrust of Theorem 8.4 is that we can always use the Box Calculus, and we always have the elegant differential representation (8.29). Never again will we need to repeat those long, ugly chain rule calculations.

Because the proof of Theorem 8.4 follows the same general pattern that we used in the proof of Theorem 8.1, the simplest Itô formula, we will not give a detailed argument. Still, we should note that this more general formula has a couple of new wrinkles. For example, the dX_s integral is really two integrals, and we need to do some work to get a proper understanding of the second ds integral. We will return to the outline of the proof of Theorem 8.4 after the discussion of quadratic variation that is given in the Section 8.6, but first we will look at an even more general version of Itô's formula.

FUNCTIONS OF SEVERAL PROCESSES

We found considerable value in the examination of functions of several independent Brownian motions, and we naturally want to extend our box calculus to functions of several processes. Formally, we just need to extend our box calculus multiplication table by one row and one column.

TABLE 8.3. EXTENDED BOX ALGEBRA

$\cdot$	dt	dB_t^1	dB_t^2
dt	0	0	0
dB_t^1	0	dt	0
dB_t^2	0	0	dt

THEOREM 8.5 (Itô's Formula for Two Standard Processes).
If $f \in C^{2,2}(\mathbb{R} \times \mathbb{R})$ and both of the standard processes $\{X_t\colon 0 \le t \le T\}$ and $\{Y_t\colon 0 \le t \le T\}$ have the integral representations

$$X_t = \int_0^t a(\omega, s)\, ds + \int_0^t b(\omega, s)\, dB_s$$

and

$$Y_t = \int_0^t \alpha(\omega, s)\, ds + \int_0^t \beta(\omega, s)\, dB_s,$$

then we have

$$\begin{aligned}
f(X_t, Y_t) = f(0,0) &+ \int_0^t f_x(X_s, Y_s)\, dX_s + \int_0^t f_y(X_s, Y_s)\, dY_s \\
&+ \frac{1}{2}\int_0^t f_{xx}(X_s, Y_s)\, b^2(\omega, s)\, ds \\
&+ \int_0^t f_{xy}(X_s, Y_s)\, b(\omega, s)\beta(\omega, s)\, ds \\
&+ \frac{1}{2}\int_0^t f_{yy}(X_s, Y_s)\, \beta^2(\omega, s)\, ds.
\end{aligned} \tag{8.30}$$

In the language of the box calculus, the Itô formula for $Z_t = f(X_t, Y_t)$ can be written much more succinctly, or, at least, four lines may be reduced to two:

$$\begin{aligned}
dZ_t =& f_x(X_t, Y_t)\, dX_t + f_y(X_t, Y_t)\, dY_t \\
&+ \frac{1}{2} f_{xx}(X_t, Y_t)\, dX_t \cdot dX_t + f_{xy}(X_t, Y_t)\, dX_t \cdot dY_t + \frac{1}{2} f_{yy}(X_t, Y_t)\, dY_t \cdot dY_t.
\end{aligned}$$

USEFUL SPECIALIZATIONS

There are many useful specializations of the previous formula, but perhaps the most natural is to take $Y_t = t$ and note that we can recapture our earlier Itô formula for functions of a general process and time. A more interesting specialization is given by the choice $f(x,y) = xy$. In this case, we have $f_{xx} = 0$, $f_{yy} = 0$, and $f_{xy} = 1$, so in the end we find

$$X_tY_t - X_0Y_0 = \int_0^t Y_s\,dX_s + \int_0^t X_s\,dY_s + \int_0^t b(\omega,s)\beta(\omega,s)\,ds. \tag{8.31}$$

This is the *product rule* for stochastic integration. In the case where one of the two processes does not have a dB_t component, the last ds integral does not appear, and we find a formula that coincides with the familiar rule for integration by parts of traditional calculus. Another special situation when we get the traditional formula is when $b(\omega,s)$ and $\beta(\omega,s)$ are orthogonal in the strong sense that

$$\int_0^t b(\omega,s)\beta(\omega,s)\,ds = 0$$

for almost every ω. This special case will be important to us much later in the proof of the famous martingale representation theorem.

One of the interesting aspects of the product rule (8.31) is that it can be proved independently of Itô's formula — say by starting with the product version of telescoping. One then can prove Itô's formula first showing that the product rule (8.31) implies Itô's formula for polynomial f and by extending the polynomial formula to more general f by approximation arguments.

8.6. Quadratic Variation

We met the notion of quadratic variation informally in Chapter 6, where in the course of making a direct calculation of $\int B_s dB_s$ we found that if $t_i = t/n$ for $i = 0, 1, \ldots, n$, then

$$\sum_{i=1}^{n} (B_{t_i} - B_{t_{i-1}})^2 \xrightarrow{p} t \quad \text{as } n \to \infty. \tag{8.32}$$

Now, in order to establish the general Itô formulas described in the previous section, we need to establish the analog of formula (8.32) for more general processes. This development will also yield two important by-products. The first of these is a new perspective on a class of martingales that generalize $B_t^2 - t$. The second is a new way to think about time — a way that will eventually tell us how we can view any continuous martingales as a Brownian motion with an altered clock.

The formal introduction of the quadratic variation requires a few basic definitions. First, a finite ordered set of times $\{t_0 \le t_1 \le t_2 \le \cdots \le t_n\}$ with $t_0 = 0$ and $t_n = t$ is called a *partition* of $[0,t]$. The *mesh* $\mu(\pi)$ of a partition π is then defined to be the length of the biggest gap between any pair of successive times t_i and t_{i+1} in π.

Now, for any partition $\pi = \{t_0 \le t_1 \le t_2 \le \cdots \le t_n\}$ of $[0,t] \subset [0,T]$ and for any process $\{X_t\}$ on $[0,T]$, the π-quadratic variation of the process $\{X_t\}$ is defined to be the random variable

$$Q_\pi(X_t) = \sum_{i=1}^{n} (X_{t_i} - X_{t_{i-1}})^2. \tag{8.33}$$

If there is a process $\{V_t\}$ such that $Q_{\pi_n}(X_t)$ converges in probability to V_t for any sequence of partitions $\{\pi\}$ of $[0,t]$ such that $\mu(\pi) \to 0$ as $n \to \infty$, then we say that $\{V_t\}$ is the *quadratic variation* of $\{X_t\}$. When the quadratic variation of $\{X_t\}$ exists, it is denoted by $\langle X\rangle_t$.

The main goal of this section is to show that the quadratic variation of a standard process can be given explicitly as an integral of the instantaneous variance of the process.

THEOREM 8.6 (Quadratic Variation of a Standard Process). *If X_t is a standard process with the integral representation*

$$X_t = \int_0^t a(\omega, s)\, ds + \int_0^t b(\omega, s)\, dB_s \text{ where } 0 \le t \le T,$$

then the quadratic variation of X_t exists, and it is given by

$$\langle X \rangle_t = \int_0^t b^2(\omega, s)\, ds \quad \text{for } 0 \le t \le T.$$

The proof of this theorem will require several steps, the easiest of which is to show how to handle the first summand of the process $\{X_t\}$.

PROPOSITION 8.4. *If $a(\omega, s)$ is a measurable adapted process with*

$$\int_0^t |a(\omega, s)|\, ds < \infty \text{ almost surely,} \tag{8.34}$$

then the quadratic variation of the process

$$A_t = \int_0^t a(\omega, s)\, ds$$

exists and is equal to zero.

PROOF. For any ω, the function $u \mapsto \int_0^u a(\omega, s)\, ds$ is uniformly continuous on the interval $[0, t]$, so for any $\epsilon > 0$ there is a $\delta(\omega)$ such that for all π with $\mu(\pi) \le \delta(\omega)$ we have

$$\sum_{i=1}^n \left(\int_{t_{i-1}}^{t_i} a(\omega, s)\, ds \right)^2 \le \max_{1 \le i \le n} \left(\int_{t_{i-1}}^{t_i} |a(\omega, s)|\, ds \right) \sum_{i=1}^n \left(\int_{t_{i-1}}^{t_i} |a(\omega, s)|\, ds \right)$$

$$\le \epsilon \int_0^t |a(\omega, s)|\, ds.$$

By our assumption (8.34) on $a(\omega, t)$, the last inequality tells us that $Q_\pi(A_t)$ converges with probability one to 0 as $\mu(\pi) \to 0$. This is more than we need to complete the proof of the proposition. □

The more challenging part of our calculation of the quadratic variation of $\{X_t\}$ comes from the contribution of the dB_t integral. The analysis of this contribution will itself take a couple of steps. The first of these is to study the quadratic variation of a *bounded* continuous martingale.

LEMMA 8.1. *If $\{Z_t : 0 \le s \le t\}$ is a continuous martingale for which we have $\sup_{0 \le s \le t} |Z_s| \le B$ for a constant $B < \infty$, then there is a finite constant C such that*

$$\sup_\pi E[Q_\pi^2(Z_t)] \le C E[(Z_t - Z_0)^2] < \infty \tag{8.35}$$

and, moreover,

$$\lim_{\mu(\pi) \to 0} \sum_{i=1}^n E[(Z_{t_i} - Z_{t_{i-1}})^4] \to 0. \tag{8.36}$$

PROOF. The square of $Q_\pi(Z_t)$ can be arranged as

(8.37)
$$Q^2_\pi(Z_t) = \sum_{i=1}^n (Z_{t_i} - Z_{t_{i-1}})^4 + 2\sum_{i=1}^{n-1}\left\{(Z_{t_i} - Z_{t_{i-1}})^2 \sum_{j=i+1}^n (Z_{t_j} - Z_{t_{j-1}})^2\right\},$$

and the inside sum can be estimated by applying the martingale property and the boundedness hypothesis to obtain

$$E\left[\sum_{j=i+1}^n (Z_{t_j} - Z_{t_{j-1}})^2 \middle| \mathcal{F}_{t_i}\right] = E((Z_t - Z_{t_i})^2 | \mathcal{F}_{t_i}) \le 4B^2. \tag{8.38}$$

Now, when we apply this last bound (and the boundedness hypothesis) to the identity (8.37), we then find

$$\begin{aligned} E[Q^2_\pi(Z_t)] \le& 4B^2 \sum_{i=1}^n E[(Z_{t_i} - Z_{t_{i-1}})^2] \\ &+ 2E\left[\sum_{i=1}^{n-1}(Z_{t_i} - Z_{t_{i-1}})^2 E\left[\sum_{j=i+1}^n (Z_{t_j} - Z_{t_{j-1}})^2 \middle| \mathcal{F}_{t_i}\right]\right] \\ \le& 4B^2 E[(Z_t - Z_0)^2] + 8B^2 E\left[\sum_{i=1}^{n-1}(Z_{t_i} - Z_{t_{i-1}})^2\right] \le 12B^2 E[(Z_t - Z_0)^2], \end{aligned}$$

so we have established the bound (8.35) with $C = 12B^2$.

For the second part of the lemma, we first note that the modulus of continuity

$$\rho(\delta, \omega) \equiv \sup\{\, |Z_u - Z_v| : 0 \le u \le v \le t \text{ and } |u - v| \le \delta\,\}$$

is bounded by $2B$, and by the uniform continuity of $s \mapsto Z_s(\omega)$ on $[0, t]$ it goes to zero as $\delta \to 0$ for all ω. Therefore, if we take the expectation of the elementary inequality

$$R_\pi \equiv \sum_{i=1}^n (Z_{t_i} - Z_{t_{i-1}})^4 \le \rho^2(\mu(\pi), \omega) \cdot Q_\pi(Z_t) \tag{8.39}$$

and then apply Cauchy's inequality, we find

$$E(R_\pi) \le ||\rho^2(\mu(\pi), \omega)||_{L^2(dP)} ||Q_\pi(Z_t)||_{L^2(dP)}.$$

Finally, by the first part of the lemma, $||Q_\pi(Z_t)||_{L^2(dP)}$ is bounded, and by the dominated convergence theorem $||\rho^2(\mu(\pi), \omega)||_{L^2(dP)}$ goes to zero as $\mu(\pi) \to 0$, so the second part of the lemma is complete. □

A New Class of Martingales

If we think back to our derivation of the simplest Itô formula, we found that $Q_\pi(B_t) \xrightarrow{p} t$, and an essential part of that proof rested on the fact that $B_t^2 - t$ is a continuous martingale. We plan to follow a similar path here, and our first step is to recall the earlier suggestion that there is a large class of martingales that generalize $B_t^2 - t$. We will now explore these new martingales, and the first step is to recall that the conditional form of the Itô isometry tells us that for any $b \in \mathcal{H}^2$ we have

$$E\left[\left(\int_s^t b(\omega,u)\,dB_u\right)^2 \,\middle|\, \mathcal{F}_s\right] = E\left(\int_s^t b^2(\omega,u)\,du \,|\, \mathcal{F}_s\right). \tag{8.40}$$

We now have an easy proposition to introduce our new martingale partners.

PROPOSITION 8.5. *If $b(\omega,t) \in \mathcal{H}^2[0,T]$, then*

$$Z_t = \int_0^t b(\omega,u)\,dB_u \quad \text{and} \quad M_t = Z_t^2 - \int_0^t b^2(\omega,u)\,du \tag{8.41}$$

are continuous martingales with respect to the standard Brownian filtration.

PROOF. We already know that Z_t is a continuous martingale, so we only need to consider M_t. The definition of M_t and the martingale property of Z_t give us

$$E(M_t - M_s|\mathcal{F}_s) = E\left(Z_t^2 - Z_s^2 - \int_s^t b^2(\omega,u)\,du\,|\mathcal{F}_s\right)$$

$$= E\left((Z_t - Z_s)^2 - \int_s^t b^2(\omega,u)\,du \mid \mathcal{F}_s\right), \tag{8.42}$$

and the conditional Itô isometry (8.40) tells us that the last expectation is equal to zero, so the martingale property of M_t is established. □

PROPOSITION 8.6. *Suppose that $b(\omega,s)$ is a measurable, adapted process such that*

$$\int_0^T b^2(\omega,s)\,ds \le C$$

with probability one for some constant $C < \infty$. If the process defined by

$$Z_t = \int_0^t b(\omega,s)\,dB_s \qquad 0 \le t \le T$$

is bounded, then the quadratic variation of $\{Z_t : 0 \le t \le T\}$ exists with probability one, and it is given by

$$\langle Z\rangle_t = \int_0^t b^2(\omega,s)\,ds \qquad 0 \le t \le T.$$

PROOF. If we let

$$\Delta_\pi = Q_\pi(Z_t) - \int_0^t b^2(\omega,s)\,ds = \sum_{i=1}^n \left\{(Z_{t_i} - Z_{t_{i-1}})^2 - \int_{t_{i-1}}^{t_i} b^2(\omega,s)\,ds\right\},$$

then the summands

$$d_i = (Z_{t_i} - Z_{t_{i-1}})^2 - \int_{t_{i-1}}^{t_i} b^2(\omega,s)\,ds$$

are martingale differences (by the conditional Itô isometry), so by the orthogonality of the d_i we have

$$E(\Delta_\pi^2) = \sum_{i=1}^n E(d_i^2).$$

If we then apply the elementary bound $(x+y)^2 \le 2x^2 + 2y^2$ to the summands of d_i, we find

$$E(\Delta_\pi^2) \le 2\sum_{i=1}^{n} E((Z_{t_i} - Z_{t_{i-1}})^4) + 2\sum_{i=1}^{n} E\left[\left(\int_{t_{i-1}}^{t_i} b^2(\omega,s)\,ds\right)^2\right]. \tag{8.43}$$

The first sum in the on the right-hand side of the bound (8.43) is easy to handle because Lemma 8.1 tells us that

$$\sum_{i=1}^{n} E((Z_{t_i} - Z_{t_{i-1}})^4) \to 0.$$

To handle the second sum in the bound (8.43), we first introduce the modulus of continuity

$$\rho(\delta,\omega) \equiv \sup\left\{\int_u^v b^2(\omega,s)\,ds : 0 \le u \le v \le T \text{ and } |u-v| \le \delta\right\},$$

and note that $\rho(\delta,\omega) \le C$, so we have

$$\sum_{i=1}^{n}\left(\int_{t_{i-1}}^{t_i} b^2(\omega,s)\,ds\right)^2 \le \rho(\mu(\pi),\omega)\int_0^t b^2(\omega,s)\,ds \le \rho(\mu(\pi),\omega)C. \tag{8.44}$$

Because $\rho(\pi,\omega)$ is bounded and goes to zero for all ω as $\mu(\pi) \to 0$, the dominated convergence theorem tells us that the expectation of the left-hand side of the bound (8.44) goes to zero as $\mu(\pi) \to 0$. □

FINALLY, THE PROOF OF THEOREM 8.6

All that remains is to assemble the individual pieces under the umbrella of a nice localization. Accordingly, we let τ_M denote the smallest value of u for which we have $u \ge r$ or any of the three conditions

$$\int_0^u |a(\omega,s)|\,ds \ge M, \quad \int_0^u b^2(\omega,s)\,ds \ge M, \quad \left|\int_0^u b(\omega,s)\,dB_s\right| \ge M. \tag{8.45}$$

Now, since $Q_\pi(X_t) = Q_\pi(X_{t\wedge\tau_M})$ for all $\omega \in \{\omega : t \le \tau_M\}$, we have

$$\begin{aligned} P\Big(\Big|Q_\pi(X_t) - \int_0^t b^2(\omega,s)\,ds\Big| \ge \epsilon\Big) &\le P\Big(\Big|Q_\pi(X_t) - \int_0^t b^2(\omega,s)\,ds\Big| \ge \epsilon,\ t \le \tau_M\Big) + P(\tau_M < t) \\ &= P\Big(\Big|Q_\pi(X_{t\wedge\tau_M}) - \int_0^t b^2(\omega,s)1(t \le \tau_M)\,ds\Big| \ge \epsilon\Big) + P(\tau_M < t), \end{aligned}$$

and since X_t is a standard process, we can choose M to make $P(\tau_M < t)$ as small as we like. The bottom line is that it therefore suffices to show that

$$Q_\pi(X_{t\wedge\tau_M}) \xrightarrow{P} \int_0^t b^2(\omega,s)1(t \le \tau_M)\,ds, \tag{8.46}$$

since we now see that this immediately implies

$$Q_\pi(X_t) \xrightarrow{P} \int_0^t b^2(\omega,s)\,ds. \tag{8.47}$$

Next, if we write $X_{t\wedge\tau_M}$ as a standard process

$$\widetilde{X}_t = X_{t\wedge\tau_M} = \int_0^t a(\omega,s)1(t\le\tau_M)\,ds + \int_0^t b(\omega,s)1(t\le\tau_M)\,dB_s$$
$$= \int_0^t \widetilde{a}(\omega,s)\,ds + \int_0^t \widetilde{b}(\omega,s)\,dB_s = \widetilde{A}_t + \widetilde{Z}_t,$$

then all that remains is to calculate the quadratic variation of $\widetilde{X}_t$. We begin by noting that

$$Q_\pi(\widetilde{X}_t) = Q_\pi(\widetilde{A}_t) + Q_\pi(\widetilde{Z}_t) + 2C_\pi(\widetilde{A}_t,\widetilde{Z}_t)$$

where

$$C_\pi(\widetilde{A}_t,\widetilde{Z}_t) = \sum_{i=1}^n \left(\int_{t_{i-1}}^{t_i} \widetilde{a}(\omega,s)\,ds\right)\left(\int_{t_{i-1}}^{t_i} \widetilde{b}(\omega,s)\,dB_s\right). \tag{8.48}$$

Now, by the construction of τ_M, we know that $\widetilde{a}(\omega,s) = a(\omega,s)1(t\le\tau_M)$ more than meets the hypotheses of Proposition 8.4, so $Q_\pi(\widetilde{A}_t)\to 0$ as $\mu(\pi)\to 0$. Also, by Proposition 8.6,

$$Q_\pi(\widetilde{Z}_t) \xrightarrow{p} \int_0^t b^2(\omega,s)1(t\le\tau_M)\,ds.$$

Finally, by equation (8.48) and Cauchy's inequality we have

$$C_\pi(\widetilde{A}_t,\widetilde{Z}_t) \le Q_\pi^{\frac{1}{2}}(\widetilde{A}_t)Q_\pi^{\frac{1}{2}}(\widetilde{Z}_t), \tag{8.49}$$

so from the fact that $Q_\pi(\widetilde{Z}_t)$ converges in probability to a bounded random variable and the fact that $Q_\pi(\widetilde{A}_t)$ converges to zero in probability, we deduce that $C_\pi(\widetilde{A}_t,\widetilde{Z}_t)$ converges to zero in probability as $\mu(\pi)\to 0$. At last, we see that

$$Q_\pi(X_{t\wedge\tau_M}) \xrightarrow{p} \int_0^t b^2(\omega,s)1(t\le\tau_M)\,ds, \tag{8.50}$$

and by our first reduction (8.46), we know that the limit (8.50) is all we need to complete the proof of Theorem 8.6.

Closing the Itô Formula Loop

This brief development of quadratic variation is all that we need to complete the proof of the Itô formula for a function of time and a standard process, but for a proof of the Itô formula for a function of two processes we need one more idea. We need to know that for the standard processes

$$X_t = \int_0^t a(\omega,s)\,ds + \int_0^t b(\omega,s)\,dB_s$$

and

$$Y_t = \int_0^t \alpha(\omega,s)\,ds + \int_0^t \beta(\omega,s)\,dB_s,$$

we have that

$$Q_\pi(X_t,Y_t) \equiv \sum_{i=1}^n (X_{t_i}-X_{t_{i-1}})(Y_{t_i}-Y_{t_{i-1}})$$

satisfies

$$Q_\pi(X_t,Y_t) \xrightarrow{p} \int_0^t b(\omega,s)\beta(\omega,s)\,ds \text{ as } \mu(\pi)\to 0. \tag{8.51}$$

Fortunately for the length of this chapter, the preceding result follows immediately from our earlier work and a well-known *polarization* trick.

The key observation is that because $X_t + Y_t$ and $X_t - Y_t$ are standard processes, Theorem 8.6 tells us that the left-hand side of

$$\frac{1}{4}Q_\pi(X_t + Y_t) - \frac{1}{4}Q_\pi(X_t - Y_t) = Q_\pi(X_t, Y_t)$$

converges in probability to

$$\begin{aligned}\frac{1}{4}\int_0^t (b(\omega, s) + \beta(\omega, s))^2\, ds \;-\; \frac{1}{4}\int_0^t (b(\omega, s) - \beta(\omega, s))^2\, ds \\ = \int_0^t b(\omega, s)\beta(\omega, s)\, ds.\end{aligned}$$

Now, we have all of the tools that are needed to apply the same analysis-synthesis approach to $f(X_t, Y_t)$ that we applied to $f(B_t)$ at the beginning of the chapter.

8.7. Exercises

To some extent, all of stochastic calculus is an exercise in the application of Itô's formula. These exercises therefore focus on ideas that are particular to this chapter. The first exercise invites the reader to follow the plan of Theorem 8.1 to complete the proof of Theorem 8.2, the space-time Itô formula. This exercise should be done by everyone. The other exercises illustrate diverse topics, including integration by parts for stochastic integrals, applications of harmonic functions, and nice local martingales that fail to be honest martingales.

EXERCISE 8.1 (Very Good for the Soul). Give an absolutely complete proof of Theorem 8.2. Take this opportunity to pick every nit, to revel in every morsel of detail, and — forever after — be able to greet variations of Itô's formula with the warm confidence of "Been there; done that."

EXERCISE 8.2 (An Integration by Parts). Use Itô's formula to prove that if $h \in C^1(\mathbb{R}^+)$ then

$$\int_0^t h(s)\, dB_s = h(t)B_t - \int_0^t h'(s)B_s\, ds.$$

This formula shows that the Itô integral of a deterministic function can be calculated just in terms of traditional integrals, a fact that should be compared to formula (8.10), where a similar result is obtained for integrands that are functions of B_t alone.

EXERCISE 8.3 (Sources Harmonic Functions and Applications).
One of the most basic facts from the theory of complex variables is that the real and imaginary parts of an analytic function satisfy the Cauchy–Riemann equations; that is, if $f(z)$ is a differentiable as a function of $z = x + iy$ and if we write $f(x + iy) = u(x, y) + iv(x, y)$, then

$$\frac{\partial u}{\partial x} = \frac{\partial v}{\partial y} \quad \text{and} \quad \frac{\partial u}{\partial y} = -\frac{\partial v}{\partial x}.$$

(a) Use these equations to show that $u(x, y)$ and $v(x, y)$ are harmonic. What harmonic functions do you obtain from the real and imaginary parts of the analytic functions $\exp(z)$ and $z\exp(z)$?

(b) Consider the family of hyperbolas given by

$$H(\alpha) = \{(x, y) : x^2 - y^2 = \alpha\}.$$

What is the probability that the standard 2-dimensional Brownian motion $\vec{B}_t$ starting at $(2, 0)$ will hit $H(1)$ before hitting $H(5)$. For a hint, one might consider the harmonic functions that may be obtained from the complex function $z \mapsto z^2$.

EXERCISE 8.4 (More Charms of the Third Martingale).

(a) The martingale $M_t = \exp(\alpha B_t - \alpha^2 t/2)$ is separable in the sense that it is of form $f(t, B_t)$, where $f(t, x) = \phi(t)\psi(x)$. Solve the equation $f_t = -\frac{1}{2} f_{xx}$ under the assumption that $f(t, x) = \phi(t)\psi(x)$ and see if you can find any other separable martingales.

(b) By expanding $M_t = \exp(\alpha B_t - \alpha^2 t/2)$ as a Taylor series in α, we can write

$$M_t = \sum_{k=0}^{\infty} \alpha^k H_k(t, B_t),$$

where the $H_k(t, x)$ are polynomials. Find the first four of the $H_k(t, x)$, and show that for each $k \in \mathbb{Z}^+$ the process defined by

$$M_t(k) = H_k(t, B_t)$$

is a martingale. You will recognize the first three of these martingales as old friends.

EXERCISE 8.5 (Nice Local Martingale, But Not a Martingale).

The purpose of this exercise is to provide an example of a local martingale that is L^2-bounded (and hence uniformly integrable) but still fails to be an honest martingale.

(a) First, show that the function $f(x, y, z) = (x^2 + y^2 + z^2)^{-\frac{1}{2}}$ satisfies $\Delta f = 0$ for all $(x, y, z) \neq 0$, so that if $\vec{B}_t$ is standard Brownian motion in $\mathbb{R}^3$ (starting from zero), then for $1 \le t < \infty$ the process defined by

$$M_t = f(\vec{B}_t)$$

is a local martingale.

(b) Second, use direct integration (say, in spherical coordinates) to show

$$E(M_t^2) = \frac{1}{t} \text{ for all } 1 \le t < \infty. \tag{8.52}$$

(c) Third, use the identity (8.52) and Jensen's inequality to show that M_t is not a martingale.

EXERCISE 8.6 (A Box Calculus Verification). The purpose of this exercise is to generalize and unify the calculations we made for functions of Brownian motion with drift and geometric Brownian motion. It provides a proof of the validity of the box calculus for processes that are functions of Brownian motion and time.

(a) Let $X_t = f(t, B_t)$, and use Itô's Lemma 8.2 to calculate dX_t. Next, use the chain rule and Itô's Lemma 8.2 to calculate dY_t, where $Y_t = g(t, X_t) = g(t, f(t, B_t))$.

(b) Finally, calculate $dX_t \cdot dX_t$ by the box calculus and verify that your expression for dY_t shows that the box calculus formula (8.28) is valid for Y_t.

CHAPTER 9

Stochastic Differential Equations

Virtually all continuous stochastic processes of importance in applications satisfy an equation of the form

$$dX_t = \mu(t, X_t)\, dt + \sigma(t, X_t)\, dB_t \text{ with } X_0 = x_0. \tag{9.1}$$

Such stochastic differential equations, or SDEs, provide an exceptionally effective framework for the construction and the analysis of stochastic models. Because the equation coefficients μ and σ can be interpreted as measures of short-term growth and short-term variability, the modeler has a ready-made pattern for the construction of stochastic processes that reflect real-world behavior.

SDEs also provide a link between probability theory and the much older and more developed fields of ordinary and partial differential equations. Wonderful consequences flow in both directions. The stochastic modeler benefits from centuries of development of the physical sciences, and many classic results of mathematical physics (and even pure mathematics) can be given new intuitive interpretations.

Our first task will be to introduce three specific SDEs of great practical importance and to show how explicit solutions for these equations can be found by systematic methods. We will then develop the basic existence and uniqueness theorem for SDEs. This theorem provides an initial qualitative understanding for most of the the SDEs that one meets in practice.

9.1. Matching Itô's Coefficients

One of the most natural, and most important, stochastic differential equations is given by

$$dX_t = \mu X_t\, dt + \sigma X_t\, dB_t \text{ with } X_0 = x_0 > 0, \tag{9.2}$$

where the equation coefficients $-\infty < \mu < \infty$ and $\sigma > 0$ are constants. The intuition behind this equation is that X_t should exhibit short-term growth and short-term variability in proportion to the level of the process. Such percentage-based changes are just what one might expect from the price of a security, and the SDE (9.2) is often used in financial models.

How can we solve an equation such as (9.2)? Because of our earlier work, we can recognize equation (9.2) as the SDE for geometric Brownian motion, but we can also proceed more systematically. Since Itô's formula is the only tool that is even remotely relevant, the obvious idea is to try to solve equation (9.2) by hunting for a solution of the form $X_t = f(t, B_t)$. When we apply Itô's formula to this candidate we see

$$dX_t = \left\{ f_t(t, B_t) + \frac{1}{2} f_{xx}(t, B_t) \right\} dt + f_x(t, B_t)\, dB_t, \tag{9.3}$$

so in order to make the coefficients of equation (9.3) match those of equation (9.2), we only need to find an $f(t,x)$ that satisfies the two equations

$$\mu f(t,x) = f_t(t,x) + \frac{1}{2} f_{xx}(t,x) \text{ and } \sigma f(t,x) = f_x(t,x). \tag{9.4}$$

The second equation can be written as $f_x/f = \sigma$, and a solution of this equation is given by $f(t,x) = \exp(\sigma x + g(t))$, where $g(\cdot)$ is an arbitrary function. When we plug this expression for f into the first equation of the system (9.4), we see that g needs to satisfy $g'(t) = \mu - \frac{1}{2}\sigma^2$. The bottom line is that a particular solution of equation (9.2) is given by

$$X_t = x_0 \exp\left((\mu - \frac{1}{2}\sigma^2)t + \sigma B_t \right). \tag{9.5}$$

For the moment, we must admit the possibility that there may be other solutions to the SDE (9.2), but we will soon prove a general result that will tell us that formula (9.5) actually gives the unique solution. The process defined by either (9.2) or (9.5) is called geometric Brownian motion, and it is one of the real workhorses of stochastic calculus. No other process is used more widely in finance and economics.

Looking Back: Dashed Expectations

The solution (9.5) of the SDE (9.2) could not be simpler, yet it already has a practical message. We think of equation (9.2) as a model for the price of a risky asset with an expected rate of return μ, and, when we take the expectation in equation (9.5), we find

$$E(X_t) = x_0 \exp(\mu t).$$

Perhaps we should not be surprised that this expectation does not depend on σ, but, if we consider the case when $\sigma^2/2 > \mu > 0$, there is some shocking news. The law of large numbers tells us that $P(B_t/t \to 0) = 1$, and, when we apply this fact to equation (9.5), we find

$$P\left(\lim_{t\to\infty} X_t = 0 \right) = 1.$$

In other words, the price of our risky asset goes to zero with probability one, despite the fact that its expected value goes to plus infinity at an exponential rate. The prudent investor is invited to probe the meaning of this modest paradox of risk without the possibility of reward.

9.2. Ornstein–Uhlenbeck Processes

The coefficient matching method that we used to solve the SDE of geometric Brownian motion is often effective, but it can fail even in simple cases. Our second example illustrates the possibility of failure and shows way that the matching method can be extended to cover further cases of importance. The equation that we will investigate is given by

$$dX_t = -\alpha X_t dt + \sigma dB_t \text{ with } X_0 = x_0, \tag{9.6}$$

where α and σ are positive constants. The drift term $-\alpha X_t dt$ in this equation is negative when X_t is larger than zero, and it is positive when X_t is smaller than zero; so, even though the process will never be free of random fluctuations, we can expect X_t to be drawn back to zero whenever it has drifted away. Moreover, because the local variability of X_t is constant — not proportional to the level as in the case of geometric Brownian motion — we expect X_t to fluctuate vigorously even when it is near zero. As a consequence, we also expect X_t to make many crossings of the zero level, unlike geometric Brownian motion, which never crosses through zero.

This model became well known after 1931, when it was used by the physicists L. S. Ornstein and G. E. Uhlenbeck to study behavior of gasses, although they needed to express the model a little differently because the theory of stochastic differential equations would not be born until more than a dozen years after their work. What interested Ornstein and Uhlenbeck was the velocity of an individual molecule of gas. They were motivated by the kinetic theory of gases which states that the molecules of a gas are in constant motion and the average velocity over all of the particles is governed by the laws of pressure and temperature. Ornstein and Uhlenbeck reasoned that an individual molecule could be expected to speed up or slow down, but, to be consistent with the constant average velocity of the ensemble of molecules, the difference between the velocity of the individual particle and the ensemble mean must exhibit reversion toward zero.

Mean reversion stories occur in many fields, so there is little wonder that the Ornstein–Uhlenbeck model has been applied (or rediscovered) in a variety of contexts. In finance, the Ornstein–Uhlenbeck process was used by O. A. Vašiček in one of the first stochastic models for interest rates. In that context, X_t was intended to capture the deviation of an interest rate around a given fixed rate — say a central bank's target rate, or perhaps the equilibrium interest rate in a model for the economy.

For us, the immediate attraction of equation (9.6) comes from the fact that it has a simple solution, yet that solution is not of the form $f(t, B_t)$. If we try to repeat the coefficient matching method based on Itô's formula, we are doomed to failure. Nevertheless, there is a way to make the coefficient matching method work; we simply need a larger set of processes in which to pursue our match. To find a class of simple processes that have SDEs that include the Ornstein–Uhlenbeck SDE (9.6), we need to find a new class of processes in which to carry out the coefficient matching procedure.

9.3. Matching Product Process Coefficients

If one experiments a bit with the O–U equation (9.6), it is easy to come to suspect that the solution X_t might be a Gaussian process, and because of our earlier work we might decide to look for a solution of equation (9.6) as an Itô integral of a deterministic function. This idea does not quite succeed, but it puts us on the right track.

The product of a Gaussian process and a deterministic process is a Gaussian process, so we can also look for a solution to equation (9.6) in the larger class of processes that can be written as

$$X_t = a(t)\left\{x_0 + \int_0^t b(s)\,dB_s\right\}, \tag{9.7}$$

where $a(\cdot)$ and $b(\cdot)$ are differentiable functions. If we now apply the product rule (or the box calculus), we find that

$$dX_t = a'(t)\left\{x_0 + \int_0^t b(s)\,dB_s\right\}dt + a(t)b(t)\,dB_t.$$

In other words, if we assume that $a(0) = 1$ and $a(t) > 0$ for all $t \geq 0$, then the process defined by formula (9.7) is a solution of the SDE:

$$dX_t = \frac{a'(t)}{a(t)} X_t\, dt + a(t)b(t)\, dB_t \text{ with } X_0 = x_0. \tag{9.8}$$

The relationship between equations (9.7) and (9.8) now gives us a straightforward plan for solving almost any SDE with a drift coefficient that is linear in X_t and that has a volatility coefficient that depends only on t. We will illustrate this plan by solving the SDEs for the Ornstein–Uhlenbeck process and the Brownian bridge.

Solving the Ornstein–Uhlenbeck SDE

For the coefficients of equation (9.8) to match those of the Ornstein–Uhlenbeck equation (9.6), we only need to satisfy the simple equations

$$\frac{a'(t)}{a(t)} = -\alpha \text{ and } a(t)b(t) = \sigma.$$

The unique solution of the first equation with $a(0) = 1$ is just $a(t) = \exp(-\alpha t)$, so the second equation tells us $b(t) = \sigma \exp(\alpha t)$. The bottom line is that we see that a solution to equation (9.6) is given by

$$X_t = e^{-\alpha t}\left\{ x_0 + \sigma \int_0^t e^{\alpha s}\, dB_s \right\} = x_0 e^{-\alpha t} + \sigma \int_0^t e^{-\alpha(t-s)}\, dB_s. \tag{9.9}$$

When we look back on the solution of the Ornstein–Uhlenbeck given by formula (9.9), we find several interesting features. First, we see that as $t \to \infty$ the influence of the initial value $X_0 = x_0$ decays exponentially. We also have $E(X_t) = x_0 e^{-\alpha t}$ since the dB_s integral has mean zero, so the mean $E(X_t)$ also goes to zero rapidly.

Finally, when we use the Itô isometry to compute the variance of X_t, we find

$$\operatorname{Var}(X_t) = \sigma^2 \int_0^t e^{-2\alpha(t-s)}\, ds = \frac{\sigma^2}{2\alpha}\left\{1 - e^{-2\alpha t}\right\}, \tag{9.10}$$

and from this variance calculation we see that

$$\operatorname{Var}(X_t) \to \frac{\sigma^2}{2\alpha}.$$

Way back in Proposition 7.6, we learned that a dB_t integral of a deterministic process will give us a Gaussian process, so as $t \to \infty$ we see that the distribution of X_t converges exponentially to a Gaussian distribution with mean zero and variance $\sigma^2/2\alpha$. As it happens, we can think of the limiting distribution $N(0, \sigma^2/2\alpha)$ as a kind of equilibrium distribution, and if we take the starting value X_0 to be a random variable with the $N(0, \sigma^2/2\alpha)$ distribution, then one can check that the distribution of X_t is equal to the $N(0, \sigma^2/2\alpha)$ distribution for all $t \geq 0$.

Solving the Brownian Bridge SDE

After our construction of Brownian motion, we observed in the exercises that there is a Gaussian process $\{X_t\}$ defined on $[0, 1]$ such that $\operatorname{Cov}(X_s, X_t) = s(1-t)$ for $0 \leq s \leq t$. We also saw that we could represent the process in terms of standard Brownian motion as

$$X_t = B_t - tB_1 \text{ for } 0 \leq t \leq 1. \tag{9.11}$$

We also suggested that one could think of this Brownian bridge process as a Brownian motion that is constrained to return to 0 at time 1. Now, we will find that Itô integration gives us a method for representing the Brownian bridge in a way that helps us understand the nature of the forces that bring the process back to zero at the correct time.

Take $t \in (0,1)$ and consider the value of X_t at time t. What drift might one reasonably require in order for X_t to be coerced to hit 0 at time 1? Because we need to go from X_t to 0 in the remaining time of $1-t$, the most natural suggestion is to consider a drift rate $-X_t/(1-t)$, and if we write the simplest genuine SDE with such a drift, we come to the equation

$$dX_t = -\frac{X_t}{1-t}\,dt + dB_t \quad \text{with } X_0 = 0. \tag{9.12}$$

We will find that this equation has a unique solution with exactly the required covariance. We will also see that the solution is a Gaussian process so the SDE (9.12) does indeed characterize the Brownian bridge.

The coefficient matching method for product processes tells us how to solve the Brownian bridge SDE (9.12) with delightful ease. When we match the coefficients of (9.12) with those from the product process (9.8) we find

$$\frac{a'(t)}{a(t)} = -\frac{1}{1-t} \quad \text{and } a(t)b(t) = 1,$$

so we find $a(t) = 1-t$ and $b(t) = 1/(1-t)$. We therefore find that a solution to the SDE (9.12) is given by

$$X_t = (1-t)\int_0^t \frac{1}{1-s}\,dB_s \qquad \text{for } t \in [0,1). \tag{9.13}$$

Since (9.13) expresses X_t as a nonrandom multiple of a dB_t integral of a nonrandom function, Proposition 7.6 again tells us that X_t is a Gaussian process. Also, we can easily confirm that the two processes have the same covariance function. First, we note that

$$\sigma(\omega, s) \stackrel{\text{def}}{=} \frac{1}{1-s} \in \mathcal{H}^2[0,t] \text{ for all } t \in [0,1),$$

so $E(X_t) = 0$ for $t \in [0,1)$. To complete the covariance calculation for $0 \le s \le t$, we only need to compute

$$E(X_s X_t) = (1-s)(1-t)E\left[\int_0^s \frac{1}{1-u}\,dB_u \int_0^t \frac{1}{1-v}\,dB_v\right].$$

But we have

$$E\left[\int_0^s \frac{1}{1-u}\,dB_u \int_s^t \frac{1}{1-v}\,dB_v\right] = 0$$

since the last two integrals are independent, and by Itô's isometry we also have

$$E\left[\left(\int_0^s \frac{1}{1-u}\,dB_u\right)^2\right] = \int_0^s \frac{1}{(1-u)^2}\,du = s/(1-s).$$

When we assemble the pieces, we find $\mathrm{Cov}(X_s, X_t) = s(1-t)$, as we hoped.

LOOKING BACK — FINDING A PARADOX

When we look back on the Brownian bridge representations as a stochastic integral (9.13), or as a function of standard Brownian motion (9.11), we see the seeds of a paradox. In the integral representation (9.13), we see that X_t depends on the Brownian path up to time t, but in the function representation (9.11) we seem to see that X_t depends on B_1.

This may seem strange, but there is nothing wrong. The processes

$$X_t = (1-t)\int_0^t \frac{1}{1-s}\,dB_s \text{ and } X_t' = B_t - tB_1$$

are both honest Brownian bridges. They can even be viewed as elements of the same probability space; nevertheless, the processes are not equal, and there is no reason to be surprised that the two filtrations

$$\mathcal{F}_t = \sigma\{\, X_s\colon 0 \le s \le t \,\} \text{ and } \mathcal{F}_t' = \sigma\{\, X_s'\colon 0 \le s \le t \,\}$$

are also different. For an even more compelling example, consider the process X_t'' defined by taking $X_t'' = X_{1-t}'$. The process X_t'' is just one more Brownian bridge, yet its natural filtration is completely tangled up with that of X_t'!

9.4. Existence and Uniqueness Theorems

In parallel with the theory of ordinary differential equations, the theory of SDEs has existence theorems that tell us when an SDE must have a solution, and there are uniqueness theorems that tell us when there is at most one solution. Here, we will consider only the most basic results.

THEOREM 9.1 (Existence and Uniqueness). *If the coefficients of the stochastic differential equation*

$$dX_t = \mu(t, X_t)\,dt + \sigma(t, X_t)\,dB_t, \quad \textit{with } X_0 = x_0 \textit{ and } 0 \le t \le T, \tag{9.14}$$

satisfy a space-variable Lipschitz condition

$$|\mu(t,x) - \mu(t,y)|^2 + |\sigma(t,x) - \sigma(t,y)|^2 \le K|x-y|^2 \tag{9.15}$$

and the spatial growth condition

$$|\mu(t,x)|^2 + |\sigma(t,x)|^2 \le K(1+|x|^2), \tag{9.16}$$

then there exists a continuous adapted solution X_t of equation (9.14) that is uniformly bounded in $L^2(dP)$:

$$\sup_{0\le t\le T} E(X_t^2) < \infty.$$

Moreover, if X_t and Y_t are both continuous L^2 bounded solutions of equation (9.14), then

$$P(\, X_t = Y_t \textit{ for all } t \in [0,T]\,) = 1. \tag{9.17}$$

NATURE OF THE COEFFICIENT CONDITION

A little experimentation with ODEs is sufficient to show that some condition such as (9.16) is needed in order to guarantee the existence of a solution of the SDE (9.14). For example, if we take

$$\sigma(t,x) \equiv 0 \text{ and } \mu(t,x) = (\beta-1)^{-1}x^\beta \text{ with } \beta > 1,$$

then our target equation (9.14) is really an ODE dressed up in SDE clothing:

$$dX_t = (\beta - 1)^{-1} X_t^\beta \, dt \quad \text{with } X_0 = 1.$$

The unique solution to this equation on the interval $[0, 1)$ is given by

$$X_t = (1-t)^{-1/(\beta-1)},$$

so there is no continuous solution to the equation on $[0, T]$ when $T > 1$. What runs amuck in this example is that the coefficient $\mu(t, x) = (\beta - 1)^{-1} x^\beta$ grows faster than x, a circumstance that is banned by the coefficient condition (9.16).

Proofs of Existence and Uniqueness

The proof of the uniqueness part of Theorem 9.1 is a bit quicker than the existence part, so we will tackle it first. The plan for the proof of the uniqueness theorem for SDEs can be patterned after one of traditional uniqueness proofs from the theory of ODEs. The basic idea is that the difference between two solutions can be shown to satisfy an integral inequality that has zero as its only solution.

From the representation given by the SDE, we know that the difference of the solutions can be written as

$$X_t - Y_t = \int_0^t [\mu(s, X_s) - \mu(s, Y_s)]\, ds + \int_0^t [\sigma(s, X_s) - \sigma(s, Y_s)]\, dB_s,$$

so by the elementary bound $(u+v)^2 \leq 2u^2 + 2v^2$ we find

$$E(|X_t - Y_t|^2) \leq 2E\left[\left(\int_0^t \mu(s, X_s) - \mu(s, Y_s)\, ds\right)^2\right] + 2E\left[\left(\int_0^t \sigma(s, X_s) - \sigma(s, Y_s)\, dB_s\right)^2\right].$$

By Cauchy's inequality (and the 1-trick), the first summand is bounded by

$$2tE\left[\int_0^t |\mu(s, X_s) - \mu(s, Y_s)|^2\, ds\right], \tag{9.18}$$

whereas by the Itô isometry the second summand simply equals

$$2E\left[\int_0^t |\sigma(s, X_s) - \sigma(s, Y_s)|^2\, ds\right]. \tag{9.19}$$

Here we should note that the use of the Itô isometry is indeed legitimate because the Lipschitz condition (9.15) on σ together with the L^2-boundedness of X_t and Y_t will guarantee that

$$|\sigma(s, X_s) - \sigma(s, Y_s)| \in \mathcal{H}^2[0, T]. \tag{9.20}$$

When the estimates (9.18) and (9.19) are combined with the coefficient condition (9.15), we find that for $C = 2K \max(T, 1)$ we have

$$E(|X_t - Y_t|^2) \leq C \int_0^t E(|X_s - Y_s|^2)\, ds < \infty. \tag{9.21}$$

If we set $g(t) = E(|X_t - Y_t|^2)$, the last equation tells us that

$$0 \leq g(t) \leq C \int_0^t g(s)\, ds \text{ for all } 0 \leq t \leq T, \tag{9.22}$$

and we will shortly find that this is just the sort of integral inequality that is good enough to show that $g(t)$ is identically zero.

To complete the argument, we first let $M = \sup\{g(t) : 0 \le t \le T\}$ and then note that (9.22) tells us $g(t) \le MCt$. By n successive applications of (9.22) we see that $g(t) \le MC^n t^n/n!$. Now, since $n!$ grows faster than $C^n t^n$, the arbitrariness of n finally tells us that $g(t) = 0$ for all $t \in [0,T]$. This brings us almost all of the way home. If we simply use the fact that $g(t) = 0$ for each rational t, we see by the countability of the rationals that we also have

$$P(X_t = Y_t \text{ for all rational } t \in [0,T]) = 1.$$

The strong uniqueness condition (9.17) that we wanted to prove now follows immediately from the continuity of X_t and Y_t.

Iteration Solutions for SDEs

Our proof of the existence theorem for SDEs is also based on an idea from the theory of ODEs — Picard's iteration. For SDEs the iterative scheme is given by taking $X_t^{(0)} \equiv x_0$ and by defining a sequence of processes by

$$X_t^{(n+1)} = x_0 + \int_0^t \mu(s, X_s^{(n)})\, ds + \int_0^t \sigma(s, X_s^{(n)})\, dB_s. \tag{9.23}$$

Before we get too carried away with this fine idea, we should check that the iterations actually make sense; that is, we should verify that the integrals on the right-hand side of equation (9.23) are always well defined.

Lemma 9.1. *If $X_t^{(n)}$ is L^2-bounded on $[0,T]$, then*

$$\sigma(t, X_t^{(n)}) \in \mathcal{H}^2[0,T] \text{ and } \mu(t, X_t^{(n)}) \in L^2\left[[0,T]\times\Omega\right]. \tag{9.24}$$

Moreover, the process $X_t^{(n+1)}$ defined by equation (9.23) is L^2-bounded on $[0,T]$.

Proof. By the L^2-boundedness of $X_t^{(n)}$, we can set

$$\sup_{t\in[0,T]} E(|X_t^{(n)}|^2) = B < \infty,$$

and if we invoke the bound (9.16) in the simplified form $|\sigma(t,x)|^2 \le K(1+x^2)$, we immediately find

$$E\left[\int_0^T |\sigma(t, X_t^{(n)})|^2\, dt\right] \le TK(1+B).$$

We therefore have $\sigma(t, X_t^{(n)}) \in \mathcal{H}^2[0,T]$, so by the Itô isometry we also have

$$E\left[\left(\int_0^t \sigma(s, X_s^{(n)})\, dB_s\right)^2\right] = E\left[\int_0^t |\sigma(s, X_s^{(n)})|^2\, ds\right] \le TK(1+B).$$

Finally, the growth condition (9.16) also gives us $|\mu(s,x)|^2 \le K(1+x^2)$, so we similarly find by Cauchy's inequality that for all $0 \le t \le T$ one has

$$E\left[\left(\int_0^t \mu(s, X_s^{(n)})\, ds\right)^2\right] \le TK(1+B).$$

The bottom line is that both integrals in the iteration (9.23) are well defined and L^2-bounded on $[0,T]$, so the proof of the lemma is complete. □

Plan for Convergence

The next step is to show that the sequence of processes $X_t^{(n)}$ defined by the recursion (9.23) actually converges to a limit. We will then need to verify that this limit is indeed a genuine solution of the SDE (9.14). Our first technical challenge is to prove that the approximations generated by the recursion (9.23) will be a Cauchy sequence in the uniform norm on $C[0,T]$ with probability one. The next lemma provides the essential step in the derivation of this fact. Curiously, the proof of the lemma closely parallels the argument that we used to obtain the uniqueness theorem.

LEMMA 9.2. *If μ and σ satisfy the coefficient condition (9.15), then there is a constant C such that the processes defined by the iteration (9.23) satisfy the inequality:*

$$E\left[\sup_{0\le s\le t} |X_s^{(n+1)} - X_s^{(n)}|^2\right] \le C\int_0^t E(|X_s^{(n)} - X_s^{(n-1)}|^2)\, ds. \tag{9.25}$$

PROOF. To prove the lemma, we first note that the difference $X_s^{(n+1)} - X_s^{(n)}$ can be written as the sum of a drift term

$$D_t = \int_0^t [\mu(s, X_s^{(n)}) - \mu(s, X_s^{(n-1)})]\, ds \tag{9.26}$$

and a second "noise" term,

$$M_t = \int_0^t [\sigma(s, X_s^{(n)}) - \sigma(s, X_s^{(n-1)})]\, dB_s. \tag{9.27}$$

Here, we should note that the process $\{M_t\}$ is in fact a martingale because we have already seen that

$$s \mapsto \sigma(s, X_s^{(n)}) \in \mathcal{H}^2[0,T] \text{ for all } n \ge 0.$$

If again we apply the elementary bound $(u+v)^2 \le 2u^2 + 2v^2$, we then see that

$$\sup_{0\le s\le t} |X_s^{(n+1)} - X_s^{(n)}|^2 \le 2\sup_{0\le s\le t} M_s^2 + 2\sup_{0\le s\le t} D_s^2, \tag{9.28}$$

so we can pursue the two terms individually.

To estimate the drift term, we simply use Cauchy's inequality to see that

$$\sup_{0\le s\le t} D_s^2 \le t\int_0^t |\mu(s, X_s^{(n)}) - \mu(s, X_s^{(n-1)})|^2\, ds. \tag{9.29}$$

To deal with the martingale term, we first call on Doob's L^2 inequality to get

$$E\Big(\sup_{0\le s\le t} M_s^2\Big) \le 4E(M_t^2),$$

and then we calculate $E(M_t^2)$ by the Itô isometry to give

$$E\Big(\sup_{0\le s\le t} M_s^2\Big) \le 4E\left[\int_0^t |\sigma(s, X_s^{(n)}) - \sigma(s, X_s^{(n-1)})|^2\, ds\right]. \tag{9.30}$$

For the final step, we tie the package together with the coefficient condition (9.15) to find

$$E\left[\sup_{0\leq s\leq t}|X_s^{(n+1)}-X_s^{(n)}|^2\right]\leq 8E\left[\int_0^t|\sigma(s,X_s^{(n)})-\sigma(s,X_s^{(n-1)})|^2\,ds\right]$$
$$+\,2tE\left[\int_0^t|\mu(s,X_s^{(n)})-\mu(s,X_s^{(n-1)})|^2\,ds\right]$$
$$\leq C\int_0^t E(|X_s^{(n)}-X_s^{(n-1)}|^2)\,ds, \tag{9.31}$$

where it suffices to take $C = 8K\max(1,T)$ to justify the last inequality and to complete the proof of the lemma. □

Existence of the Limit

We can now show that the processes $X_t^{(n)}$ converge with probability one to a continuous process X_t. If we let

$$g_n(t)=E\left[\sup_{0\leq s\leq t}|X_s^{(n+1)}-X_s^{(n)}|^2\right],$$

Lemma 9.2 then tells us that

$$g_n(t)\leq C\int_0^t g_{n-1}(s)\,ds,\quad n\geq 1. \tag{9.32}$$

Except for the appearance of the subscripts, this is precisely the inequality we used in the uniqueness argument. The rest of the proof follows a familiar pattern.

To start the ball rolling, we first note that coefficient conditions imply that $g_0(t) \leq M$ for some constant M and all $t \in [0,T]$. Now, by one application of inequality (9.32), we also see that $g_1(t) \leq MCt$, and by induction we find more generally that

$$0\leq g_n(t)\leq MC^nt^n/n!. \tag{9.33}$$

From the last bound and Markov's inequality, we see

$$P\left(\sup_{0\leq s\leq T}|X_s^{(n+1)}-X_s^{(n)}|\geq 2^{-n}\right)\leq MC^nT^n2^{2n}/n!,$$

so the familiar Borel–Cantelli argument now tells us that there is a set Ω_0 with probability one such that for all $\omega \in \Omega_0$ the continuous functions $t \mapsto X_t^{(n)}(\omega)$ form a Cauchy sequence in the supremum norm on $C[0,T]$. The bottom line is that for all $\omega \in \Omega_0$ there is a continuous function $X_t(\omega)$ such that $X_t^{(n)}(\omega)$ converges uniformly to $X_t(\omega)$ on $[0,T]$.

L^2 Boundedness

We now know that we have a well-defined continuous process $\{X_t\colon 0 \leq t \leq T\}$, and after a few simple checks we will be able to show that it is the desired solution of the SDE (9.14). The first order of business is to show that X_t is L^2-bounded on $[0,T]$, and that we have

$$X_t^{(n)}\to X_t \text{ in } L^2(dP) \text{ for all } t\in[0,T].$$

In fact, both properties turn out to be easy consequences of our earlier estimates.

Specifically, by the definition of g_n, we have

$$||X_t^{(n+1)} - X_t^{(n)}||_{L^2(dP)} \le T^{\frac{1}{2}} g_n(T)^{\frac{1}{2}} \text{ for all } t \in [0,T], \tag{9.34}$$

and, since the sequence $\{g_n(T)^{\frac{1}{2}}\}$ is summable, we see that $\{X_t^{(n)}\}$ is a Cauchy sequence in $L^2(dP)$. As a consequence, $X_t^{(n)}$ also converges in $L^2(dP)$ and by uniqueness of limits it in fact converges to X_t in $L^2(dP)$. Finally, by summing the bound (9.34), we obtain the required $L^2(dP)$ boundedness together, and we even have an explicit bound:

$$||X_t||_{L^2(dP)} \le g_0(T)^{\frac{1}{2}} + \sum_{n=0}^{\infty} g_n(T)^{\frac{1}{2}} \text{ for all } t \in [0,T].$$

L^2 Convergence

The next step is to note that by the second inequality of (9.31), we also have

$$E\left(\int_0^T |\sigma(s, X_s^{(n)}) - \sigma(s, X_s^{(n-1)})|^2\, ds\right)$$
$$\le C \int_0^T E(|X_s^{(n)} - X_s^{(n-1)}|^2)\, ds \le C \int_0^T g_{n-1}(s)\, ds,$$

and, since $g_{n-1}(s)$ is monotone increasing in s, the last inequality gives us the norm bound

$$||\sigma(t, X_t^{(n)}) - \sigma(t, X_t^{(n-1)})||_{L^2(dP\times dt)} \le C^{\frac{1}{2}} T^{\frac{1}{2}} g_{n-1}^{\frac{1}{2}}(T). \tag{9.35}$$

In exactly the same way, the second inequality of (9.31) also gives us the bound

$$||\mu(t, X_t^{(n)}) - \mu(t, X_t^{(n-1)})||_{L^2(dP\times dt)} \le C^{\frac{1}{2}} T^{\frac{1}{2}} g_{n-1}^{\frac{1}{2}}(T). \tag{9.36}$$

Now, from the bound (9.35) and the Itô isometry, the summability of the $g_n^{\frac{1}{2}}(T)$ implies

$$\int_0^t \sigma(s, X_s^{(n)})\, dB_s \to \int_0^t \sigma(s, X_s)\, dB_s \quad \text{in } L^2(dP). \tag{9.37}$$

Finally, the bound (9.36) plus the summability of $g_n^{\frac{1}{2}}(T)$ gives us

$$\int_0^t \mu(s, X_s^{(n)})\, ds \to \int_0^t \mu(s, X_s)\, ds \quad \text{in } L^2(dP). \tag{9.38}$$

Solution of the SDE

Now, we need to check that $X_t(\omega)$ is an honest solution of the SDE (9.14). When we let $n \to \infty$ in the recurrence (9.23), we see on the left-hand side that with probability one we have

$$X_t^{(n)} \to X_t \text{ uniformly on } [0,T].$$

To handle the right-hand side, we just note that by the limits (9.38) and (9.37) there is a subsequence $\{n_k\}$ such that as $k \to \infty$

$$\int_0^t \mu(s, X_s^{(n_k)})\, ds \to \int_0^t \mu(s, X_s)\, ds \quad \text{a.s. for all } t \in [0,T] \cap \mathbb{Q} \tag{9.39}$$

and

$$\int_0^t \sigma(s, X_s^{(n_k)})\, dB_s \to \int_0^t \sigma(s, X_s)\, dB_s \quad \text{a.s. for all } t \in [0,T] \cap \mathbb{Q}. \tag{9.40}$$

Therefore, when we take the limit in the recurrence (9.23) as $n_k \to \infty$, we find that there is a set Ω_0 with $P(\Omega_0) = 1$ such that

(9.41)

$$X_t = x_0 + \int_0^t \mu(s, X_s)\, ds + \int_0^t \sigma(t, X_s)\, dB_s \text{ for all } t \in [0,T] \cap \mathbb{Q} \text{ and } \omega \in \Omega_0.$$

Finally, since both sides of equation (9.41) are continuous, the fact that it holds on $[0,T] \cap \mathbb{Q}$ implies that it holds on all of $[0,T]$ for all $\omega \in \Omega_0$.

9.5. Systems of SDEs

If the theory of differential equations were confined to one dimension, we would not have airplanes, radio, television, guided missiles, or much else for that matter. Fortunately, the theory of differential equations carries over to systems of equations, and the same is naturally true for SDEs.

To keep our notation close to that of our one-dimensional problems, we often find it useful to write systems of SDEs as

$$d\vec{X}_t = \vec{\mu}(t, \vec{X}_t)\, dt + \sigma(t, \vec{X}_t)\, d\vec{B}_t, \text{ with } X_0 = x_0, \tag{9.42}$$

where we have

$$\vec{\mu}(t, \vec{X}_t) = \begin{bmatrix} \mu_1(t, \vec{X}_t) \\ \mu_2(t, \vec{X}_t) \\ \cdot \\ \cdot \\ \cdot \\ \mu_d(t, \vec{X}_t) \end{bmatrix} \quad \text{and} \quad d\vec{B}_t = \begin{bmatrix} dB_t^1 \\ dB_t^2 \\ \cdot \\ \cdot \\ \cdot \\ dB_t^d \end{bmatrix} \tag{9.43}$$

together with

$$\sigma(t, \vec{X}_t) = \begin{bmatrix} \sigma_{11} & \sigma_{12} & \cdots & \sigma_{1d} \\ \sigma_{21} & \sigma_{22} & \cdots & \sigma_{2d} \\ \cdot & \cdot & \cdots & \cdot \\ \cdot & \cdot & \cdots & \cdot \\ \cdot & \cdot & \cdots & \cdot \\ \sigma_{d1} & \sigma_{d2} & \cdots & \sigma_{dd} \end{bmatrix}, \tag{9.44}$$

where we use the shorthand $\sigma_{ij} = \sigma_{ij}(t, \vec{X}_t)$. The basic existence and uniqueness theorem for SDEs extends to the case of SDE systems with only cosmetic changes.

Example: Brownian Motion on a Circle

If we let $X_t = \cos(B_t)$ and $Y_t = \sin(B_t)$, the two dimensional process (X_t, Y_t) is known as Brownian motion on a circle. The associated system of SDEs is given by

$$dX_t = -\sin(B_t)\, dB_t - \frac{1}{2}\cos(B_t)\, dt = -Y_t\, dB_t - \frac{1}{2} X_t\, dt$$

and

$$dY_t = \cos(B_t)\,dB_t - \frac{1}{2}\sin(B_t)\,dt = X_t\,dB_t - \frac{1}{2}Y_t\,dt,$$

or in vector notation

$$d\begin{bmatrix} X_t \\ Y_t \end{bmatrix} = -\frac{1}{2}\begin{bmatrix} X_t \\ Y_t \end{bmatrix} dt + \begin{bmatrix} -Y_t & 0 \\ X_t & 0 \end{bmatrix}\begin{bmatrix} dB_t^1 \\ dB_t^2 \end{bmatrix}.$$

In this case, we see that the second Brownian motion does not have an honest role in the process. This degeneracy is reasonable because Brownian motion on the circle is essentially a one-dimensional process that just happens to live in $\mathbb{R}^2$.

EXAMPLE: SYSTEMS OF LOCATION AND MOTION

Some of the most important systems of SDEs arise in telemetry where the task is to observe a moving object from a distance and to make the best possible estimate of its location and velocity. To give a simple illustration of such a problem, consider a satellite that is designed to maintain a fixed location over a point on the Earth's surface — the typical situation of a communications satellite in geosynchronous orbit. If we focus on just one component X_t of the location vector and the corresponding component of the velocity vector, then perhaps the simplest model for the change in the location variable uses the velocity variable and a random error:

$$dX_t = V_t\,dt + \sigma_1\,dB_t^1.$$

Also, if the system is designed to maintain a zero velocity then the usual mean reversion argument suggests that we might model the evolution of the velocity component by an Ornstein–Uhlenbeck process:

$$dV_t = -\alpha V_t\,dt + \sigma_2\,dB_t^2.$$

In vector notation, this system can be written as

$$d\begin{bmatrix} X_t \\ V_t \end{bmatrix} = \begin{bmatrix} V_t \\ -\alpha V_t \end{bmatrix} dt + \begin{bmatrix} \sigma_1 & 0 \\ 0 & \sigma_2 \end{bmatrix}\begin{bmatrix} dB_t^1 \\ dB_t^2 \end{bmatrix}.$$

This system is too simple to be taken very seriously, but it is a stepping stone to a truly major enterprise. When we couple these *system equations* together with *observation equations* that model the relationship between observables and the underlying system variables, we then have the basis for an estimation of position and velocity. Such estimates can then be used for the purpose of control — either of the observed systems, as in the case of a satellite, or of an independent system, such as a heat-seeking anti-aircraft missile.

Engineering systems built on appropriate elaborations of these ideas are ubiquitous. They were first used in military hardware, and their descendants are now critical elements of all major systems that are designed to fly, orbit, or shoot. More recently, these models have served as essential components in automated manufacturing systems and anti-lock brakes for automobiles.

9.6. Exercises

The first four exercises provide essential practice in the solution of SDEs. The fifth exercise then gives an extension of the recurrence technique that was at the heart of our existence and uniqueness proofs. Finally, the sixth exercise points out a common misunderstanding about the diffusion parameter σ of geometric Brownian motion.

EXERCISE 9.1. Solve the SDE

$$dX_t = (-\alpha X_t + \beta)\,dt + \sigma\,dB_t \text{ where } X_0 = x_0 \text{ and } \alpha > 0,$$

and verify that the solution can be written as

$$X_t = e^{-\alpha t}\left(x_0 + \frac{\beta}{\alpha}(e^{\alpha t} - 1) + \sigma \int_0^t e^{\alpha s}\,dB_s\right).$$

Use the representation to show that X_t converges in distribution as $t \to \infty$, and find the limiting distribution. Finally, find the covariance $\mathrm{Cov}(X_s, X_t)$.

EXERCISE 9.2. Solve the SDE

$$dX_t = tX_t\,dt + e^{t^2/2}\,dB_t \text{ with } X_0 = 1.$$

EXERCISE 9.3. Use appropriate coefficient matching to solve the SDE

$$X_0 = 0 \quad dX_t = -2\frac{X_t}{1-t}\,dt + \sqrt{2t(1-t)}\,dB_t \quad 0 \le t < 1.$$

Show that the solution X_t is a Gaussian process. Find the covariance function $\mathrm{Cov}(X_s, X_t)$. Compare this covariance function to the covariance function for the Brownian bridge.

EXERCISE 9.4. Show that if X_t is a process that satisfies $X_0 = 0$ and

$$dX_t = a(X_t)\,dt + \sigma(X_t)\,dB_t,$$

where $a(\cdot)$ and $\sigma(\cdot)$ are smooth functions and $\sigma(\cdot) \ge \epsilon > 0$, then there is a monotone increasing function $f(\cdot)$ and a smooth function $b(\cdot)$ such that $Y_t = f(X_t)$ satisfies

$$Y_t = Y_0 + \int_0^t b(Y_s)\,ds + B_t.$$

The benefit of this observation is that it shows that many one-dimensional SDEs can be recast as an integral equation where there is no stochastic integral (other than a standard Brownian motion). Anyone who wants a hint might consider applying Itô's formula to $f(X_t)$ and making appropriate coefficient matches.

EXERCISE 9.5. The argument we used at the conclusion of Theorem 9.1 can be applied to prove more general results. Suppose $f \ge 0$ is continuous and $g \ge 0$ nondecreasing on $[a, b]$. Show that if

$$f(t) \le g(t) + c\int_a^t f(s)\,ds \text{ for all } t \in [a, b] \tag{9.45}$$

for some $c > 0$, then

$$f(t) \le g(t)e^{c(t-a)} \text{ for all } t \in [a, b].$$

There are many variations on this theme, any one of which may be called *Gronwall's lemma*.

EXERCISE 9.6 (Estimation of Real-world σ's). If X_t is a stochastic process that models the price of a security at time t, then the random variable

$$R_k(h) = \frac{X_{kh}}{X_{(k-1)h}} - 1$$

is called the *kth period return*. It expresses in percentage terms the profit that one makes by holding the security from time $(k-1)t$ to time kt.

When we use geometric Brownian motion

$$dX_t = \mu X_t \, dt + \sigma X_t \, dB_t \tag{9.46}$$

as a model for the price of a security, one common misunderstanding is that σ can be interpreted as a normalized standard deviation of sample returns that more properly estimate

$$s = \sqrt{\mathrm{Var}[R_k(h)]/h}.$$

Sort out this confusion by calculating $E[R_k(h)]$ and $\mathrm{Var}[R_k(h)]$ in terms of μ and σ. Also, use these results to show that an honest formula for σ^2 is

$$\sigma^2 = \frac{1}{h} \log \left(1 + \frac{\mathrm{Var}[R_k(h)]}{\left(1 + E[R_k(h)]\right)^2} \right),$$

and suggest how you might estimate σ^2 from the data $R_1(h), R_2(h), \ldots, R_n(h)$. Finally, make a brief table of σ versus s, where $E[R_k(h)]$ and $\sqrt{\mathrm{Var}[R_k(h)]}$ are assigned values that would be reasonable for the S&P500 Index with sampling period h of one month. Do the differences that you find seem as though they would be of economic significance?

CHAPTER 10

Arbitrage and SDEs

The heart of the modern financial aspirant never beats so quickly as in the presence of an *arbitrage opportunity*, a circumstance where the simultaneous purchase and sale of related securities is guaranteed to produce a riskless profit. Such opportunities may be thought of as rare, but on a worldwide basis they are a daily occurrence, although perhaps not quite as riskless as one might hope.

In financial theory, arbitrage stands as an *éminence grise* — a force that lurks in the background as a barely visible referee of market behavior. In many cases, market prices are what they are precisely because if they were something else, there would be an arbitrage opportunity. The possibility of arbitrage enforces the price of most derivative securities, and technology for the calculation of arbitrage possibilities has brought the markets for derivative securities into a Golden Age.

10.1. Replication and Three Examples of Arbitrage

The basis of arbitrage is that any two investments with identical payout streams must have the same price. If this were not so, we could simultaneously sell the more expensive instrument and buy the cheaper one; we would make an immediate profit and the payment stream from our purchase could be used to meet the obligations from our sale. There would be no net cash flows after the initial action so we would have secured our arbitrage profit. What adds force to this simple observation is the relatively recent discovery of methods that tell us how to replicate the payout streams of important financial instruments such as stock options.

FINANCIAL FRICTIONS

The theory of arbitrage shares with the theory of mechanical systems the fact that it is most easily understood in a frictionless world. In finance, frictions are those irritating realities like the fact that it is easier to buy a sailboat than to sell one. For individuals, stock commissions are a friction in the transaction of equity shares, and even institutions that do not pay commissions still face important frictions like the bid–ask spread or the impact of large sales on market prices. Frictions are of great importance in financial markets; they are in many ways the krill that feed the financial Leviathans.

Nevertheless, the first step in either finance or mechanics is to consider models that are free of frictions. In our case, this means that first we study models where there are no transaction costs in the purchase or sale of shares, or in the borrowing or lending of funds. This restriction may seem harmless enough in the case of shares, but the stringency of the assumption may be less clear for bonds. We are all familiar with asymmetry in interest rates — witness the worn-out joke about the traditional banker who borrows at 6, lends at 9, and leaves at 4.

In our initial financial models, we will often assume that there is a continuous compounding interest rate r at which we can equally well borrow or lend funds, a *two-way* interest rate. Still, if we do not make this assumption, all is not lost. We will see later that our analysis may be informative even if the borrowing and lending rates are different. In the typical situation, an identity that is derived under the assumption that we have a two-way rate can be replaced by a pair of inequalities when that assumption is dropped.

Forward Contracts

Forward contracts provide an example of arbitrage pricing that has been honored for centuries. A typical forward contract is an agreement to buy a commodity — say 10,000 ounces of gold — at time T for a price K. If the current time is t and the current gold price is S, then in a world where there are economic agents who stand ready to borrow or lend at the continuous compounding rate r, the arbitrage price F of the forward contract is given by

$$F = S - e^{-r(T-t)}K.$$

In other words, if the forward contract were to be available at a different price, one would have an arbitrage opportunity.

The key observation is that there is an easy way to replicate the financial consequences of a forward contract. Specifically, one could buy the gold right now and borrow $e^{-r(T-t)}K$ dollars for a net cash outlay of $S - e^{-r(T-t)}K$, then at time T pay off the loan (with the accrued interest) and keep the gold. At the end of this process, one makes a payment of K dollars and gets ownership of the gold, so the payout of the forward contract is perfectly replicated, both with respect to cash and the commodity. The cash required to initiate the immediate purchase strategy is $S - e^{-r(T-t)}K$ and the cost of the forward contract is F, so the arbitrage argument tells us that these two quantities must be equal.

Table 10.1. Replication of a Forward Contract

	Cash Paid Out (Time=t)	Commodity and Cash (Time=T)
Forward Contract	F	Gold owned, K\$ Cash paid
Replication	$S - e^{-r(T-t)}K$	Gold owned, K\$ Cash paid

We do not want to get too distracted by side variations on our hypotheses, but we still should note that if the economic agents in our model who borrow money at r^* could only lend at r_*, where $r_* < r^*$, then our arbitrage argument no longer gives us an identity. Instead, we find that the absence of arbitrage profits would only guarantee the double inequality

$$S - e^{-r_*(T-t)}K \le F \le S - e^{-r^*(T-t)}K.$$

Put–Call Parity

A European call option on a stock is the right, but not the obligation, to buy the stock for the price K at time T. The European put option is the corresponding right to sell the stock at time T at a price of K. Our second illustration of arbitrage pricing will tell us that the arbitrage price of a European put is a simple function of the price of the call, the price of the stock, and the two-way interest rate.

First, we consider a question in the geometry of puts and calls. What is the effect of buying a call and selling a put, each with the same strike price K? Some funds will flow from the initiation of this position, then we will find at time T that

- if the stock price is above K, we will realize a profit of that price minus K;
- if the stock price is below K, we will realize a loss equal to K minus the stock price.

A moment's reflection will tell us this payout is exactly what we would get from a contract for the purchase of the stock at time T for a price K. Because we already know what the price of such a contract must be, we see that the price C of the call and the price P of the put must satisfy

$$C - P = S - e^{-r(T-t)}K. \tag{10.1}$$

This relationship is often referred to as the *put–call parity formula*, and it tells us how to price the European put if we know how to price the European call, or vice versa. One might expect this put call parity formula to be every bit as ancient as the formula for pricing a forward contract, but such an expectation would be misplaced. This elegant formula is barely thirty years old.

The Binomial Arbitrage

It does not take a rocket scientist to replicate a forward contract, or to value a put in terms of a call, but the idea of replication can be pushed much further, and, before long, some of the techniques familiar to rocket scientists start to show their value. Before we come to a mathematically challenging problem, however, there is one further question that deserves serious examination — even though it requires only elementary algebra.

For the sake of argument, we first consider an absurdly simple world with one stock, one bond, and two times — time 0 and time 1. The stock has a price of \$2 at time 0, and its price at time 1 is either equal to \$1 or \$4. The bond has a price of \$1 at time 0 and is also \$1 at time 1. People in this thought-experiment world are so kind that they borrow or lend at a zero interest rate.

Now, consider a contract that pays \$3 if the stock moves to \$4 and pays nothing if the stock moves to \$1. This contract is a new security that derives its value from the value of the stock, a toy example of a *derivative security.* The natural question is to determine the arbitrage price X of this security.

From our earlier analysis, we know that to solve this problem we only need to find a replicating portfolio. In other words, we only need to find α and β such that the portfolio consisting of α units of the stock and β units of the bond will exactly replicate the payout of the contract. The possibility of such a replication is made perfectly clear when we consider a table that spells out what is required under the two possible contingencies — the stock goes up, or the stock goes down.

Table 10.2. Replication of a Derivative Security

	Portfolio	Derivative Security
Original cost	$\alpha S + \beta B$	X
Payout if stock goes up	$4\alpha + \beta$	3
Payout if stock goes down	$\alpha + \beta$	0

When we require that the portfolio must replicate the payout of the derivative security, we get the two equations

$$4\alpha + \beta = 3 \text{ and } \alpha + \beta = 0.$$

We can solve these equations to find $\alpha = 1$ and $\beta = -1$, so by the purchase of one share of stock and the short sale of one bond, we produce a portfolio that perfectly replicates the derived security. This replicating portfolio requires an initial investment of one dollar to be created, so the arbitrage price of the derived security must also equal one dollar.

REPLICATION IS THE KEY

The preceding example is simple, but it rewards careful consideration. Before the 1970s everyone on the planet would have approached the valuation of this security by first forming a view of the probability of the stock going up. We now see that this may be a shaky way to start. The price of the security is uniquely determined by arbitrage without consideration of such probabilities. In fact, even if we are told upfront that the stock has a 50 % chance of going up to \$4, it would not change the fact that the only price for the security that keeps the world free of arbitrage is \$1. Since the expected value of the claim's payout would then be \$1.50, we see that the trader who bases his valuations on expected values may become grist for the arbitrage mill, even if he gets his expected values exactly right.

10.2. The Black–Scholes Model

The world of binomial price changes does a fine job isolating some very important ideas, but it cannot make a strong claim on realism. Nevertheless, considerations like those used for binomial changes can be applied in the much more realistic context where prices change continuously through time. Even this situation is not perfectly realistic, but it provides a palpable sense of progress.

We will now follow a remarkable continuous-time arbitrage argument that will lead us to the famous Black–Scholes formula for the pricing of European call options. We let S_t denote the price at time t of a stock and let β_t denote the price at time t of a bond. We then take the time dynamics of these two processes to be given by the SDEs

$$\text{(10.2)} \qquad \text{Stock model: } dS_t = \mu S_t\,dt + \sigma S_t\,dB_t \qquad \text{Bond model: } d\beta_t = r\beta_t\,dt;$$

that is, we assume that the stock price is given by a geometric Brownian motion, and the bond price is given by a deterministic process with exponential growth.

For a European call option with strike price K at termination time T, the payout is given by $h(S_T) = (S_T - K)_+$. To find the arbitrage price for this security, we need to find a way to replicate this payout. The new idea is to build a dynamic portfolio where the quantities of stocks and bonds are continuously readjusted as time passes.

ARBITRAGE AND REPLICATION

If we let a_t denote the number of units of stock that we hold in the replicating portfolio at time t and let b_t denote the corresponding number of units of the bond, then the total value of the portfolio at time t is

$$V_t = a_t S_t + b_t \beta_t.$$

The condition where the portfolio replicates the contingent claim at time T is simply

$$\text{(10.3)} \qquad \text{terminal replication constraint: } V_T = h(S_T).$$

In the one-period model of the binomial arbitrage, we only needed to solve a simple linear system to determine the stock and bond positions of our replicating portfolio, but in the continuous-time model we face a more difficult task. Because the prices of the stock and bond change continuously, we have the opportunity to continuously rebalance our portfolio — that is, at each instant we may sell some of the stock to buy more bonds, or vice versa. This possibility of continuous rebalancing gives us the flexibility we need to replicate the cash flow of the call option.

Because the option has no cash flow until the terminal time, the replicating portfolio must be continuously rebalanced in such a way that there is no cash flowing into or out of the portfolio until the terminal time T. This means that at all intermediate times we require that the restructuring of the portfolio be *self-financing* in the sense that any change in the value of the portfolio value must equal the profit or loss due to changes in the price of the stock or the price of the bond. In terms of stochastic differentials, this requirement is given by the equation

$$\text{(10.4)} \qquad \text{self-financing condition: } dV_t = a_t\, dS_t + b_t\, d\beta_t.$$

This equation imposes a strong constraint on the possible values for a_t and b_t. When coupled with the termination constraint $V_t = h(S_T)$, the self-financing condition (10.4) turns out to be enough to determine a_t and b_t uniquely.

Coefficient Matching

In order to avail ourselves of the Itô calculus, we now suppose that the portfolio value V_t can be written as $V_t = f(t, S_t)$ for an appropriately smooth f. Under this hypothesis, we will then be able to use the self-financing condition to get expressions for a_t and b_t in terms of f and its derivatives. The replication identity can then be used to turn these expressions into a PDE for f. The solution of this PDE will in turn provide formulas for a_t and b_t as functions of the time and the stock price.

To provide the coefficient matching equation, we need to turn our two expressions for V_t into SDEs. First, from the self-financing condition and the models for the stock and bond, we have

$$\begin{aligned} dV_t &= a_t dS_t + b_t d\beta_t = a_t\{\mu S_t dt + \sigma S_t dB_t\} + b_t\{r\beta_t dt\} \\ &= \{a_t\mu S_t + b_t r\beta_t\}\, dt + a_t\sigma S_t\, dB_t. \end{aligned} \tag{10.5}$$

From our assumption that $V_t = f(t, S_t)$ and the Itô formula for geometric Brownian motion (or the box calculus), we then find

$$\begin{aligned} dV_t &= f_t(t, S_t)\, dt + \frac{1}{2} f_{xx}(t, S_t)\, dS_t \cdot dS_t + f_x(t, S_t)\, dS_t \\ &= \left\{ f_t(t, S_t) + \frac{1}{2} f_{xx}(t, S_t)\, \sigma^2 S_t^2 + f_x(t, S_t)\mu S_t \right\} dt + f_x(t, S_t)\sigma S_t\, dB_t. \end{aligned} \tag{10.6}$$

When we equate the dB_t coefficients from (10.5) and (10.6), we find a delightfully simple expression for the size of the stock portion of our replicating portfolio:

$$a_t = f_x(t, S_t).$$

Now, to determine the size of the bond portion, we only need to equate the dt coefficients from equations (10.5) and (10.6) to find

$$\mu S_t f_x(t, S_t) + r b_t \beta_t = f_t(t, S_t) + \frac{1}{2} f_{xx}(t, S_t)\sigma^2 S_t^2 + f_x(t, S_t)\mu S_t.$$

The $\mu S_t f_x(t, S_t)$ terms cancel, and we can then solve for b_t to find

$$\text{(10.7)} \qquad b_t = \frac{1}{r\beta_t}\left\{ f_t(t, S_t) + \frac{1}{2} f_{xx}(t, S_t)\sigma^2 S_t^2 \right\}.$$

Because V_t is equal to both $f(t, S_t)$ and $a_t S_t + b_t \beta_t$, the values for a_t and b_t give us a PDE for $f(t, S_t)$:

$$\begin{aligned} f(t, S_t) &= V_t = a_t S_t + b_t \beta_t \\ &= f_x(t, S_t) S_t + \frac{1}{r\beta_t}\{ f_t(t, S_t) + \frac{1}{2} f_{xx}(t, S_t)\sigma^2 S_t^2 \}\beta_t. \end{aligned}$$

Now, when we cancel β_t from the last term and replace S_t by x, we arrive at the justly famous *Black–Scholes PDE*:

$$\text{(10.8)} \qquad f_t(t, x) = -\frac{1}{2}\sigma^2 x^2 f_{xx}(t, x) - r x f_x(t, x) + r f(t, x),$$

with its terminal boundary condition

$$f(T, x) = h(x) \quad \text{for all } x \in \mathbb{R}.$$

10.3. The Black–Scholes Formula

When we solve the Black-Scholes PDE, we find a function $f(t, x)$ such that $f(t, S_t)$ is the arbitrage price of a European call option under the stock and bond model given by equation (10.2). In subsequent chapters, we will give several methods for solving equation (10.8), but for the moment we will concentrate on the lessons that can be learned from that solution.

Nature of the Solution

In the first place, the solution is simpler than one has any right to expect. The only special function needed to represent the solution is the Gaussian integral Φ, and all of the parameters of our model appear in a thoroughly intuitive way. Specifically, the arbitrage price of the European call option at time t with a current stock price of S, exercise time T, strike price K, and residual time $\tau = T - t$ is given by

$$\text{(10.9)} \qquad S\Phi\left(\frac{\log(S/K) + (r + \frac{1}{2}\sigma^2)\tau}{\sigma\sqrt{\tau}}\right) - Ke^{-r\tau}\Phi\left(\frac{\log(S/K) + (r - \frac{1}{2}\sigma^2)\tau}{\sigma\sqrt{\tau}}\right).$$

There are several remarkable features of this famous *Black–Scholes formula*, but perhaps the most notable of these is the absence of the μ — the growth rate parameter of the stock model. Prior to the 1970s this absence would have been so unintuitive that it would have cast doubt on the validity of the pricing formula — or the credibility of the underlying model. Today, we have a much clearer understanding of the absence of μ, although even now there may be *some* room for mystery.

In a mechanical sense, we know that μ does not appear in the Black–Scholes formula simply because it was cancelled away before we came to the Black–Scholes

PDE (10.8). This is not a very insightful observation, and eventually we will do much better. For the moment, perhaps the best way to understand the disappearance of μ is by thinking about the binomial arbitrage. In that example, we found that the probability of the stock going up or down was irrelevant to the calculation of the arbitrage price. We would have found the same phenomenon if we had looked at binomial models with more than two time periods.

Multiperiod models lead quickly to random walk, and from random walk it is a short stroll to Brownian motion and the Black–Scholes model. This is all to argue, far too quickly, that since the expected growth rate of the stock does not matter in the binomial model, neither should it matter in the continuous-time model.

Naturally, this waving of the hands does not fully resolve the μ problem, or μ Koan.[1] Still, it should at least lend some intuition to the absence of μ in the Black–Scholes formula. In a later chapter, we will revisit the option pricing formula from an (almost) entirely new perspective, and eventually we will come to a much richer understanding of the μ problem.

The Formula and the Portfolio

Perhaps the most important practical feature of the Black–Scholes formula is that it tells us very simply and exactly how to build the replicating portfolio. Thus, the Black–Scholes formula not only specifies the arbitrage price, but it even tells us how to enforce that price — if we are agile in the execution of our transactions and if our model honestly reflects reality.

We recall from the derivation of the Black–Scholes PDE that the amount of stock a_t that we hold in the replicating portfolio at time t is given by

$$a_t = f_x(t, S_t),$$

and, now that we know f, we can calculate a_t. If we differentiate the Black–Scholes formula (10.9) with respect to the stock price S, then with some algebraic care we find that the number of units of stock in the replicating portfolio is given by

$$a_t = \Phi\left(\frac{\log(S_t/K) + (r + \frac{1}{2}\sigma^2)\tau}{\sigma\sqrt{\tau}}\right). \tag{10.10}$$

The positivity of a_t tells us that the replicating portfolio always has a long position in the stock, and the fact that $a_t < 1$ tells us that the replicating portfolio never holds more than one unit of the stock for each unit of the call option.

When we then look at the bond position, we also find from the Black–Scholes formula (10.9) that the amount of money that is invested in bonds is given by

$$b_t\beta_t = -Ke^{-r\tau}\Phi\left(\frac{\log(S_t/K) + (r - \frac{1}{2}\sigma^2)\tau}{\sigma\sqrt{\tau}}\right). \tag{10.11}$$

Because $b_t\beta_t$ is negative we see that in the replicating portfolio we are always short the bond. Moreover, the formula for the bond position (10.11) tells us that dollar size of that short position, $|b_t\beta|$, is never larger than the strike price K.

---

[1]Not to be confused with the *Mu Koan* of Jōshū from the *Mumonkan*, a famous Sung Dynasty (960-1279 A.D.) collection of philosophical problems (or koans): A monk asked Jōshū, "Has a dog the Buddha Nature?" Jōshū answered "Mu." The meaning of the Mu koan has created an evolving conversation for almost a millennium. One cannot expect the μ Koan to be so lucky.

10.4. Two Original Derivations

Few activities are more consistently rewarding than the study of the original sources of important results, and the 1973 paper by F. Black and M. Scholes should be examined by anyone who wishes to understand how the theory of option pricing came into being. In that paper, Black and Scholes give two derivations of their PDE, and, although both arguments have rough spots, they are highly instructive for the unique insight they provide into the creative interaction of calculation and modeling. The main goal of this section is to review these arguments in a way that is faithful to the original exposition yet provides mathematical clarifications at several points.

The Original Hedged Portfolio Argument

In their first argument, Black and Scholes consider a portfolio that consists of a long position of one share of stock and a short position that consists of a continuously changing quantity of options. The size of the option position is chosen so that the option position tracks the stock position as closely as possible, and, in our earlier notation, the value of this portfolio can be written as

$$X_t = S_t - f(t, S_t)/f_x(t, S_t) \qquad \text{[BS-eqn-2]}, \tag{10.12}$$

where $f(t, S_t)$ denotes the (unknown) value of the call option. Also, the redundant right-hand equation label [BS-eqn-2] is given so that one can make a line-by-line comparison between this discussion and the original paper of Black and Scholes. Thus, our equation (10.12) is just a transcription into our notation of the original Black–Scholes equation (2).

Next, we need to introduce the evocative notation Δ, and one naturally thinks of ΔY_t as something like $Y_{t+h} - Y_t$ for some small value of h. Black and Scholes then calculate

$$\Delta X_t = \Delta S_t - \{\Delta f(t, S_t)\}/f_x(t, S_t) \qquad \text{[BS-eqn-3]}, \tag{10.13}$$

and they do not comment on the process by which one goes from the defining equation [BS-eqn-2] to the interesting [BS-eqn-3], even though a straightforward Δ calculation would have given a much messier right-hand side in equation (10.13). This step is an important one which we will examine in detail after we have seen the rest of the argument, but, for the moment, we are content to note that [BS-eqn-3] is self-evident, provided that we think of $f_x(t, S_t)$ as "held fixed" while the change $\Delta f(t, S_t)$ "takes place."

Black and Scholes continue their argument by saying[2] "Assuming the short position is changed continuously, we can use the stochastic calculus to expand $\Delta f(t, S_t)$, which is $f(t + \Delta t, S_t + \Delta S_t) - f(t, S_t)$, as follows":

(10.14)

$$\Delta f(t, S_t) = f_x(t, S_t)\Delta S_t + \frac{1}{2} f_{xx}(t, S_t)\sigma^2 S_t^2 \Delta t + f_t(t, S_t)\Delta t \qquad \text{[BS-eqn-4]},$$

and for their next step Black and Scholes say "Substituting from equation [BS-eqn-4] into expression [BS-eqn-3], we find that the change in value of the equity in the

[2]The *notation* within the quotes has been changed to conform to the standard notations.

hedged position is":

$$(10.15)\qquad -\left(\frac{1}{2}f_{xx}(t,S_t)\sigma^2 S_t^2 + f_t(t,S_t)\right)\Delta t / f_x(t,S_t) \qquad \text{[BS-eqn-5]}.$$

Now, Black and Scholes make a basic arbitrage argument that we will amplify a tiny bit. Since the constructed portfolio is a tradable instrument and since the return on the hedged portfolio in the time period Δt is certain, the usual arbitrage argument tells us that this return must equal the return that one would receive from an investment of the same size in a risk-free bond for a comparable period. At time t, the amount invested in the hedged position is $X_t = S_t - f(t,S_t)/f_x(t,S_t)$, so during the time interval Δt the return on the hedged portfolio should be $X_t \cdot r \cdot \Delta t$. As a consequence, one should be able to equate these two rates to find [BS-eqn-6]:

$$-\left(\frac{1}{2}f_{xx}(t,S_t)\sigma^2 S_t^2 + f_t(t,S_t)\right)\Delta t / f_x(t,S_t) = \left(S_t - f(t,S_t)/f_x(t,S_t)\right) r\Delta t.$$

Now, the Δt can be dropped, and the last equation can be rearranged to provide

$$f_t(t,S_t) = rf(t,S_t) - rS_t f_x(t,S_t) - \frac{1}{2}f_{xx}(t,S_t)\sigma^2 S_t^2 \qquad \text{[BS-eqn-7]}.$$

Naturally, we recognize this equation as the classic Black–Scholes PDE that we derived earlier in the chapter by rather different means.

What Do We Learn from This Argument?

This argument is of considerable historical importance, and it deserves to be analyzed to see what it can teach us about mathematics and modeling. One way we may work toward these goals is to try to put ourselves into the time and place where the work was done and ask what we would have done, or where we might have become stuck.

In this case, we have to imagine the ideas that were in the air at the time. Fortunately, Black and Scholes share some of this background with us in their paper, and they point out that one of their sources of inspiration came from *Beat the Market*, a popular 1967 book by Thorp and Kassouf. One of the ideas in this book was to use empirical methods to estimate the size of a short position in a derivative security that would minimize the risk of holding a long position in the underlying security. When Black and Scholes looked for a theoretical analog to this process, they were naturally led to consider the portfolio consisting of a long position of one share of stock and a short position of $1/f_x(t,S_t)$ options, so, in symbols, they came to consider a portfolio whose value at time t is given by

$$(10.16)\qquad X_t = S_t - f(t,S_t)/f_x(t,S_t) \qquad \text{[BS-eqn-2]}.$$

At this point, many of us may be tempted to say that the modeling idea is on the table and all that is needed from this point is simply honest calculation. In fact, the most impressive modeling step is just about to take place.

If we apply the Δ operation to both sides of [BS-eqn-2], we definitely *do not* get [BS-eqn-3], or even an approximation to [BS-eqn-3]. Nevertheless, the argument that follows after [BS-eqn-3] is completely straightforward, and it leads directly to the Black–Scholes equation [BS-eqn-7], a result that we have every reason to believe to be valid. Therefore, we know that there must be some way to justify

$$\Delta X_t = \Delta S_t - \Delta f(t, S_t)/f_x(t, S_t) \qquad \text{[BS-eqn-3]}, \tag{10.17}$$

even if this identity is not a mechanical consequence of [BS-eqn-2].

One way to look for a justification is to consider how one would actually implement the management of a portfolio such as that given by [BS-eqn-2]. In practice, we cannot do continuous rebalancing, so, the natural process would be to choose a discrete sequence of times $t_1, t_2, \ldots t_n$ at which to do our buying and selling. If we take this point of view, and if we interpret the Δ's as the change over the interval $[t_i, t_{i+1}]$ based on the investments that are held at time t_i, then the meaning of [BS-eqn-3] is simply

$$X_{t_{i+1}} - X_{t_i} = S_{t_{i+1}} - S_{t_i} - [f(t, S_{t_{i+1}}) - f(t, S_{t_i})]/f_x(t, S_{t_i}). \tag{10.18}$$

With this interpretation, [BS-eqn-3] is as clear as a bell; but, unfortunately, the new interpretation brings a new problem.

Is X_t defined by the portfolio equation [BS-eqn-3], or is it defined by the current value of a portfolio that evolves according to some discrete strategy based on rebalancing at a time skeleton $t_1, t_2, \ldots, t_n$? In pure mathematics, one cannot have it both ways, but in applied mathematics one can and *should.* The whole issue is one of objective.

In the down-and-dirty world of applied mathematics, what matters is that one eventually comes to a model that can be tested against reality. In this case, we come to a PDE that can be solved and has solutions that are empirically informative, even if imperfect. This outcome and the process that leads to it are 100% satisfactory, even if we must be a bit schizophrenic in our interpretation of X_t. Schizophrenia is not an easy condition to live with, but sometimes it is the price of creativity.

The Original CAPM Argument

In their 1973 paper, Black and Scholes were not content with only one argument for their PDE. They also gave a second derivation which was based on the market equilibrium ideas of the capital asset pricing model, or CAPM. For the purpose of this derivation, the main assertion of the CAPM is that if Y_t is the price process of an asset Y and M_t is the price process of the "market asset" M, then there is a quantity β_Y such that the Δt-period return $\Delta Y_t/Y_t$ on the Y asset has an expected value that can be written as

$$E(\Delta Y_t/Y_t) = r\Delta t + a\beta_Y \Delta t, \tag{10.19}$$

where r is the risk-free interest rate and a is understood (a bit informally) as the rate of expected return on the market minus the risk-free interest rate r. Moreover, in the theory of the CAPM, the constant β_Y is given explicitly in terms of the covariance of the asset and market returns:

$$\beta_Y \stackrel{\text{def}}{=} \text{Cov}(\,\Delta Y_t/Y_t,\ \Delta M_t/M_t\,)/\text{Var}(\Delta M_t/M_t).$$

Now, in order to meld the CAPM point of view with our earlier notation, we just need to think of $f(t, S_t)$ as the price process of asset O, the option, and to think of S_t as the price process of the asset S, the stock. The first step is to relate the β's of the option and the stock, and we begin by using Itô's formula to write

$$\Delta f(t, S_t) = f_x(t, S_t)\Delta S_t + \frac{1}{2} f_{xx}(t, S_t)\sigma^2 S_t^2 \Delta t + f_t(t, S_t)\Delta t. \tag{10.20}$$

When we recast this formula in terms of the rate of return on the O asset, the option, we see that the rate of return on the option can be viewed as an affine function of the rate of return on the stock,

$$\frac{\Delta f(t, S_t)}{f(t, S_t)} = \frac{f_x(t, S_t)}{f(t, S_t)} S_t \frac{\Delta S_t}{S_t} + \frac{1}{f(t, S_t)} \left(\frac{1}{2} f_{xx}(t, S_t)\sigma^2 S_t^2 \Delta t + f_t(t, S_t)\Delta t \right).$$

Now, Black and Scholes take a remarkable step. They proceed to calculate the covariance of $\Delta f(t, S_t)/f(t, S_t)$ and $\Delta S_t/S_t$ *as if* all of the other quantities in the preceding formula were *constants*. Once one makes this ansatz, the definition of β_O and the preceding formula for $\Delta f(t, S_t)/f(t, S_t)$ combine in an easy calculation to give

$$\beta_O = (S_t f_x(t, S_t)/f(t, S_t))\, \beta_S. \qquad \text{[BS-eqn-15]} \tag{10.21}$$

When one looks back on this equation, it certainly seems at least a little strange. For one thing, the left-hand side is a constant, whereas the right-hand side is a random variable. Still, applied mathematics always takes courage, so we press forward with the optimistic hope that an appropriate interpretation of this step will emerge once we have the whole argument in view.

The next two equations offer a brief respite from the stress of modeling since they only require us to express two basic assertions of the CAPM in our notation; specifically, we have a model for the S asset returns,

$$E\left[\frac{\Delta S_t}{S_t}\right] = r\Delta t + a\beta_S \Delta t \qquad \text{[BS-eqn-16]}, \tag{10.22}$$

and a model for the O asset returns,

$$E\left[\frac{\Delta f(t, S_t)}{f(t, S_t)}\right] = r\Delta t + a\beta_O \Delta t \qquad \text{[BS-eqn-17]}. \tag{10.23}$$

Now, Black and Scholes are poised to take another remarkable step. In their words (but our notation), they continue: "Multiplying [BS-eqn-17] by $f(t, S_t)$, and substituting for β_O from equation [BS-eqn-15], we find"

$$E[\Delta f(t, S_t)] = r f(t, S_t)\Delta t + a S_t f_x(t, S_t)\beta_S \Delta t \qquad \text{[BS-eqn-18]}. \tag{10.24}$$

If we hope to follow this step, we need to note that Black and Scholes seem to be counting on certain interpretations of their equations that might confuse notationally fastidious readers. First, they surely intend for us to interpret Δ as a forward difference, so, for example, $\Delta Y_t = Y_{t+h} - Y_t$. The more critical point is that they evidently also intend for us to interpret the expectations as *conditional* expectations such as those that one more commonly renders as $E(\cdot|\mathcal{F}_t)$. With these interpretations, there is no difficulty with equation [BS-eqn-18] or with the corresponding multiplication of [BS-eqn-16] by S_t to conclude that

$$E(\Delta S_t) = r S_t \Delta t + a\beta_S S_t \Delta t. \tag{10.25}$$

The next step in the CAPM derivation of the Black–Scholes PDE is to use Itô's formula in the form of equation (10.20) to calculate

$$E(\Delta f(t,S_t)) = f_x(t,S_t)E(\Delta S_t) + \frac{1}{2}f_{xx}(t,S_t)\sigma^2 S_t^2 \Delta t + f_t(t,S_t)\Delta t. \tag{10.26}$$

Now, when the expression for $E(\Delta S_t)$ given by [BS-eqn-16] (by way of equation (10.25)) is substituted into equation (10.26), we find

$$\begin{aligned}E(\Delta f(t,S_t)) &= rS_t f_x(t,S_t)\Delta t + aS_t f_x(t,S_t)\beta_S \Delta t \\ &\quad + \frac{1}{2}f_{xx}(t,S_t)\sigma^2 S_t^2 \Delta t + f_t(t,S_t)\Delta t \quad \text{[BS-eqn-20]},\end{aligned}$$

and, with this equation in hand, we are essentially done. We have two expressions for $E(\Delta f(t,S_t))$ given by [BS-eqn-18] and [B-eqn-20], so we can simply equate them, cancel the Δt's, and simplify the result to find

$$f_t(t,S_t) = rf(t,S_t) - rS_t f_x(t,S_t) - \frac{1}{2}\sigma^2 S_t^2 f_{xx}(t,S_t).$$

Perspective on the CAPM Derivation

The most amazing aspect of the second derivation given by Black and Scholes may be that it is so different from the first derivation. There is no arbitrage argument at all. The CAPM derivation obtains the Black–Scholes PDE from purely structural relations in a way that almost seems magical.

Perhaps the trickiest step of the argument takes place with the derivation of the covariance formula given in [BS-eqn-15]: $\beta_O = S_t\{f_x(t,S_t)/f(t,S_t)\}\beta_S$. Even though we whined a bit when this formula first appeared, both the formula and its derivation make perfect sense provided that Δ is interpreted as a forward difference and all the expectations, variances, and covariances are interpreted as conditional expectations, variances, and covariances given $\mathcal{F}_t$.

Nevertheless, there is some crafty applied mathematical insight that lurks under the surface. The formulas of the CAPM were written in the unconditional form in which they were widely understood at the time, yet these formulas were subsequently used in a conditional form. This is a situation that a purist might view as unsportsmanlike. More informatively, one can view the calculations of Black and Scholes as offering a reinterpretation of the traditional formulas of the CAPM. In essence these calculations suggest the use of the stronger conditional form of the CAPM.

Even now a small puzzle remains. If Black and Scholes had this reinterpretation in mind, why did they not comment on such an interesting theoretical idea? One possible answer was that there is no real need to go as far as positing an untested interpretation of the CAPM. The point of both of the derivations was not one of proving the validity of the Black–Scholes PDE as a mathematical theorem. The intention was rather to motivate a specific PDE model as a model for option prices.

The ultimate test of such a PDE model is always an empirical one, and there is no compelling reason for one to draw a firm distinction between the use of the CAPM and the use of an *analogy* where conditional formulas stand in for more traditional CAPM formulas. In both cases, one comes to the same point. Either by extension or by analogy, Black and Scholes show how widely studied equilibrium principles of finance motivate a concrete PDE model. The true test of the model then rests with its empirical performance.

10.5. The Perplexing Power of a Formula

How much should one worry about the assumptions that lead to the Black–Scholes option pricing formula? This depends entirely on how seriously the model is to be taken. As it turns out, models for derivative pricing and arbitrage engineering have had more impact on financial markets than any other development of that industry in recent memory. The market for risk-sharing financial products would be a mere trickle of its present flood if market makers and other financial institutions did not have confidence in their ability to design portfolios that permit them to control the risk in selling or buying derivative securities. At this stage, every nuance of replication theory is fair game for detailed examination.

When we examine the model that leads us to the Black–Scholes formula, we find many assumptions that one might want to modify. To the probabilistic modeler, the boldest assumption is the use of geometric Brownian motion (GBM) for the underlying stock model. Then, even if one accepts GBM as a reasonable proxy for the stock price process, one still has to question whether we should believe that μ and σ may be honestly viewed as constant, except perhaps for very short periods of time. The same concerns apply to the interest rate r. In a world where interest rates do not change, the fixed-income markets would lose a considerable amount of their excitement, yet the trading pit for bond futures is not yet an island of calm.

There is also a structural objection to the options contracts that we have analyzed thus far; the European option is a very rare bird compared to American options that allow the holder to exercise his purchase or sale at any time up until the terminal time. In a later chapter, we will find that this problem can be nicely resolved, at least in the case of call options.

Finally, there is the issue of transaction costs. One can assume that if the portfolios involved are small enough then market impact costs are not important. There is also no harm in assuming that brokerage costs will not matter, provided that we take the point of view of a large institution or a member firm of a stock exchange. Nevertheless, one cost — the bid–ask spread — cannot be avoided, and even one unavoidable cost is enough to destroy the fundamental logic of our derivations. After all, to trade continuously implies that we are *continuously trading*, and, with a bit of work, one can show that in the typical case where the stock price S_t is a process with unbounded variation then the total trading volume — and the associated transaction cost — is also infinite.

Popularity of a Formula

If the Black–Scholes model faces all of these objections, why has it become so popular — and so genuinely important? In the first place, the model leads to a concrete formula. In those lucky instances when sustained modeling and reasoning manage to jell into a single line, there is tremendous psychological power in that line. To the world beyond professional scientists, a formula is at the heart of what one means by the *solution* to a problem.

The Black–Scholes formula also has the benefit of being easy to use and understand. The only troubling parameter in the formula is the volatility σ, and, despite the fact that σ is harder to estimate than μ, the disappearance of μ is a celebrated blessing. All of the parameters that are present in the Black–Scholes formula appear in physically logical ways.

Finally, a formula that engages real phenomena does not need to be perfectly accurate to provide insight. As it turns out, the empirical performance of the Black–Scholes formula is reasonably good. To be sure, there are important discrepancies between observed prices and those predicted by the Black–Scholes formula, but for options with a strike price that is not too far from the current stock price the Black–Scholes formula anticipates observed prices rather well. The biggest empirical defect is that the market price for an option that is deep out of the money is almost always higher than the price suggested by the Black–Scholes formula.

Redundancy and Utility

Finally, we must reflect for a moment on the absence of any influence of utility theory on the Black–Scholes formula. Utility theory is at the core of almost all economic modeling, yet it makes no appearance here. One has to ask why.

To get a feeling for this issue, suppose for a moment that someone has offered to sell you a random prize X that is to be given to you at time T in exchange for a payment that you are to make now at time $t < T$. Also, for the moment, you can suppose that the payout X is not contingent on the price of a tradable entity such as a stock, so there is not any issue of arbitrage pricing for X. How much should you be willing to pay for X?

To keep life pinchingly simple, we will even assume that the time horizon T is sufficiently small that we can ignore the time value of money, so there is no need to worry about discounting the value of the time T prize to get its present value. This would be more or less reasonable for short time horizons and low interest rates, but our only concern is simplicity; model refinements can come later.

Now, at last, if we let $\mathcal{F}_t$ summarize our knowledge of the world at time t, then, in order to work out a value for the random prize, we almost have to consider the benchmark given by the conditional expectation:

$$V_t = E(X|\mathcal{F}_t). \tag{10.27}$$

If the random prize is small with respect to our wealth, this is indeed a decent candidate for a fair price for the prize. On the other hand, if the random prize is large compared to our wealth, then this expectation does not represent what we would be willing to pay. For example, if W is our net worth, we are not willing to pay W for a random prize that is worth $2W$ with probability one-half and 0 with probability one-half. This reflects the fact that our utility for money is not linear; twice as much is not twice as good, unless the amounts involved are small. The issues involved in the valuation of our random prize are psychological, and there is no reason to suppose that different individuals would not come to different valuations. This is quite a contrast to the enforced prices provided by arbitrage theory.

Now, here is the last puzzle. Why did we hear nothing about the nonlinearity of utility or about our aversion to risk during the development of the theory of arbitrage prices? The simple observation is that arbitrage pricing theory is a *certainty theory.* To be sure, we did begin with a probability model, but at the critical moment probability is cut out of the loop. The possibility of arbitrage removed all concern with probability, with one small exception. We did need to know which events were *possible*, even though we never needed to know their probabilities. Finally, the absence of chance reduces utility theory to the trivial; one needs no finer understanding of money than to know that more is better.

10.6. Exercises

The first exercise should be done by everyone. It provides basic practice with the most important argument of the chapter and provides an important perspective on the range and limitation of the PDE approach to option pricing. The next two problems are easy but amusing. They explore special solutions of the basic Black–Scholes PDE, and they suggest how these solutions may be used to obtain a second derivation of the put–call parity formula. The last problem outlines how the Black–Scholes PDE (and its derivation) may be modified to account for the payment of stock dividends.

EXERCISE 10.1 (A More General Black–Scholes PDE). Consider the stock and bond model given by

$$dS_t = \mu(t, S_t)\, dt + \sigma(t, S_t)\, dB_t \quad \text{and} \quad d\beta_t = r(t, S_t)\beta_t\, dt, \tag{10.28}$$

where all of the model coefficients $\mu(t, S_t), \sigma(t, S_t)$, and $r(t, S_t)$ are given by explicit functions of the current time and current stock price.

(a) Use the coefficient matching method of Section 10.2 to show that arbitrage price at time t of a European option with terminal time T and payout $h(S_T)$ is given by $f(t, S_t)$ where $f(t, x)$ is the solution of the terminal value problem

$$f_t(t,x) = -\tfrac{1}{2}\sigma^2(t,x) f_{xx}(t,x) - r(t,x)x f_x(t,x) + r(t,x) f(t,x),$$
$$f(T,x) = h(x).$$

(b) Show that $f(t, x)$ and its derivatives may be used to provide explicit formulas for the portfolio weights a_t and b_t for the self-financing portfolio $a_t S_t + b_t \beta_t$ that replicates $h(S_T)$.

EXERCISE 10.2 (Exploration of Separable Solutions). (a) Find all solutions of the basic Black–Scholes PDE that are of the form

$$f(x,t) = \phi(x) \quad \text{or} \quad f(x,t) = \psi(t).$$

If you consider the just linear combinations of these solutions, what terminal boundary value problems can you solve?

(b) Now find all of the solutions of the form $\phi(x)\psi(t)$. What terminal-value problems can you solve with these basic solutions?

EXERCISE 10.3 (Black–Scholes PDE and Put–Call Parity). In the basic Black–Scholes model, the time t arbitrage price of a European style contingent claim that pays $h(S_T)$ at time T can be written as $f(t, S_t)$ where $f(t, x)$ satisfies the terminal-value problem

$$f_t(t,x) = -\frac{1}{2}\sigma^2 x^2 f_{xx}(t,x) - rxf_x(t,x) + rf(t,x) \text{ and } f(T,x) = h(x). \tag{10.29}$$

(a) Consider the collective contingent claim that corresponds to being long one call and short one put, where each option has the strike price K and expiration time T. What is the $h(\cdot)$ that corresponds to this collective claim?

(b) Show by direct substitution that $f(t,x) = x - e^{-r(T-t)}K$ is a solution to the Black–Scholes PDE. Use this observation and your answer to part (a) to give an alternative proof of the put–call parity formula. Is this derivation more or less general than the one given at the beginning of the chapter?

EXERCISE 10.4 (Dealing with Dividends). Suppose that we have a stock that pays a dividend at a constant rate that is proportional to the stock price, so that if you hold a_t units of the stock during the time period $[\alpha, \beta]$, then your total dividend receipts will be given by

$$r_D \int_\alpha^\beta a_t S_t \, dt,$$

where the constant r_D is called the *dividend yield.* Suppose we start with a stock that pays no dividend that satisfies the SDE

$$dS_t = \mu S_t \, dt + \sigma S_t \, dB_t \quad \text{and} \quad d\beta_t = r\beta_t \, dt.$$

If the dividend policy of the stock is then changed so that the stock pays a continuous dividend (with constant dividend yield r_D), then, with a bit of thought, one can see that the SDE for S_t should be changed to

$$dS_t = (\mu - r_D)S_t \, dt + \sigma S_t \, dB_t,$$

since one must change the drift coefficient in order to compensate for the cash dividends that one receives.

If we assume that we have the usual bond model $d\beta_t = r\beta_t \, dt$, then, by analogy with other cases we have studied, one might guess that the change in dividend policy would not lead to a change in the arbitrage pricing PDE. Surprisingly enough, the change in dividend policy does call for a change in the PDE, and it turns out that the appropriate terminal value problem for a European option with time T payout $h(S_T)$ is then given by the modified equation:

$$f_t(t,x) = -\frac{1}{2}\sigma^2 x^2 f_{xx}(t,x) - (r - r_D)xf_x(t,x) + rf(t,x),$$

with the usual terminal condition $f(T,x) = h(x)$.

Use the coefficient matching method of Section 10.2 to show that this assertion is correct. To solve this problem you will need to make an appropriate modification of the self-financing condition before you start matching coefficients.

CHAPTER 11

The Diffusion Equation

In its protean form, the diffusion equation is one of the simplest partial differential equations:

$$\frac{\partial u}{\partial t} = \lambda \frac{\partial^2 u}{\partial x^2}. \tag{11.1}$$

Applications of this equation are now found throughout science, engineering, and even finance. The equation first became well known through Fourier's 1822 study of heat, and, for this reason, it is often referred to as the heat equation. Since the time of Fourier's work, the diffusion equation has become central to many areas, and, most relevant for us, the theory of the diffusion equation is inextricably intertwined with Brownian motion.

Further afield, the diffusion equation gave us one of the first physics-based estimations of the age of the Earth, and it still provides the basis for the calculation of the distribution of slowly moving neutrons in graphite — useful information if you want to build an atomic reactor. It has even been used to catch criminals and to measure the brightness of stars since the diffusion equation guides us to methods that enhance images from both surveillance and astronomical cameras.

We begin our study of the diffusion equation by first observing how it pops up naturally in physical systems that obey simple conservation and constitutive laws. We then give three ways to solve the initial-value problem for the diffusion equation: a Fourier method, a series method, and a similarity method. Each of these approaches puts a new face on the diffusion equation, and each illustrates a mathematical idea with far reaching consequences. Finally, we use the solution of the diffusion equation to complete our investigation of the Black–Scholes PDE.

11.1. The Diffusion of Mice

Each of the fields that leans on the diffusion equation has its own intuition, technology, and vocabulary, yet the equation retains a simple logic that can stand by itself. Perhaps the easiest way to master this logic is to consider a model that is technologically remote, or even modestly absurd. For example, we might consider the world of a dense population of mice living on the real line.

We imagine these mice running around according to their murine preferences, and we describe the population dynamics in terms of two basic functions: the density $u(t, x)$ of mice at location x at time t and the rate $q(t, x)$ at which mice cross the point x at time t (moving in the positive direction of the x-axis). The natural problem is to predict the population distribution at a future time given the distribution at the present.

Conservation of Mice

As a first approximation to the population dynamics, we will suppose that mice are neither created nor destroyed. In other words, we consider a world where mice are *conserved*. If we can calculate the rate at which mice *leave* the interval $[a, b]$ in terms of $q(t, x)$, and if we can make an independent calculation of this rate in terms of $u(t, x)$, then we will obtain an equation that relates q and u.

We first note that the rate at which mice leave the interval $[a, b]$ is given by $q(t, b) - q(t, a)$, and we can naturally recast this difference as an integral:

$$q(t,b) - q(t,a) = \int_a^b \frac{\partial}{\partial x} q(t,x)\,dx. \tag{11.2}$$

We then note that the total population of mice in the interval $[a, b]$ is also equal to the integral of the population density $u(t, x)$, so the rate at which mice leave the interval $[a, b]$ has a second representation which is given by

$$-\frac{\partial}{\partial t}\int_a^b u(t,x)\,dx = -\int_a^b \frac{\partial}{\partial t} u(t,x)\,dx. \tag{11.3}$$

By the equality of the two rates (11.2) and (11.3) for all intervals $[a, b]$, we see the corresponding integrands must be equal, so at last we find a first order partial differential equation

$$\frac{\partial u}{\partial t} = -\frac{\partial q}{\partial x}. \tag{11.4}$$

Such an equation is traditionally called a *conservation law*, or, in this case, the conservation law for mice.

Constitution of Mice

The movements of mice are determined by their preferences, and, in the austere language of our model, this boils down to saying that there is some function F for which we have

$$q(t,x) = F(u, u_x, x).$$

If we hope to find a simple model for our population, we need to find a simple choice of F.

Although it is difficult to speak for our fellow mammals, we might reasonably suppose that mice prefer to live in less crowded areas. We might also suppose that as they snoop around the mice can sense the spatial rate of change u_x of the population density, even though the nominal levels of u and x would be irrelevant to their actions. This brings us to consider an F that depends only on u_x, and, if we boldly approximate F as a linear function, we come to the hypothesis

$$q(t,x) = -\lambda \frac{\partial u}{\partial x}, \tag{11.5}$$

where we take λ to be a positive constant so that the minus sign reflects the assumption that the mice prefer to move to less crowded conditions.

Any relationship that expresses $q(t, x)$ in terms of x, u, and u_x is called a *constitutive law*, and equation (11.5) is the simplest and most widely used of the constitutive laws. It is not the only possibility, however, and in every instance we must regard the constitutive law as an approximation to be judged by experience.

At first we might only require that the constitutive law (11.5) be qualitatively reasonable, but, as one comes to rely on the model, the shape of the constitutive law must be formed by experimental evidence. Rather miraculously, simple constitutive laws such as equation (11.5) have been found to serve reliably in many important contexts.

The Diffusion Equation

The diffusion equation instantly boils out of any pot that contains a conservation law and a constitutive law. When we substitute equation (11.5) into equation (11.4), we immediately find the basic diffusion equation:

$$\frac{\partial u}{\partial t} = \lambda \frac{\partial^2 u}{\partial x^2}. \tag{11.6}$$

One of the interesting features of this derivation is that the function $q(t,x)$ has left the stage after serving as a trustworthy intermediary. Actually, this process is quite common. Many of the second order equations that one finds in science and engineering come about as the combination of two first order equations that have more direct physical interpretations.

Diffusions in $\mathbb{R}^d$

Equation (11.6) is just the one-dimensional member of a d-dimensional family, and the physical interpretation of the d-dimensional diffusion equation is perfectly parallel to the case of $d = 1$. New issues do emerge, however, and, at a minimum, we must be a bit more precise in the definition of some of our terms. Also, we will cater to the usual conventions and speak of material rather than mice. In $\mathbb{R}^3$, one might think of the diffusion of a drop of iodine in a glass of water.

As before, we can begin with a density function $u : \mathbb{R}^+ \times \mathbb{R}^d \mapsto \mathbb{R}$, but for a diffusion in $\mathbb{R}^d$ we can no longer take q to be simply a real valued function of time and location; now it must be a time-dependent vector field, $q : \mathbb{R}^+ \times \mathbb{R}^d \mapsto \mathbb{R}^d$. We think of the magnitude $|q|$ as the rate at which material at $x \in \mathbb{R}^d$ moves away from x, and we think of $q/|q|$ as the direction in which the material moves. Often, q is called the *flux* field, although the term may be more trouble than it is worth.

Just as before, the true charm of q is that it permits us to express the amount of material leaving the domain $D \subset \mathbb{R}^d$ as an integral, though this time we need to use the surface integral

$$\int_{\partial D} q(t,x) \cdot n \, dS, \tag{11.7}$$

where n is the outward normal to the surface ∂D of the domain D. We also have that the rate at which material leaves the domain D is given by

$$-\frac{\partial}{\partial t} \int_D u(t,x) \, dV, \tag{11.8}$$

so, as we saw before, the conservation of material tells us that these two quantities must be equal.

Because the surface integral (11.7) can be rewritten as a volume integral by Gauss's formula, and since we can take the time derivative inside the volume integral

(11.8), we therefore have

$$-\int_D \frac{\partial}{\partial t} u(t,x)\,dV = \int_D \nabla \cdot q(t,x)\,dV.$$

The validity of this identity for all D tells us that we have equality of the integrands; that is, we have

$$\frac{\partial u(t,x)}{\partial t} = -\nabla \cdot q(t,x), \tag{11.9}$$

and equation (11.9) is the d-dimensional version of the conservation law.

The generic — plain vanilla — constitutive law in $\mathbb{R}^d$ is given by

$$q(t,x) = -\lambda \nabla u(t,x) \tag{11.10}$$

where λ is a positive constant. This law is the natural analog to the one-dimensional law, and in many physical contexts it can be motivated by arguments that parallel those of $d = 1$. The bottom line is that for systems that satisfy the conservation law (11.9) and the constitutive law (11.10), we have the d-dimensional diffusion equation:

$$\frac{\partial u}{\partial t} = \lambda \Delta u. \tag{11.11}$$

11.2. Solutions of the Diffusion Equation

Here is the basic problem in the simplest case. How do we find $u(t,x)$ when we are given the equation

$$\frac{\partial u}{\partial t} = \lambda \frac{\partial^2 u}{\partial x^2} \tag{11.12}$$

and the initial condition

$$u(0,x) = f(x)\,? \tag{11.13}$$

We will examine three approaches to solving this problem. One might well think that a single solution method would be enough, and, of course, it would be. Nevertheless, each of the methods that we use to solve equation (11.12) has its advantages. Moreover, each of these methods is an acorn from which a great theory has grown.

Fourier Transform Method: Search for a Candidate

When we go hunting for the solution of an equation, we are perfectly willing to make whatever assumptions we need to justify our calculations. After all, we can always retrace our steps after we have succeeded in finding a candidate solution. There is never any real cost to bold calculation, provided that we eventually confirm that our candidate is an honest solution.

One of the great innovations of Fourier's 1822 treatise was the introduction of what we now call the Fourier transform:

$$\phi(\cdot) \mapsto \hat{\phi}(\theta) = \int_{-\infty}^{\infty} e^{i\theta x} \phi(x)\,dx.$$

When we apply this transformation to the space variable of a function of space and time, we have $g(t,\cdot) \mapsto \hat{g}(t,\theta)$, where

$$\hat{g}(t,\theta) = \int_{-\infty}^{\infty} e^{i\theta x} g(t,x)\,dx, \tag{11.14}$$

and, quite remarkably, this process converts the diffusion equation (11.12) from a PDE for $u(t, x)$ to an ODE for $\hat{u}(t, \theta)$.

To see how this works, we first note that if we take the Fourier transform (11.14) of both sides of the diffusion equation (11.12), then by interchanging integration and differentiation we can write the left-hand side as

$$\int_{-\infty}^{\infty} e^{i\theta x} u_t(t, x)\, dx = \hat{u}_t(t, \theta). \tag{11.15}$$

The Fourier transform $\hat{u}_{xx}(t, \theta)$ of $u_{xx}(t, x)$ can also be rewritten by integration by parts to give

$$\hat{u}_{xx}(t, \theta) = \int_{-\infty}^{\infty} e^{i\theta x} u_{xx}(t, x)\, dx = -\theta^2 \int_{-\infty}^{\infty} e^{i\theta x} u(t, x)\, dx = -\theta^2 \hat{u}(t, \theta), \tag{11.16}$$

so Fourier transformation automatically converts the initial-value problem for the PDE (11.12) into a much simpler initial-value problem for an ODE:

$$\hat{u}_t(t, \theta) = -\lambda\theta^2 \hat{u}(t, \theta), \tag{11.17}$$

with the initial condition

$$\hat{u}(0, \theta) = \int_{-\infty}^{\infty} e^{i\theta x} u(0, x)\, dx = \int_{-\infty}^{\infty} e^{i\theta x} f(x)\, dx = \hat{f}(\theta). \tag{11.18}$$

Now, for each fixed θ, this t variable ODE is easy to solve, and, when we take into account that the constant of integration can depend on θ, we see that the general solution of equation (11.17) is given by

$$\hat{u}(t, \theta) = h(\theta) e^{-\lambda\theta^2 t}, \tag{11.19}$$

where $h(\theta)$ is a function that remains to be determined. By the initial condition given in equation (11.18), we have $\hat{u}(0, \theta) = \hat{f}(\theta)$, and, when we let $t = 0$ in equation (11.19), we see that $h(\theta) = \hat{u}(0, \theta)$. By combining these observations, we see that $h(\theta) = \hat{f}(\theta)$, so the unique solution of equation (11.17) is given by

$$\hat{u}(t, \theta) = \hat{f}(\theta) e^{-\lambda\theta^2 t}. \tag{11.20}$$

The determination of $u(t, x)$ now boils down to the inversion of the Fourier transform.

Inverting the Transform

By our great luck, the Fourier transform $\hat{u}(t, \theta)$ is easy to invert. We just need two facts. First, the Fourier transform takes convolution into multiplication. That is, if

$$\gamma(x) = \psi * \phi(x) = \int_{-\infty}^{\infty} \psi(x - y)\phi(y)\, dy = \int_{-\infty}^{\infty} \psi(y)\phi(x - y)\, dy,$$

then

$$\hat{\gamma}(\theta) = \hat{\psi}(\theta)\hat{\phi}(\theta).$$

Second, the Fourier transform of a normal density with mean zero and variance $2\lambda t$ is $e^{-\lambda\theta^2 t}$.

To exploit these facts, we first write $\hat{f}(\theta)$ for the Fourier transform of $f(x)$ and then note that the convolution

$$\int_{-\infty}^{\infty} f(x - y) \frac{1}{2\sqrt{\pi\lambda t}} e^{-y^2/4\lambda t}\, dy = \int_{-\infty}^{\infty} f(y) \frac{1}{2\sqrt{\pi\lambda t}} e^{-(x-y)^2/4\lambda t}\, dy \tag{11.21}$$

has Fourier transform equal to $\hat{f}(\theta)e^{-\lambda\theta^2 t}$. By equation (11.20), this is also the Fourier transform of $u(t,x)$, so, by the uniqueness of the transform, we find a perfectly explicit formula for $u(t,x)$:

$$u(t,x) = \int_{-\infty}^{\infty} f(y)\frac{1}{2\sqrt{\pi\lambda t}}e^{-(x-y)^2/4\lambda t}\,dy. \tag{11.22}$$

Checking the Candidate

In the course of our derivation of equation (11.22), we happily calculated without any concern for the legitimacy of the individual steps. Now that we have a solid candidate for the solution to the initial-value problem, we should confirm that our candidate is a bona fide solution. Later, we will also need to sort out the extent to which the solution that we have found may, or may not, be unique.

The function defined by

$$k(\lambda;t,x) = \frac{1}{2\sqrt{\pi\lambda t}}e^{-x^2/4\lambda t}$$

is nothing more than the density of a Gaussian random variable with mean 0 and variance $2\lambda t$, but because of its prominent role in the world of PDEs $k(\lambda;t,x)$ is also called the *heat kernel*. In the standardized case when $\lambda = 1$, we will also write $k_t(x)$ for $k(1;t,x)$.

The main duty of the heat kernel is to serve as the kernel of the integral operator K_t that is defined by

$$(K_t f)(x) = \int_{-\infty}^{\infty} f(y)k(\lambda;t,x-y)\,dy = \int_{-\infty}^{\infty} f(x-y)k(\lambda;t,y).\,dy$$

The integral operator K_t is called the the *heat operator*, and it gives us a tidy way to write down a solution of the initial-value problem for the diffusion equation.

THEOREM 11.1. *Suppose the continuous function $f:\mathbb{R}\to\mathbb{R}$ satisfies the growth condition*

$$|f(x)| \le A\exp(B|x|^\rho) \tag{11.23}$$

for some constant A,B, and $\rho < 2$. If we define $u(t,x)$ for $t > 0$ by

$$u(t,x) = K_t(f)(x) = \int_{-\infty}^{\infty} f(y)k(\lambda;t,x-y)\,dy, \tag{11.24}$$

then $u \in C^{1,2}((0,\infty)\times\mathbb{R})$ and

$$\frac{\partial u}{\partial t} = \lambda\frac{\partial^2 u}{\partial x^2} \quad \text{for all } t>0 \text{ and } x\in\mathbb{R}. \tag{11.25}$$

Moreover, $u(t,x)$ can be extended continuously to $[0,\infty)\times\mathbb{R}$ in such a way that

$$u(0,x) = f(x) \quad \text{for all } x\in\mathbb{R}. \tag{11.26}$$

PROOF. To begin, we will check the important fact that the heat kernel $k(\lambda;t,x)$ is itself a solution to the diffusion equation. First, by differentiating $\log k(\lambda;t,x)$, we find

$$\frac{k_t(\lambda;t,x)}{k(\lambda;t,x)} = -\frac{1}{2t} + \frac{x^2}{4\lambda t^2} \quad \text{and} \quad \frac{k_x(\lambda;t,x)}{k(\lambda;t,x)} = -\frac{x}{2\lambda t}. \tag{11.27}$$

Next, if we take the x-derivative of $k_x(\lambda; t, x) = -(x/2\lambda t)k(\lambda; t, x)$, we also find

$$k_{xx}(\lambda; t, x) = \left\{ -\frac{1}{2\lambda t} + \frac{x^2}{4\lambda^2 t^2} \right\} k(\lambda; t, x).$$

When we compare the first identity of equation (11.27) with the formula for the second derivative $k_{xx}(\lambda; t, x)$ we find that $k(\lambda; t, x)$ satisfies

$$k_t(\lambda; t, x) = \lambda k_{xx}(\lambda; t, x),$$

just as we hoped.

Now, since the heat kernel satisfies the diffusion equation, we see that the convolution (11.24) is also a solution, provided that we can interchange the integration with the necessary derivatives $\partial/\partial t$ and $\partial^2/\partial x^2$. With patience, one can show that for each fixed $t > 0$ there are constants $C(t)$ and $\delta(t)$ such that the absolute value of the difference quotients

$$\frac{1}{h}\left(k(\lambda; t+h, x-y) - k(\lambda; t, x-y)\right) \tag{11.28}$$

and

$$\frac{1}{h^2}\left(k(\lambda; t, x+2h-y) - 2k(\lambda; t, x+h-y) + k(\lambda; t, x-y)\right) \tag{11.29}$$

are bounded above by $C(t)$ for all $|h| \le \delta(t)$ and all x, y. This observation and the growth condition (11.23) on f are more than we need to invoke the dominated convergence theorem to justify the required differentiation under the integral.

All that remains is to show that u may be extended to a function that satisfies the initial-value and continuity conditions. It turns out that if we take $Z \sim N(0,1)$, then the required extension can be written rather neatly as

$$u(t, x) = E[f(x + \sqrt{2\lambda t}Z)]. \tag{11.30}$$

For $t > 0$, this definition certainly coincides with the integral in equation (11.24), and by direct substitution into equation (11.30) we see that we also have $u(0, x) = f(x)$. All that remains is to show that the u defined by equation (11.30) is continuous on $\mathbb{R}^+ \times \mathbb{R}$, and this is quite easy.

If we take $0 \le t \le (2\lambda)^{-1}$, then the growth condition on f given by equation (11.23) tells us that

$$|f(x + \sqrt{2\lambda t}Z)| \le A \exp\big(B(|x| + |Z|)^p\big),$$

and the last random variable has a finite expectation. The dominated convergence theorem and the continuity of f then tell us that

$$\lim_{t \to 0} E[f(x + \sqrt{2\lambda t}Z)] = E[\lim_{t \to 0} f(x + \sqrt{2\lambda t}Z)] = f(x),$$

so the function $u(t, x)$ defined by the integral (11.24) for $t > 0$ may be extended to a continuous function on $[0, \infty) \times \mathbb{R}$ by taking $u(0, x) = f(x)$. □

Series Solution

The integral representation of the solution of the initial-value problem has many practical and theoretical benefits, but there is a radically different approach to the initial-value problem that also has its merits. Strangely, it calls on infinitely many derivatives rather than one integral, a situation that would seem to spell doom. Nevertheless, there are many instances where it prevails nicely.

If we fix x and expand $u(t,x)$ as a Taylor series about $t = 0$, we are brought to the representation

$$u(t,x) = \sum_{n=0}^{\infty} \frac{t^n}{n!} \frac{\partial^n u}{\partial t^n}(0,x). \tag{11.31}$$

What makes this formal expansion attractive is that the diffusion equation gives us a way to express all of the t derivatives of u in terms of f derivatives. If we assume that $u(t,x)$ is sufficiently smooth, then the equation

$$\frac{\partial u}{\partial t}(t,x) = \lambda \frac{\partial^2 u}{\partial x^2}(t,x)$$

can be iterated to give

$$\frac{\partial^2 u}{\partial t^2}(t,x) = \lambda \frac{\partial}{\partial t}\frac{\partial^2 u}{\partial x^2}(t,x) = \lambda \frac{\partial^2}{\partial x^2}\frac{\partial u}{\partial t}(t,x) = \lambda^2 \frac{\partial^4 u}{\partial x^4}(t,x),$$

or, more generally,

$$\frac{\partial^n u}{\partial t^n}(t,x) = \lambda^n \frac{\partial^{2n} u}{\partial x^{2n}}(t,x). \tag{11.32}$$

Now, if we heroically set $t = 0$ in equation (11.32) and then apply our initial condition $u(0,x) = f(x)$ on the right-hand side, we find

$$\frac{\partial^n u}{\partial t^n}(0,x) = \lambda^n f^{(2n)}(x).$$

Now, at last, we can use this formula for the t derivatives of u in our original Taylor expansion (11.31). The bottom line is that we obtain an almost magical formula for $u(t,x)$:

$$u(t,x) = \sum_{n=0}^{\infty} \frac{t^n \lambda^n}{n!} f^{(2n)}(x). \tag{11.33}$$

This formula can be used to get the solution of the diffusion equation in many interesting cases, often with a minimum of calculation. One of the nicest situations arises when f is a polynomial. In this case, the representation of $u(t,x)$ has only finitely many terms, and only trivial differentiations are needed to calculate u as a simple polynomial in x and t. Other nice examples include $f(x) = \exp(\alpha x)$, where we quickly find

$$u(t,x) = \sum_{n=0}^{\infty} \frac{t^n \lambda^n}{n!} \alpha^{2n} e^{\alpha x} = e^{\lambda t \alpha^2 + \alpha x}, \tag{11.34}$$

and $f(x) = \cos x$, where the fact that the $2n$'th derivative of f is $(-1)^n \cos x$ brings us immediately to

$$u(t,x) = e^{-\lambda t} \cos x.$$

In all of these instances, one can easily check that the function given by the formal series is in fact an honest solution of the initial-value problem.

Similarity Solutions

Two solutions may seem to be more than enough, but there is a third approach to the diffusion equation that helps us see more clearly the *structure* of the solutions. The big idea here is that if an equation is invariant under a group of transformations, then the functions that are invariant under that group are the natural candidates for solutions of the equation. A familiar instance of this principle is the universal habit of looking for a radial function whenever we want to solve an equation that is unchanged by rotation of the coordinate axes.

To begin our hunt for invariance, we first consider the simple linear change of variables $\tau = \alpha t$ and $y = \beta x$, and we introduce a new function $v(\tau, y)$ by requiring that $u(t,x) = v(\tau, y)$. Now, if we apply the chain rule, we find

$$u_t = \alpha v_\tau \quad u_x = \beta v_y, \text{ and } u_{xx} = \beta^2 v_{yy},$$

so the original equation $u_t = \lambda u_{xx}$ implies that $\alpha v_\tau = \beta^2 \lambda v_{yy}$.

Now we have the opportunity to make a useful observation. If we take $\alpha = \beta^2$, then $v(\tau, y)$ is again a solution of the simple diffusion equation $v_\tau = \lambda v_{yy}$. In other words, the diffusion equation is unchanged by the transformation

$$t \mapsto \alpha t \quad \text{and} \quad x \mapsto \alpha^{\frac{1}{2}} x. \tag{11.35}$$

The invariance of the diffusion equation under the one-parameter group of transformations (11.35) suggests that we should look for solutions to the diffusion equation that are also invariant. In other words, we should look for a solution $u(t,x)$ that satisfies $u(t,x) = u(\alpha t, \sqrt{\alpha} x)$ for all $\alpha > 0$. This idea greatly simplifies our search because any $u(t,x)$ with this invariance property can be written as a function of the ratio $x/\sqrt{t}$, say

$$u(t,x) = \psi(x/\sqrt{t}).$$

When we substitute this candidate into the diffusion equation, we find

$$-\psi'(x/\sqrt{t}) \frac{x}{2t^{3/2}} = \lambda \psi''(x/\sqrt{t}) \frac{1}{t},$$

and if we introduce the variable $y = x/\sqrt{t}$, we arrive at the elementary ODE

$$\frac{\psi''(y)}{\psi'(y)} = -\frac{y}{2\lambda}, \text{ or } (\log \psi'(y))' = -\frac{y}{2\lambda}.$$

Now, when we integrate once, we find $\psi'(y) = c \exp(-y^2/4\lambda)$, where c is a constant, so, when we integrate again and replace y by $x/\sqrt{t}$, we see that one solution of the equation $u_t = \lambda u_{xx}$ is given in terms of the Gaussian distribution function as $u(t,x) = \Phi(x/\sqrt{2\lambda t})$. When we let $t > 0$ decrease to zero, we find that this solution corresponds to the initial-value problem with $u(0,x) = f(x)$ equal to zero for negative x, one for positive x, and one-half for $x = 0$. More generally, we see that for any $a < b$ the difference

$$u(x,t) = \Phi\left(\frac{x-a}{\sqrt{2\lambda t}}\right) - \Phi\left(\frac{x-b}{\sqrt{2\lambda t}}\right) \tag{11.36}$$

solves $u_t = \lambda u_{xx}$ with the initial-value condition

$$u(x,0) = f(x) = \begin{cases} 0 & x \notin [a,b] \\ \frac{1}{2} & x = a \text{ or } x = b \\ 1 & x \in (a,b). \end{cases}$$

Finally, by the linearity of the equation $u_t = \lambda u_{xx}$, we also see that equation (11.36) tells us how to write down the solution to the initial-value problem for any step function f, or for any f that we are willing to approximate by step functions.

The simple idea of searching for the symmetries of the equation $u_t = \lambda u_{xx}$ under linear transformations has brought us amazingly far. Now, after only a minimal amount of calculation, we have found a rich class of problems that we can solve without any further computation. Moreover, the solutions given by equation (11.36) are useful intuition builders; one cannot beat a simple Mathematica (or Maple) animation of the formula (11.36) for showing how a diffusion *diffuses.*

11.3. Uniqueness of Solutions

One of the humbling properties of the diffusion equation is that the initial-value problem does not have a unique solution, no matter how nice our initial data f may be. In order to specify a unique solution $u(t,x)$, one must look beyond f; a guarantee of uniqueness can only be found if one imposes conditions on the set of functions that one is willing to admit as possible solutions. We will soon obtain a uniqueness theorem that serves well enough for most applications, but first we look at an example that illustrates the horrors of nonuniqueness.

An Example of Nonuniqueness

If one has a good eye for index shifting, it is easy to guess that we can get a formal solution to $u_t = u_{xx}$ just by taking

$$u(t,x) = \sum_{n=0}^{\infty} \frac{x^{2n}}{(2n)!} \frac{d^n}{dt^n}\phi(t) = \phi(t) + \frac{x^2}{2!}\phi'(t) + \frac{x^4}{4!}\phi''(t) + \cdots . \tag{11.37}$$

To see how this works, we just do our formal differentiations and see what happens to n:

$$\begin{aligned} u_{xx}(t,x) &= \sum_{n=1}^{\infty} \frac{1}{(2n)!} \cdot 2n \cdot (2n-1)\, x^{2n-2} \frac{d^n}{dt^n}\phi(t) \\ &= \sum_{n=0}^{\infty} \frac{x^{2n}}{(2n)!} \frac{d^{n+1}}{dt^{n+1}}\phi(t) = u_t(t,x). \end{aligned}$$

Now, if we were looking for pathological solutions to the diffusion equation, one idea we might try is to hunt for an interesting ϕ. This turns out to be a fine idea if we consider

$$\phi(t) = \begin{cases} e^{-1/t^2} & \text{for } t > 0 \\ 0 & \text{for t} \le 0. \end{cases} \tag{11.38}$$

This famous function is a building block for many counterexamples in analysis. What makes it so handy is that it goes to zero amazingly fast as $t \downarrow 0$. In particular, this rapid convergence permits one to show that $\phi \in C^\infty(\mathbb{R})$ and $\phi^n(0) = 0$ for all n, so $\phi(t)$ provides an example of a nontrivial smooth function that has a Taylor expansion at zero that is identically zero.

This lame Taylor series may be amusing enough to make ϕ worth remembering, but ϕ really comes into its own when it is used as an element in the construction of more complex examples. In particular, when we plug ϕ into equation (11.37), we end up with a function $u(t,x)$ such that

- $u_t = u_{xx}$ for all $(t,x) \in \mathbb{R} \times \mathbb{R}$,
- $u(0,x) = 0$ for all $x \in \mathbb{R}$, and
- $u(t,x) \not\equiv 0$.

These properties of u are easy to check if one is content with formal calculations, and, with honest toil, they can be rigorously justified. Such a justification would take us too far off our trail, so we must be content with our formal arguments.

Nevertheless, we should underscore the bottom line. Because there are nonzero solutions to the initial-value problem with $f(x) \equiv 0$, *any* initial-value problem for the diffusion equation will fail to have a unique solution. We can only talk about uniqueness within subclasses of solutions, like nonnegative solutions, bounded solutions, or solutions that satisfy some more liberal growth condition.

The Idea of a Maximum Principle

Many of the most useful function classes satisfy a *maximum principle* of one sort or another. Two of the most important of these are the class of harmonic functions and the class of analytic functions. Each of these classes has several varieties of maximum principles, but the simplest are those that deal with bounded domains. In this case, the maximum principle for harmonic functions says that any harmonic function h on a bounded domain $D \subset \mathbb{R}^d$ satisfies

$$\max_{x \in \partial D} h(x) = \max_{x \in \bar{D}} h(x), \tag{11.39}$$

where $\bar{D}$ is the closure of the domain and ∂D is the boundary of D. Similarly, for any analytic function f on a bounded domain $D \subset \mathbb{C}$, the maximum *modulus* principle tells us

$$\max_{x \in \partial D} |f(z)| = \max_{x \in \bar{D}} |f(z)|. \tag{11.40}$$

The reason we mention these results is that maximum principles are the key to many uniqueness theorems. For example, if D is a bounded domain in $\mathbb{R}^d$ and if u and v are both solutions of the boundary-value problem

$$\Delta h(x) = 0 \text{ for } x \in D \text{ and } h(x) = b(x) \text{ for } x \in \partial D, \tag{11.41}$$

then the maximum principle for harmonic functions gives us an easy way to prove that $u(x) = v(x)$ for all $x \in \bar{D}$. All we need to do is note that $h(x) = u(x) - v(x)$ is a harmonic function on D that is zero on ∂D, so the maximum principle tells us $h(x) \le 0$ on D, and, since the same argument applies to $-h(x)$, we actually have $h(x) = 0$ on D. In other words, $u(x) = v(x)$ on $\bar{D}$, and — *voilá* — the solution of the boundary value problem (11.41) is unique.

This well-worn pattern suggests that if we want to find a uniqueness theorem for the diffusion equation, then we might first think about an appropriate maximum principle. This suggestion turns out to be a sound one.

Parabolic Maximum Principle

In the diffusion equation, the time variable and the space variable play different roles, and these differences must be taken into account in order to frame an appropriate maximum principle. The simplest natural domain in which we might study the diffusion equation is surely the *space-time slab* defined by

$$D = \{(t, x) : 0 < t < T, \, A < x < B\}$$

for some constants A and B. We will shortly find that in many circumstances a solution of the equation $u_t = u_{xx}$ will take on its maximum value in $\bar{D}$ at a point in the set

$$\partial_0 D = \partial D \setminus \{(t, x) : t = T \text{ and } A < x < B\}.$$

The set $\partial_0 D$ is called the *parabolic boundary* of D, and it would be equal to the usual boundary of D except that the open segment between the points (T, A) and (T, B) has been excluded.

We could now state a maximum principle for the solutions of the diffusion equation that directly parallels the maximum principle for harmonic functions, but, for reasons that will be revealed shortly, we will actually have an easier time proving a "one-sided" version of the natural result. To understand why such an apparently more general result turns out to be easier to prove, one only needs recall why we often prefer to work with submartingales rather than martingales. You can do a lot to a submartingale and still have a submartingale, whereas a martingale is more fragile. A square or maximum will destroy the martingale property, but such operations leave the submartingale property unruffled. Proofs of maximum principles and uniqueness theorems for PDEs often require us to build modifications of our original functions, and we often need these modifications to satisfy our original hypotheses. A one-sided condition such as that used in the next theorem provides just the required flexibility.

THEOREM 11.2 (Parabolic Maximum Principle). *Suppose D is a space-time slab and $u \in C^{1,2}(D) \cap C(\bar{D})$. If u satisfies the inequality*

$$u_t(t,x) \leq u_{xx}(t,x) \text{ for all } (t,x) \in D, \tag{11.42}$$

then the maximum of u on $\bar{D}$ is assumed on the parabolic boundary; that is,

$$\max_{(t,x)\in\bar{D}} u(t,x) = \max_{(t,x)\in\partial_0 D} u(t,x). \tag{11.43}$$

PROOF. If we first fix $\delta > 0$ and let $v(t,x) = u(t,x) - \delta t$, then by the hypothesis on u we have

$$v_t(t,x) < v_{xx}(t,x) \text{ for all } (t,x) \in D. \tag{11.44}$$

Now, if we can prove that

$$\max_{(t,x)\in\bar{D}} v(t,x) = \max_{(t,x)\in\partial_0 D} v(t,x), \tag{11.45}$$

then we would have for all $\delta > 0$ that

$$\begin{aligned} -\delta T + \max_{(t,x)\in\bar{D}} u(t,x) &\leq \max_{(t,x)\in\bar{D}} \{u(t,x) - \delta t\} \\ &= \max_{(t,x)\in\partial_0 D} \{u(t,x) - \delta t\} \leq \max_{(t,x)\in\partial_0 D} u(t,x). \end{aligned}$$

The arbitrariness of δ would give us the inequality (11.43) that we are after.

Now, take any $0 < \epsilon < T$ and consider the slightly smaller slab $D(\epsilon)$ defined by

$$D(\epsilon) = \{(t,x) : 0 < t < T - \epsilon, A < x < B\}.$$

If the maximum value of $v(t,x)$ in $\bar{D}(\epsilon)$ takes place at a point (t_0, x_0) in the interior of $D(\epsilon)$, then we would have $v_{xx}(t_0,x_0) \leq 0$ by maximality of the function $x \mapsto v(t_0, x)$ at x_0. Similarly, we have $v_t(t_0, x_0) = 0$, since the function $t \mapsto v(t, x_0)$ has an extreme point at t_0. These two inequalities then tell us $v_{xx}(t_0,x_0) \leq v_t(t_0,x_0)$, and this inequality contradicts equation (11.44). Therefore, the maximum value of $v(t,x)$ in $\bar{D}(\epsilon)$ takes place on the boundary.

Next, consider the possibility of a maximum at a point $(T - \epsilon, x_0)$ in the open line segment $L = \{(T-\epsilon, x) : x \in (A, B)\}$. Again, by maximality we have the inequality $v_{xx}(T-\epsilon, x_0) \leq 0$, but this time we just have $v_t(T-\epsilon, x_0) \geq 0$, since otherwise we would have some $\epsilon' > \epsilon$ such that $v(T-\epsilon', x_0) > v(T-\epsilon, x_0)$. These two inequalities tell us that $v_{xx}(T-\epsilon, x_0) \le v_t(T-\epsilon, x_0)$. This gives us a contradiction to equation (11.44), and consequently we have

$$\max_{(t,x)\in \bar{D}(\epsilon)} v(t,x) = \max_{(t,x)\in \partial_0 D(\epsilon)} v(t,x). \tag{11.46}$$

Because $u(t,x)$ is continuous on the compact set $\bar{D}$, the functions

$$\epsilon \mapsto \max_{(t,x)\in \bar{D}(\epsilon)} v(t,x) \text{ and } \epsilon \mapsto \max_{(t,x)\in \partial_0 D(\epsilon)} v(t,x)$$

are continuous for all $\epsilon \in [0,T]$, so letting $\epsilon \to 0$ in equation (11.46) gives us the target inequality (11.45). By our earlier reductions, this completes the proof of the theorem. □

A Uniqueness Theorem

We now have the tools we need to state and prove a general uniqueness theorem for the heat equation. A bit later, we will see that this result provides the key to proving a basic uniqueness theorem for the Black–Scholes terminal-value problem.

Theorem 11.3 (Uniqueness Theorem). *Suppose that $u \in C^{1,2}$ on $(0,\infty)\times\mathbb{R}$ and that u is continuous on $[0,\infty)\times\mathbb{R}$. Suppose also that there are constants C and B such that*

$$|u(t,x)| \le Ce^{Bx^2} \text{ for all } (t,x) \in [0,\infty)\times\mathbb{R}. \tag{11.47}$$

If u satisfies the inequality

$$u_t(t,x) \le u_{xx}(t,x) \text{ for all } (t,x) \in (0,\infty)\times\mathbb{R}, \tag{11.48}$$

and if $u(0,x) \le 0$ for all $x \in \mathbb{R}$, then

$$u(t,x) \le 0 \text{ for all } (t,x) \in [0,\infty)\times\mathbb{R}. \tag{11.49}$$

Proof. We first observe that it suffices to show that there is some fixed T such that the hypotheses imply $u(t,x) \le 0$ for all $(t,x) \in [0,T]\times\mathbb{R}$. To see why this is so, we only need to note that if we take $t' = t - T$ the hypotheses of the theorem continue to hold, and we can conclude that $u(t,x) \le 0$ on $[0,T]\times\mathbb{R}$. This argument can be repeated as many times as we like, so eventually we find for any $(t,x) \in [0,T]\times\mathbb{R}$ that $u(t,x) \le 0$.

We therefore choose a small fixed T (any $T < 1/(8B)$ will suffice), and we consider the auxiliary function defined by

$$v(\delta, y; t, x) \equiv v(t,x) \equiv u(x,t) - \delta(2T-t)^{-\frac{1}{2}} \exp[(x-y)^2/4(2T-t)], \tag{11.50}$$

where $\delta > 0$ and $y \in \mathbb{R}$ are arbitrary fixed parameters. One can use the same calculation that we used during our investigation of the heat kernel to show that the difference $v - u$ satisfies the diffusion equation, so by equation (11.48) we also have

$$v_t(t,x) \le v_{xx}(t,x) \text{ for all } (t,x) \in D,$$

where $D = D(y,h)$ is the space-time slab

$$D(y,h) = \{\ (t,x) : 0 < t < T,\ y-h < x < y+h\ \}$$

and $h > 0$ is arbitrary. We can now apply the parabolic maximum principle to v on $D(y,h)$ to find

$$\max_{(t,x)\in\bar{D}} v(t,x) = \max_{(t,x)\in\partial_0 D} v(t,x), \tag{11.51}$$

so our task eventually boils down to understanding the behavior of v on $\partial_0 D$.

On the upper part of the parabolic boundary, we have $x = y+h$ and $0 \le t \le T$, so

$$\begin{aligned} v(\delta, y; t, x) &\le C\exp[B(y+h)^2] - \delta(2T-t)^{-\frac{1}{2}}\exp[h^2/4(2T-t)] \\ &\le C\exp[B(y+h)^2] - \delta T^{-\frac{1}{2}}\exp(h^2/(8T)). \end{aligned}$$

Now, with our choice of T, we have $B < 1/(8T)$, so the last equation tells us that we can choose h_0 such that for all $h \ge h_0$ the right-hand side is negative. The same estimates also show that $v(t,x)$ is negative on the lower part of the parabolic boundary, and the definition of $v(t,x)$ immediately implies that

$$v(0,x) \le u(0,x) \le 0,$$

so $v(t,x)$ is nonpositive on the whole parabolic boundary $D(h,y)$ when $h \ge h_0$. The maximum principle then tells us that

$$v(t,x) \le 0 \text{ for all } (t,x) \in D = D(y,h) \text{ and } h \ge h_0.$$

This actually tells us that

$$u(x,t) - \delta(2T-t)^{-\frac{1}{2}}\exp[(x-y)^2/4(2T-t)] \le 0 \tag{11.52}$$

for all $0 \le t \le T$ and $x \in \mathbb{R}$. Because the region where inequality (11.52) holds does not depend on δ, we can let $\delta \to 0$ in equation (11.52) to obtain

$$u(x,t) \le 0 \text{ for all } 0 \le t \le T \text{ and } x \in \mathbb{R}.$$

By our first remarks, the last inequality suffices to imply the theorem. □

Auxiliary comparison functions such as the v defined by equation (11.50) are widely used throughout the theory of partial differential equations, and the flexibility we gain by using only 'one-sided' conditions is almost always essential for their construction. In our case, the construction of the auxiliary function $v(\delta, y; t, x)$ was not difficult, but in some PDE problems the construction of the right comparison function can be extremely challenging.

11.4. How to Solve the Black–Scholes PDE

The Black–Scholes PDE may look like a long stride from the heat equation, but there are a couple of standard manipulations that quickly span the gap. The first of these is the PDE analog to a familiar trick for solving the ODE $f'(x)+\lambda f(x) = g(x)$. This equation differs from the trivial ODE $f'(x) = g(x)$ by the addition of the linear term $\lambda f(x)$, and a way to deal with this minor complication has been known since the dawn of calculus. If we multiply our original equation by $e^{\lambda x}$, then the resulting equation can be rewritten as $(e^{\lambda x} f(x))' = e^{\lambda x} g(x)$, an equation that is of the trivial variety. For PDEs there is an analogous multiplier technique that shows us how to solve versions of the heat equation that have been complicated by the addition of constant multiples of u and u_x.

A Multiplier Method for PDEs

Suppose that $u(t,x)$ satisfies the equation $u_t = u_{xx}$, and consider the function $v(t,x)$ that lets us reexpress $u(t,x)$ as

$$u(t,x) = e^{\alpha t+\beta x}v(t,x). \tag{11.53}$$

We then have

$$u_t(t,x) = \alpha e^{\alpha t+\beta x}v(t,x) + e^{\alpha t+\beta x}v_t(t,x),$$
$$u_x(t,x) = \beta e^{\alpha t+\beta x}v(t,x) + e^{\alpha t+\beta x}v_x(t,x),$$

and

$$u_{xx}(t,x) = \beta^2 e^{\alpha t+\beta x}v(t,x) + 2\beta e^{\alpha t+\beta x}v_x(t,x) + e^{\alpha t+\beta x}v_{xx}(t,x),$$

so the equation $u_t = u_{xx}$ tells us that we have

$$v_t = v_{xx} + 2\beta v_x + (\beta^2 - \alpha)v \text{ and } v(0,x) = e^{-\beta x}u(0,x). \tag{11.54}$$

This equation is our ticket to the solution of a nice class of initial-value problems, and it brings us one step closer to the solution of the Black–Scholes equation. If we combine this trick with a natural scaling argument, we will be able to reduce any constant-coefficient parabolic PDE to the heat equation.

Solving Constant-Coefficient Diffusion Equations

The most natural extension of the initial-value problem for the basic diffusion equation is the problem with constant coefficients:

$$v_t = av_{xx} + bv_x + cv \text{ and } v(0,x) = \psi(x). \tag{11.55}$$

This problem shows up in a great variety of contexts, and in many of these situations the term bv_x can be interpreted as a *transport term* and the term cv can be interpreted as a *source term*. The intuitive content of these evocative names is explored in Exercise 11.1, which revisits the world of one-dimensional mice to show how source and transport terms may make a natural appearance in models that are built from first principles.

By equation (11.54), we know how to solve the initial-value problem equation for the standardized case $a = 1$, so, to write down a general solution for the initial-value problem (11.55), we only need to make an appropriate transformation. If we introduce the new variable $y = \rho x$ and write our original function $v(t,x)$ as $w(t,y)$, then we have

$$v_x(t,x) = w_y(t,y)\rho \quad \text{and} \quad v_{xx}(t,x) = w_{yy}(t,y)\rho^2.$$

When we substitute into our original equation, we find the new initial-value problem:

$$w_t(t,y) = \rho^2 aw_{yy}(t,y) + \rho bw_y(t,y) + cw(t,y) \quad \text{and} \quad w(0,y) = \psi(y/\rho),$$

and, if we take $\rho^2 = 1/a$, the lead term of equation (11.56) becomes simply w_{yy}. At last, we come to an initial-value problem in the desired form:

$$w_t(t,y) = w_{yy}(t,y) + \frac{b}{\sqrt{a}}w_y(t,y) + cw(t,y) \text{ and } w(0,y) = \psi(y\sqrt{a}). \tag{11.56}$$

Because of equation (11.55), we know how to solve this problem, so only arithmetic stands between us and and the general solution of equation (11.55).

THEOREM 11.4 (A Solution Formula). *If a, b, and c are constants, then the initial-value problem given by*

$$v_t = av_{xx} + bv_x + cv \text{ and } v(0,x) = \psi(x) \tag{11.57}$$

has a solution that can be written as

$$v(t,x) = \exp(-t(b^2 - 4ac)/4a - xb/2a) \int_{-\infty}^{\infty} k_t(x/\sqrt{a} - s) e^{sb/2\sqrt{a}} \psi(s\sqrt{a})\, ds, \tag{11.58}$$

provided that $a > 0$ and that the initial data satisfy the exponential bound

$$|\psi(x)| \le A \exp(B|x|^\rho)$$

for some constants A,B, and $\rho < 2$.

PROOF. We already know that $v(t,x)$ satisfies equation (11.57) if and only if the function defined by $w(t,y) = v(t,x)$ and $y = x/\sqrt{a}$ satisfies the initial-value problem (11.56). Moreover, by equation (11.54) and Theorem 11.1, we know how to write down the solution of the last equation. It is simply

$$w(t,y) = e^{-\alpha t - \beta y} \int_{-\infty}^{\infty} k_t(y-s) e^{\beta s} \psi(s\sqrt{a})\, ds,$$

where α and β are given by $2\beta = b/\sqrt{a}$ and $\beta^2 - \alpha = c$, or

$$\beta = \frac{b}{2\sqrt{a}} \text{ and } \alpha = (b^2 - 4ac)/4a.$$

When we make these substitutions and use the fact that $v(t,x) = w(t, y/\sqrt{a})$, we come directly to the stated solution (11.58). □

UNIQUENESS THEOREM FOR CONSTANT COEFFICIENT DIFFUSIONS

There is a natural temptation to press on immediately to the solution of the Black–Scholes PDE, but we do best to take care of another important task while the calculations that brought us to Theorem 11.4 are still fresh in mind. We will soon need a uniqueness theorem for initial-value problems for the constant-coefficient diffusion equation, and the key link is given by the correspondence between solutions of equation (11.57) and those of the simple diffusion equation for which we have the basic uniqueness result of Theorem 11.3.

THEOREM 11.5 (Uniqueness of Solutions). *Suppose that $v \in C^{1,2}((0,T] \times \mathbb{R})$ and $v \in C([0,T] \times \mathbb{R})$. If v satisfies the constant-coefficient PDE*

$$v_t = av_{xx} + bv_x + cv \text{ for } t \in (0,T] \text{ and } x \in \mathbb{R}, \tag{11.59}$$

the initial condition

$$v(0,x) = 0 \quad \text{for all} \quad x \in \mathbb{R},$$

and the growth condition

$$|v(t,x)| \le Ae^{Bx^2} \tag{11.60}$$

for some constants A and B and all $(t,x) \in [0,T] \times \mathbb{R}$, then

$$v(t,x) \equiv 0 \quad \text{for all } (t,x) \in [0,T] \times R.$$

PROOF. The theorem is almost longer to state than to prove. By the multiplier method calculations and the change of variables that gave us Theorem 11.4, we know there are constants α_0, β_0, γ_0 such that the function

$$u(t,x) = e^{\alpha_0 x + \beta_0 t} v(t, \gamma_0 x) \tag{11.61}$$

satisfies the basic diffusion equation $u_t = u_{xx}$. The representation also tells us that u satisfies the initial condition that $u(0,x) = 0$ for all $x \in \mathbb{R}$.

Now, by the growth constraint on v and boundedness of the time interval $[0,T]$ we also see that there exist constants A_0 and B_0 such that u satisfies the growth condition

$$|u(t,x)| \leq A_0 e^{B_0 x^2}$$

All of the conditions of Theorem 11.5 are met for u, and we can conclude that $u(t,x) = 0$ for all $(t,x) \in [0,T] \times R$. By the correspondence (11.61) between u and v, we also have the same conclusion for v. □

EULER'S EQUIDIMENSIONAL EQUATION

We could solve the Black–Scholes equation right now, but the solution will seem a lot simpler if we first take advantage of an old observation of Euler about the so-called *equidimensional* equation:

$$ax^2 f''(x) + bxf'(x) + cf(x) = 0. \tag{11.62}$$

If we introduce the new variable $y = \alpha x$ and write $f(x) = g(y)$, then the usual chain rule calculations give us

$$f'(x) = g'(y)\alpha \text{ and } f''(x) = g''(y)\alpha^2$$

and, since $x = y/\alpha$, we see that

$$xf'(x) = (y/\alpha) \cdot (g'(y)\alpha) = yg'(y) \quad \text{and} \quad x^2 f''(x) = (y/\alpha)^2 \cdot (g''(y)\alpha^2) = y^2 g''(y).$$

These equations tell us that the change of variables $y = \alpha x$ leaves Euler's equidimensional equation (11.62) unchanged; that is, changes of scale do not change the solutions of equation (11.62) — just as changes of location $y = x + c$ do not change the class of solutions of an ODE with constant coefficients.

These observations suggest that we may be able to convert equation (11.62) to an ODE with constant coefficients if we make the change of variables $y = \log x$ and introduce the new function $g(y) = f(x)$. In this case, the chain rule calculations give us

$$f'(x) = g'(y)y_x = g'(y)(1/x) \text{ and } f''(x) = g''(y)(1/x^2) - g'(y)(1/x^2),$$

so Euler's equation (11.62) becomes

$$ag''(y) + (b-a)g'(y) + cg(y) = 0.$$

The bottom line is that a logarithmic change of variables gives us a surefire method for reducing an equidimensional equation to an ODE with constant coefficients.

Looking Back: Organizing Calculations

There is nothing particularly difficult in the process of transforming one PDE to another, but there is an element of complexity that sometimes slows us down. There is no universal remedy for this molasses effect, but the calculations do seem to go more quickly if one follows a well-defined plan.

If we know that $u(t,x)$ satisfies an equation, we are guaranteed that we can make good use of that equation in the derivation of an equation for a new function $v(\tau, y)$ if we write the old u as a function of the new v and write the new τ and y as functions of the old t and x. This order of things puts everything in the direct line of fire of the chain rule; the partial derivatives u_t, u_x, u_{xx} are easy to compute, and, at the end, the original equation stands ready for immediate use.

To be sure, one can make correct and useful transformations of equations without following this rule, or any other. Still, with a little practice, many people find that calculations that are guided by this suggestion tend to flow more quickly (and with more certainty) than calculations that are organized only by the will of the moment.

First Simplification of the Black–Scholes PDE

Now that we have seen that the diffusion equation is not as narrow as it seemed at first, we will go a step further to show that we can obtain the Black–Scholes formula by means of the diffusion equation. First, we recall that in the last chapter we saw that under a reasonable model for stock and bond prices, the arbitrage price for a European call option was given as the solution to the Black–Scholes PDE

$$f_t(t,x) = -\frac{1}{2}\sigma^2 x^2 f_{xx}(t,x) - rxf_x(t,x) + rf(t,x), \tag{11.63}$$

together with its terminal boundary condition

$$f(T,x) = (x-K)_+ \quad \text{for all } x \in \mathbb{R}.$$

Because we face an equation with a terminal condition instead of an initial condition, our first thought is to make a change of the time variable to reverse time, say by defining a new variable

$$\tau = T - t$$

so that $\tau = 0$ corresponds to $t = T$. We could now rewrite the Black–Scholes PDE as a function of τ and x, but we can save a line or two if we first transform the x variable.

Final Transformation of the Black–Scholes PDE

Since the right-hand side of equation (11.63) is equidimensional, our experience with Euler's equation screams out to us that we should introduce the new variable

$$y = \log x.$$

When we write $f(t,x)$ as $g(\tau, y)$, we find that

$$f_t = g_\tau \tau_t = -g_\tau,\ f_x = g_y y_x = g_y(1/x), \text{ and } f_{xx} = g_{yy}(1/x^2) - g_y(1/x^2),$$

so equation (11.63) gives us a new initial-value problem for g:

$$g_\tau = \frac{1}{2}\sigma^2 g_{yy} + \left(r - \frac{1}{2}\sigma^2\right) g_y - rg \text{ and } g(0,y) = (e^y - K)_+. \tag{11.64}$$

This problem is precisely of form (11.59), so after we identify the coefficients

$$a = \frac{1}{2}\sigma^2,\ b = r - \frac{1}{2}\sigma^2, \text{ and } c = -r, \tag{11.65}$$

we see that $f(t,x) = g(\tau, y)$ is given by the product of the exponential factor

$$\exp(-\tau(b^2 - 4ac)/4a - yb/2a) \tag{11.66}$$

and the corresponding integral term of (11.58):

$$I = \int_{-\infty}^{\infty} k_\tau(y/\sqrt{a} - s)e^{sb/2\sqrt{a}}(e^{s\sqrt{a}} - K)_+\, ds. \tag{11.67}$$

Two Integrals and the Black–Scholes Formula

To compute this integral, we first make the change of variables $u = y/\sqrt{a} - s$ and restrict our integration to the domain D where the integrand is nonzero,

$$D = \{\, u\colon y - u\sqrt{a} \geq \log K \,\} = \{\, u\colon u \leq (y - \log K)/\sqrt{a} \,\}, \tag{11.68}$$

so

$$\begin{aligned} I = \exp(y + yb/2a) &\int_D k_\tau(u)\exp(-u(b/2\sqrt{a} + \sqrt{a}))\, du \\ - K\exp(yb/2a) &\int_D k_\tau(u)\exp\left(-ub/2\sqrt{a}\right)\, du. \end{aligned} \tag{11.69}$$

Now, by the familiar completion of the square in the exponent, we can compute the general integral

$$\int_{-\infty}^{\alpha} k_\tau(s)e^{-\beta s}\, ds = e^{\tau\beta^2}\Phi\left(\frac{\alpha}{\sqrt{2\tau}} + \beta\sqrt{2\tau}\right), \tag{11.70}$$

and we can also check the accuracy of equation (11.70) with a quick differentiation by α. Finally, since both integrals of equation (11.69) are of the same type as equation (11.70), we see that nothing remains but a bit of arithmetic.

If we recall that $x = e^y$ and patiently collect our several terms, we find that the formula for $f(t,x)$ emerges from (11.58) like a butterfly from a chrysalis. The solution $f(t,x)$ of the Black–Scholes terminal-value problem is indeed given by the Black–Scholes formula:

$$x\Phi\left(\frac{\log(x/K) + (r + \frac{1}{2}\sigma^2)\tau}{\sigma\sqrt{\tau}}\right) - Ke^{-r\tau}\Phi\left(\frac{\log(x/K) + (r - \frac{1}{2}\sigma^2)\tau}{\sigma\sqrt{\tau}}\right).$$

11.5. Uniqueness and the Black–Scholes PDE

Now that we have found a solution for the Black–Scholes terminal-value problem for the call option, we would like to be reassured that there are no other solutions. From our experience with the diffusion equation, we know that this is a false hope unless we restrict the class of acceptable solutions in some way. Fortunately, there is a natural growth rate restriction that is liberal enough to include any economically reasonable solutions.

THEOREM 11.6 (Uniqueness in the Black–Scholes Problem). *Suppose that*

$$f \in C^{1,2}[(0,T] \times (0,\infty)] \cap C([0,T] \times (0,\infty))$$

satisfies the Black–Scholes PDE

$$f_t(t,x) = -\frac{1}{2}\sigma^2 x^2 f_{xx}(t,x) - rxf_x(t,x) + rf(t,x) \qquad (t,x) \in (0,T] \times (0,\infty), \tag{11.71}$$

and the terminal condition

$$f(T,x) = 0 \qquad x > 0.$$

If there exist constants A and B such that

$$|f(t,x)| \leq A \exp(B \log^2(1 + |x|)) \qquad (t,x) \in [0,T] \times [0,\infty), \tag{11.72}$$

then f is identically zero on $[0,T] \times [0,\infty)$.

PROOF. This result only requires that we examine the transformations that we found in the course of solving the Black–Scholes terminal-value problem for the call option. We found that any solution $f(t,x)$ of the PDE (11.71) can be written as $g(\tau, y)$, where $\tau = T - t$, $y = \log x$, and where $g(\tau, y)$ satisfies a diffusion equation with constant coefficients. By the assumption that $f(T,x) = 0$ for all $x > 0$, we have $g(0,y)$ for all $y \in \mathbb{R}$. Also, since $f(t,x)$ satisfies the growth condition (11.72), we see that there are constants A_0 and B_0 such that $g(\tau, y)$ satisfies

$$|g(\tau, y)| \leq A_0 e^{B_0 y^2} \quad \text{for all } \tau \in [0,T] \text{ and all } y \in \mathbb{R}.$$

By Theorem 11.5 on the uniqueness solutions of diffusion equations with constant coefficients, we see that $g(\tau, y) = 0$ for all $\tau \in [0,T]$ and $y \in \mathbb{R}$. As a consequence, we see that $f(t,x)$ is also identically zero, and the proof of the theorem is complete. □

Here we should remark on the generosity of the growth bound. The function $A \exp(B \log^2(1+|x|))$ grows more slowly than an exponential function, but it grows faster than any power of x.

A GLANCE AROUND ... AND AHEAD

The theory of partial differential equations is intimately intertwined with the theory and application of stochastic calculus. The interplay between the two fields is one of the most active areas in mathematics, and new developments are a regular occurrence. This chapter has provided only a taste of the service that classical PDE provides to the users of stochastic calculus. Nevertheless, we have obtained the solution of the Black–Scholes PDE and have found that under economically realistic conditions the solution of the Black–Scholes terminal-value problem is unique. This is a worthy achievement.

In later chapters, we will find two alternative solutions to the option pricing problem, each of which has a direct probabilistic interpretation and each of which tells us something about the solution of PDEs. The first of these is the exceptionally powerful martingale method that guides us to the valuation of a large class of contingent claims. The second solution is based on the remarkable Feynman–Kac formula. This formula is a true piece of rocket science that repays part of probability's debt to the theory of PDE by providing probabilistic representations for a rich class of important equations, including the notorious Schrödinger equation of quantum mechanics.

11.6. Exercises

The first exercise adds to our intuition by showing how basic modeling principles can be used to derive diffusion equations that contain source terms and transport terms. The next exercise then shows how martingales and integral representations can also be used to prove maximum principles. The integral representation technique is a useful complement to the more elementary local extremum arguments that gave us the parabolic maximum principle.

The two remaining exercises deal with the Black–Scholes equation. The first of these completes the last details of our derivation of the Black–Scholes formula. The last problem is more exciting. It suggests how the superposition principle can be used to design an algorithm that prices a large class of options.

EXERCISE 11.1 (Intuition for Source Terms and Transport Terms).

(a) Suppose that at time t additional one-dimensional mice are born at the location x at a rate that is a multiple μ of the population density $u(t,x)$. How does this change the conservation law (11.4)? How does this change in the conservation law change the basic diffusion equation (11.6)? Does your argument go through even if μ is a function of x? What if it is a function of x and t?

(b) Further, suppose that there is a steady cold wind blowing along the x-axis from $-\infty$ to $+\infty$, and suppose that this gives the mice a steady tendency to drift in the positive direction in addition to their continuing desire to find less crowded living conditions. How does this change the constitutive law (11.5)? How does this change the diffusion equation found in part (a).

EXERCISE 11.2 (Integral Representations and Maximum Principles).

Suppose A is a connected open set in $\mathbb{R}^2$ that is bounded, and suppose that $h : A \mapsto \mathbb{R}$ is a harmonic function on A. It is known that for any disk D of radius r and center (x,y) such that $D \subset A$, we have

$$h(x,y) = \frac{1}{2\pi r}\int_{\partial D} h(u,v)\,d\gamma \text{ and } h(x,y) = \frac{1}{\pi r^2}\int_D h(u,v)\,dudv; \tag{11.73}$$

that is, h both equals its average over circle ∂D around (x,y) and equals its average over the disk D.

(a) Use either of these representations to show that if h has a maximum in the interior of A then h is a constant. This is a version of the maximum principle for harmonic functions.

(b) Prove the first formula of (11.73) by exploiting the fact that if (B_t^1, B_t^2) is two-dimensional Brownian motion starting at (x,y), and τ is the hitting time of ∂D, then $M_t = h(B^1_{t\wedge\tau}, B^2_{t\wedge\tau})$ is a bounded martingale.

(c) Prove the second formula of (11.73) from the first by using polar coordinates and the fact that the first formula holds for any disk contained in A.

EXERCISE 11.3 (A Rite of Passage).

Complete the last step in the computation of the solution of the Black–Scholes PDE. In particular, compute the integrals in equation (11.69) by exploiting equation (11.70), then collecting the exponential terms (including those from (11.66), and finally make the necessary parameter identifications. Try to organize the calculation in a way that makes the final check as clear as possible. The task is a little dreary but it is a rite of passage, and one can take refreshment by noticing how informative elements such as $(r + \frac{1}{2}\sigma^2)\tau)/\sigma\sqrt{\tau}$ pop out of the assembled bits.

EXERCISE 11.4 (Superposition, Wavelets, and a General Algorithm).

(a) Assuming the Black–Scholes stock and bond model, find the time t arbitrage price of a European style option with payout $h(S_T)$ at terminal time T, where $K > 0$, $D > 0$, and

$$h(x) = \begin{cases} 0 & x \le K \\ x - K & x \in [K, K + D] \\ D & x \ge K + D. \end{cases}$$

Hint: You should be able to write down the answer almost immediately after you have graphed the payout function h.

(b) Suppose instead that the function h is given by

$$h(x) = \begin{cases} 0 & x \le K - D \\ H(x - K + D)/D & x \in [K - D, K] \\ H & x \in [K, L] \\ H - H(x - L)/D & x \in [L, L + D] \\ 0 & x \ge L + D, \end{cases}$$

where $0 \le K \le L$, $H \ge 0$, and $D > 0$. Find a formula for the option price now.

(c) As a special case of part (b), can you now price the European style option with payout $h(S_T)$ where $h(x)$ equals the Mother wavelet $\Delta_0(x)$ of Chapter 3? Suggest how one may use this connection to create an algorithm for pricing a large class of options.

CHAPTER 12

Representation Theorems

One of the everyday miracles of mathematics is that some objects have two (or more) representations that somehow manage to reveal different features of the object. Our constant use of Taylor expansions and Fourier series tends to blunt their surprise, but in many cases the effectiveness of the right representation can be magical. Consider the representation for $\min(s,t)$ that we found from Parceval's theorem for the wavelet basis and which formed the cornerstone of our construction of Brownian motion; or, on a more modest level, consider the representation of a convex function as the upper envelope of its tangents and the automatic proof it provides for Jensen's inequality.

This chapter focuses on three important representation theorems of stochastic calculus. The first of these is Dudley's theorem, which tells us that any suitably measurable function can be represented as the stochastic integral of an $\mathcal{L}^2_{\text{LOC}}$ function. The second is the famous martingale representation theorem, which plays an important role in the arbitrage pricing of derivative securities by telling us when we can represent a continuous martingale as the stochastic integral of an $\mathcal{H}^2$ function. Finally, we establish an important characterization of the stochastic integral as a time change of Brownian motion. This magnificent theorem leads to the wholesale translation of results for Brownian motion into corresponding results for stochastic integrals and Brownian martingales.

12.1. Stochastic Integral Representation Theorem

Sometimes, a simple observation opens the door to a world of new possibilities, and the next proposition is a perfect illustration of this phenomenon. It is also a very optimistic result in that it shows that even the tiniest bit of a Brownian motion's future is good enough to recapture any part of its past.

PROPOSITION 12.1 (Capturing the Past). *For any $0 \le a < b$ and any finite random variable $X \in \mathcal{F}_a$, there is a stopping time τ with $a \le \tau < b$, such that*

$$X = \int_a^\tau \frac{1}{b-t}\, dB_t. \tag{12.1}$$

PROOF. In order to understand the right-hand side of equation (12.1) a little more generally, we first consider the Gaussian process Y_t defined by taking

$$Y_t = \begin{cases} 0 & \text{for } 0 \le t \le a \\ \int_a^t (b-u)^{-1}\, dB_u & \text{for } a \le t < b. \end{cases}$$

Clearly, we have $E(Y_t) = 0$ for all $0 \le t < b$, and by our earlier work we know $E(Y_s Y_t) = \min\{h(s), h(t)\}$, where the function $h(\cdot)$ is defined by taking $h(t) = 0$

for $0 \le t \le a$ and by taking

$$h(t) = \int_a^t (b-s)^{-2}\, ds \text{ for } a \le t < b.$$

The Gaussian process $\{Y_t\}$ therefore has the same covariance function as the time-changed Brownian motion $\{B_{h(t)}\}$, so in fact the two processes are equivalent. Now, since we have assumed that $X \in \mathcal{F}_a$, we see that the random variable defined by

$$\tau = \min\{t \ge a \colon Y_t = X\} \tag{12.2}$$

serves as the required stopping time provided that we can show $P(\tau < b) = 1$.

Fortunately, the required bound on τ is quite easy. First, we just need to note that $h(t) \to \infty$ as $t \to b$, so the familiar properties of Brownian motion tell us that with probability one we have

$$\limsup_{t\to b} B_{h(t)} = \infty \text{ and } \liminf_{t\to b} B_{h(t)} = -\infty.$$

Now, since the process $\{Y_t\}$ is equivalent to the process $\{B_{h(t)}\}$, we also have

$$\limsup_{t\to b} Y_t = \infty \quad \text{and} \quad \liminf_{t\to b} Y_t = -\infty \tag{12.3}$$

with probability one, and the equations of (12.3) are more than we need to confirm $P(\tau < b) = 1$. □

The last proposition provides us a simple example of a representation theorem. It tells us that any random variable in a certain measurability class can be written as a stochastic integral. The principal task before us is to generalize the last lemma to random variables that satisfy a more modest measurability assumption.

A Tail Bound

Our first step in this process will be to exploit the equivalence of $\{Y_t\}$ and $\{B_{h(t)}\}$ to get quantitative information on the tail probability $P(h(\tau) \ge t)$. To begin, we take any α and note that

$$P(h(\tau) \ge t) \le P(h(\tau) \ge t,\ |X| < |\alpha|) + P(|X| \ge |\alpha|), \tag{12.4}$$

so, if we introduce the parametric family of hitting times

$$\nu_\alpha = \min\{t : Y_t = \alpha,\ t \ge a\},$$

then by (12.3) we have $P(\nu_\alpha < b) = 1$. We also have the trivial bounds

$$1(\tau \ge s, |X| < \alpha) \le 1(\nu_\alpha \ge s) + 1(\nu_{-\alpha} \ge s) \quad \text{for all } s \ge 0,$$

so by inequality (12.4) and the monotonicity of $h(\cdot)$ we find

$$P(h(\tau) \ge t) \le 2P(h(\nu_\alpha) \ge t) + P(|X| \ge |\alpha|). \tag{12.5}$$

The equivalence of the processes $\{Y_t\}$ and $\{B_{h(t)}\}$ then tells us that for any α we have

$$\begin{aligned} h(\nu_\alpha) &= \min\{h(t) : Y_t = \alpha\} \\ &\overset{\mathrm{d}}{=} \min\{h(t) : B_{h(t)} = \alpha\} \\ &= \min\{t : B_t = \alpha\ \}, \end{aligned}$$

where in the last step we exploited the monotonicity of h. Because we already know that the density of the first time that Brownian motion reaches a level α is given by $|\alpha|\phi(|\alpha|/\sqrt{s})/s^{3/2}$ for $s \geq 0$, we find from inequality (12.5) that for all $t \geq 0$ we have

$$P(h(\tau) \geq t) \leq 2P(h(\nu_\alpha) \geq t) + P(|X| \geq |\alpha|) \tag{12.6}$$
$$\leq 2\int_t^\infty |\alpha|\phi(|\alpha|/\sqrt{s})/s^{3/2}\,ds + P(|X| \geq \alpha)$$
$$\leq 2|\alpha|/\sqrt{t} + P(|X| \geq |\alpha|),$$

where in the last step we used the trivial bound $\phi(u) \leq \frac{1}{2}$ to help simplify the integral.

We will confirm shortly that inequality (12.6) provides us with just enough control over the size of $h(\tau)$ to build a far-reaching extension of Proposition 12.1, but first we need to derive a simple result that tells us how any $\mathcal{F}_T$-measurable random variable can be written as a nice sum of random variables such as those given by the integral of Proposition 12.1.

Representation by Informative Increments

Our intuition about the filtration of Brownian motion suggests that if $X \in \mathcal{F}_T$, and if we successively come to know $\mathcal{F}_{t_n}$ where $t_n \uparrow T$, then we should also come to know X more and more precisely as $n \to \infty$. There are several ways to make this intuition precise, and one of the most pleasing ways is to represent X as a sum of terms that are $\mathcal{F}_{t_n}$-measurable. We can even exercise considerable control over the size of the summands.

Proposition 12.2 (Representation by Informative Increments). *For any random variable $X \in \mathcal{F}_T$ and any pair of summable sequences of decreasing real numbers $\{a_n\}$ and $\{b_n\}$, there exists an increasing real sequence $\{t_n\}$ such that $t_n \to T$ and a sequence of random variables Δ_n such that*

$$\Delta_n \in \mathcal{F}_{t_n} \text{ for all } n, \quad X = \sum_{n=0}^{\infty} \Delta_n \text{ with probability one,} \tag{12.7}$$

and

$$P(|\Delta_n| \geq a_n) \leq b_n \text{ for all } n \geq 1. \tag{12.8}$$

Proof. To begin, we let $Y = \Phi(X)$ where $\Phi(\cdot)$ is the standard Gaussian distribution function. If we then set $s_n = T - 1/n$ and $\mathcal{G}_n = \sigma\{B_t \colon t \leq s_n\}$, then by the boundedness of Y and the martingale convergence theorem, the conditional expectations $Y_n = E(Y|\mathcal{G}_n)$ converge with probability one to $E(Y|\mathcal{F}_T) = \Phi(X)$. By the strict monotonicity of $\Phi(\cdot)$, we therefore have

$$X_n \stackrel{\text{def}}{=} \Phi^{-1}(Y_n) \to X \text{ with probability one,}$$

so we can choose a sequence of integers $\{k_n : 0 \leq n < \infty\}$ such that

$$P(\,|X - X_{k_n}| \geq a_{n+1}/2\,) \leq b_{n+1}/2 \text{ for all } n \geq 0.$$

We then take $\Delta_0 = X_{k_0}$, and for $n \geq 1$ we take $\Delta_n = X_{k_n} - X_{k_{n-1}}$, so, for $t_n = T - 1/k_n$, we have $\Delta_n \in \mathcal{F}_{t_n}$ for all $n \geq 0$. This establishes the first requirement of (12.7).

Now, we also have

$$P(|\Delta_n| \ge a_n) \le P(|X - X_{k_n}| \ge a_n/2) + P(|X - X_{k_{n-1}}| \ge a_n/2)$$
$$\le b_n/2 + b_n/2 = b_n,$$

and the last inequality is just what we needed for (12.8). Finally, we note that by the construction of the Δ_n, we have

$$\sum_{n=0}^{N} \Delta_n = X_{k_N},$$

so the representation (12.7) follows from the fact that X_n converges to X with probability one. □

A Construction to Perplex Zeno

Zeno, the old Greek philosopher, found it paradoxical that a fast runner can catch a slow one. After all, he reasoned, the fast runner would have to cover one half of the distance between himself and the slow runner an *infinite* number of times. Zeno strained to understand how this could be possible. Nowadays we are comfortable with the execution of an infinite number of actions in a finite amount of time, and Zeno's concerns are hard to take seriously. Nevertheless, in some constructions — such as the one that follows — we may find room to sympathize with Zeno's forgotten torment.

Theorem 12.1 (Dudley's Representation Theorem). *If $X \in \mathcal{F}_T$, then there is a $\phi \in \mathcal{L}^2_{\text{LOC}}[0,T]$ such that*

$$X = \int_0^T \phi(\omega, s)\, dB_s. \tag{12.9}$$

Proof. To begin the construction, we take any two decreasing sequences $\{a_n\}$ and $\{b_n\}$ that satisfy

$$\sum_{n=0}^{\infty} n a_n < \infty \text{ and } \sum_{n=0}^{\infty} b_n < \infty. \tag{12.10}$$

We then let $\{t_n\}$ be the sequence whose existence is guaranteed by the Proposition 12.2, so that we have random variables $\Delta_n \in \mathcal{F}_{t_n}$ such that

$$P(|\Delta_n| \ge a_n) \le b_n \tag{12.11}$$

and

$$X = \lim_{N\to\infty} \sum_{n=0}^{N} \Delta_n \quad \text{with probability one.} \tag{12.12}$$

Now, for each Δ_n in the convergent sum for X, we can apply Proposition 12.1 with $a = t_n$ and $b = t_{n+1}$ to find a stopping time $\tau_n \in [t_n, t_{n+1})$ such that

$$\Delta_n = \int_{t_n}^{\tau_n} (t_{n+1} - s)^{-1}\, dB_s = \int_0^T \phi_n(\omega, s)\, dB_s$$

where we have defined $\phi_n(\omega, s)$ by

$$\phi_n(\omega, s) = \begin{cases} (t_{n+1} - s)^{-1} & \text{for } s \in [t_n, \tau_n) \\ 0 & \text{otherwise.} \end{cases}$$

Our candidate for the integrand ϕ in the representation (12.9) for X is then given by

$$\phi(\omega, s) = \sum_{n=0}^{\infty} \phi_n(\omega, s), \tag{12.13}$$

and the first order of business is to check that ϕ is actually an element of $\mathcal{L}^2_{\text{LOC}}[0, T]$.

Because the functions $\phi_n(\omega, s)$ have disjoint supports, we can simply square $\phi(\omega, s)$ and take the integral over $[0, T]$ to find

$$\int_0^T \phi^2(\omega, s)\, ds = \sum_{n=0}^{\infty} \int_0^T \phi_n^2(\omega, s)\, ds. \tag{12.14}$$

Now, by the tail bound (12.6) applied to τ_n (where $a = t_n$ and $b = t_{n+1}$), we have for any $\lambda > 0$ and any $\alpha > 0$ that

$$P\left(\int_0^T \phi_n^2(\omega, s)\, ds \geq \lambda\right) \leq \frac{2\alpha}{\sqrt{\lambda}} + P(|\Delta_n| \geq \alpha),$$

so when we take $\alpha = a_n$ and apply the bound (12.11), we find

$$P\left(\int_0^T \phi_n^2(\omega, s)\, ds \geq \lambda\right) \leq \frac{2a_n}{\sqrt{\lambda}} + b_n.$$

Finally, if we take $\lambda = n^{-2}$, the summability of the sequences $\{na_n\}$ and $\{b_n\}$ teams up with the Borel–Cantelli lemma to imply that the sum in (12.14) is finite with probability one. In other words, we have $\phi \in \mathcal{L}^2_{\text{LOC}}$, just as we hoped.

Our only remaining task is to check the representation (12.9), and this step is almost immediate from our construction. First, we note that we have

$$\sum_{n=0}^{N} \Delta_n = \sum_{n=0}^{N} \int_{t_n}^{\tau_n} (t_{n+1} - s)^{-1}\, dB_s = \int_0^{t_{N+1}} \phi(\omega, s)\, dB_s. \tag{12.15}$$

Now, since $\phi \in \mathcal{L}^2_{\text{LOC}}$, we also know that the process

$$t \mapsto \int_0^t \phi(\omega, s)\, dB_s \tag{12.16}$$

is continuous, so when we let $N \to \infty$ we have by (12.12) and (12.16) that each side of the identity (12.15) converges to the respective side of equation (12.9). □

Now that we are done with the construction, we can see what might have upset Zeno. Each Δ_n is $\mathcal{F}_{t_n}$-measurable, yet the integral summand that represents Δ_n is never $\mathcal{F}_{t_n}$-measurable; the representing integral must always peek ahead at least bit past t_n. Nevertheless, in finite time —but infinite n — the two sums end up in exactly the same place. For us, the proof is clear and complete, yet more philosophical spirits may still sympathize with Zeno and sense a hint of paradox.

NONUNIQUENESS OF THE REPRESENTATION

Because of the many flexible choices that were made during the construction of the integrand $\phi(\omega, s) \in \mathcal{L}^2_{\text{LOC}}$ that represents X in equation (12.9), one would not expect the representation to be unique. In fact, the nonuniqueness is quite extreme in the sense that we can construct a bewildering variety of $\phi(\omega, s) \in \mathcal{L}^2_{\text{LOC}}$ such that

$$\int_0^T \phi(\omega, s)\, dB_s = 0. \tag{12.17}$$

To give one example that leverages our earlier construction, just consider any real number $0 < a < T$ and set

$$\psi(\omega, s) = \begin{cases} 1 & \text{for } 0 \leq s < a \\ (T-s)^{-1} & \text{for } a \leq s < \tau \\ 0 & \text{for } \tau \leq s \leq T, \end{cases}$$

where

$$\tau = \min\left\{ t \geq a : \int_a^t \frac{1}{T-u}\, dB_u = -B_a \right\}.$$

From our discussion of "capturing the past" and Proposition 12.1, we know that $P(\tau < T) = 1$ and that $\psi(\omega, s)$ is indeed an element of $\mathcal{L}^2_{\text{LOC}}$. Also, just by integration and the definition of τ, we see that

$$\int_0^a \psi(\omega, s)\, dB_s = B_a = -\int_a^T \psi(\omega, s)\, dB_s,$$

so we also find that (12.17) holds. In other words, ψ is a nontrivial integrand whose Itô integral is identically zero. Now, given any ϕ with a stochastic integral that represents X as in (12.9), we see that $\phi + \psi$ also provides a representation of X, so the representation of equation (12.9) is certainly not unique.

12.2. The Martingale Representation Theorem

Dudley's representation theorem is inspirational, but it is not all that easy to use because integrands in $\mathcal{L}^2_{\text{LOC}}$ can be terribly wild. Fortunately, there is another representation theorem that provides us with nice $\mathcal{H}^2$ integrands.

THEOREM 12.2 ($\mathcal{H}^2$ Representation Theorem). *Suppose that X is a random variable that is $\mathcal{F}_T$-measurable where $\{\mathcal{F}_t\}$ is the standard Brownian filtration. If X has mean zero and $E(X^2) < \infty$, then there is a $\phi(\omega, s) \in \mathcal{H}^2[0, T]$ such that*

$$X = \int_0^T \phi(\omega, s)\, dB_s. \tag{12.18}$$

Moreover, the representation in equation (12.18) is unique in the sense that if $\psi(\omega, s)$ is another element of $\mathcal{H}^2[0, T]$ that satisfies equation (12.18), then we have $\psi(\omega, s) = \phi(\omega, s)$ for all $(\omega, s) \in \Omega \times [0, T]$ except a set of $dP \times dt$ measure zero.

The phrasing of Theorem 12.2 has been chosen to show the parallel to Dudley's theorem, but most applications of Theorem 12.2 actually call on a corollary that makes the connection to martingales more explicit. We also state this result as a theorem because of its importance.

THEOREM 12.3 (Martingale Representation Theorem). *Suppose that X_t is an $\{\mathcal{F}_t\}$ martingale, where $\{\mathcal{F}_t\}$ is the standard Brownian filtration. If there is a T such that $E(X_T^2) < \infty$ and if $X_0 = 0$, then there is a $\phi(\omega, s) \in \mathcal{H}^2[0,T]$ such that*

$$X_t = \int_0^t \phi(\omega, s)\, dB_s \quad \text{for all } 0 \le t \le T. \tag{12.19}$$

Moreover, the representation in equation (12.19) is unique up to a set of $dP \times dt$ measure zero.

This is one of the most useful theorems in the theory of stochastic integration, and, more than any other result, it shapes our thinking about martingales with respect to the Brownian filtration. To check that Theorem 12.3 does indeed follow from Theorem 12.2, we just note that $X = X_T$ satisfies the hypotheses of Theorem 12.2, so there is a $\phi(\omega, s) \in \mathcal{H}^2[0,T]$ such that

$$X_T = \int_0^T \phi(\omega, s)\, dB_s. \tag{12.20}$$

Now, given any $t \in [0,T]$, we can take the conditional expectation $E(\,\cdot\,|\mathcal{F}_t)$ of both sides of equation (12.20) and use the fact that $\phi \in \mathcal{H}^2$ to get equation (12.19). The uniqueness of ϕ in equation (12.19) is an immediate consequence of the uniqueness in Theorem 12.2, so we see that Theorem 12.3 is indeed an immediate corollary of Theorem 12.2.

We now take up the proof of Theorem 12.2. This turns out to be a challenging project, although one should not think of the proof as difficult. It simply calls on techniques that are a bit different from those we have been using. These new techniques are important for many purposes, so we will engage them rather fully, even when they take us off the customary trail.

IN $\mathcal{H}^2$ UNIQUENESS IS EASY

First of all we should note that the uniqueness in Theorem 12.2 is almost trivial. We only need to note that for any two representing functions ψ and ϕ in $\mathcal{H}^2[0,T]$ the difference $\psi - \phi$ is again an element of $\mathcal{H}^2[0,T]$. Since the Itô integral of the difference $\psi - \phi$ represents the zero random variable, Itô's isometry tells us that

$$0 = \int_0^T E[\,(\psi(\omega, s) - \phi(\omega, s))^2\,]\, ds.$$

Finally, the integrand $(\psi(\omega, s) - \phi(\omega, s))^2$ is nonnegative, so it must equal zero almost surely with respect to the measure $dP \times dt$. This is precisely the uniqueness that we sought.

While we are here, we should note the sharp contrast between the nonuniqueness in Dudley's $\mathcal{L}^2_{\text{LOC}}$ representation theorem and the easy proof of uniqueness of the $\mathcal{H}^2$ representation. This difference is due to the fact that Itô's isometry is available to us in $\mathcal{H}^2$, but it does not extend to $\mathcal{L}^2_{\text{LOC}}$. This point is developed a bit further in Exercise 12.1.

A SPECIAL CASE IS ALREADY KNOWN

Whenever we are trying to prove a new result, we always benefit from the recognition of any special cases where the conjectured result is already known, and we happen to be in one of those lucky situations right now. During our investigation of geometric Brownian motion, we found that the process

$$X_t = x_0 \exp[(\mu - \sigma^2/2)t + \sigma B_t]$$

solves the equation

$$dX_t = \mu X_t\, dt + \sigma X_t\, dB_t \qquad X_0 = x_0.$$

If we take $\mu = 0$, the previous equation reduces to $dX_t = \sigma X_t dB_t$, so if we also set $x_0 = 1$ we see that

$$X_t = \exp(-\sigma^2 t/2 + \sigma B_t)$$

solves the integral equation

$$X_t = 1 + \int_0^t \sigma X_u\, dB_u.$$

Now, if we substitute the expression $X_t = \exp(-\sigma^2 t/2 + \sigma B_t)$ back into the integral equation, then a tiny bit of arithmetic leads us to an intriguing self-referential representation for $\exp(\sigma B_t)$:

$$\exp(\sigma B_t) = \exp(\sigma^2 t/2) + \int_0^t \sigma \exp\left(-\sigma^2(u-t)/2 + \sigma B_u\right) dB_u. \tag{12.21}$$

This is a very useful and informative formula. In particular, it tells us that every random variable Y that can be written as

$$Y = \sum_{k=1}^n \alpha_k \exp(\sigma_k B_{t_k})$$

for some α_k, σ_k, and t_k can be written as an $\mathcal{H}^2$ stochastic integral. The class of random variables defined by such sums is large, but it is not large enough to be dense in $L^2(\Omega, \mathcal{F}_T, P)$. We need to find some way to build an even richer class of representable functions.

A Modest Generalization

There are many times when we can extract new information from an old formula by replacing one of the real parameters of the formula by a complex parameter. In this case, one of the ideas we can try is to substitute $i\theta$ for σ in equation (12.21). This substitution suggests that $\exp(i\theta B_t)$ will have the representation

$$\exp(i\theta B_t) = \exp(-\theta^2 t/2) + \int_0^t i\theta \exp\left(-\theta^2(t-u)/2 + i\theta B_u\right) dB_u. \tag{12.22}$$

This derivation by complex substitution is not 100 % honest; in essence, it calls for an imaginary standard deviation. Nevertheless, once one writes down an identity such as (12.22), Itô's formula and a little patience are all one needs to verify its truth, although here — if we are picky— we would need to make separate checks of the identities given by the real and imaginary parts. Nevertheless, twice an easy calculation is still an easy calculation.

The higher art is that of guessing the existence of such an identity — and of guessing how the new identity may help with our project. In our case, there are two features of equation (12.22) that offer new hope. First, the random variable $\exp(i\theta B_t)$ is bounded, and this can offer a technical benefit over the unbounded variable $\exp(\sigma B_t)$. More critically, $\exp(i\theta B_t)$ creates a connection to Fourier transforms, and such connections are often desirable.

In order to squeeze the benefits out of equation (12.22), we will first generalize it a bit. In particular, we note that if the standard Brownian motion $\{B_t\}$ is replaced by the shifted (but still standard) Brownian motion $\{ B_{t+s} - B_s : 0 \le t < \infty \}$, we find that the random variable $\exp[i\theta(B_{t+s} - B_s)]$ has the representation

$$\exp(-\theta^2 t/2) + \int_s^{s+t} i\theta \exp\left(-\theta^2(s+t-u)/2 + i\theta(B_u - B_s)\right) dB_u. \tag{12.23}$$

What makes this representation so attractive is that it expresses the rather general random variable $\exp[i\theta(B_{s+t} - B_s)]$ as a stochastic integral *with support in the interval* $[s, s+t]$. When we think back to the product rule for stochastic integrals, this small observation opens the possibility of generating a huge class of random variables that have nice integral representations.

A Product Rule Application

We have seen before that the rule for computing the SDE of a product of two standard processes is particularly simple when the cross-variation is zero. The idea of the next proposition is to exploit this fact by isolating an important case where the product of random variables with integral representations will have an integral representation.

PROPOSITION 12.3. *Suppose that X and Y are bounded random variables with the representations*

$$X = x_0 + \int_0^T \phi(\omega,t)\,dB_t \text{ and } Y = y_0 + \int_0^T \psi(\omega,t)\,dB_t,$$

where ϕ and ψ are elements of $\mathcal{H}^2$. If

$$\int_0^T \phi(\omega,t)\psi(\omega,t)\,dt = 0, \tag{12.24}$$

then the product XY has the representation

$$XY = x_0 y_0 + \int_0^T X_s\psi(\omega,s) + Y_s\phi(\omega,s)\,dB_s,$$

where the processes $\{X_t\}$ and $\{Y_t\}$ are the bounded martingales defined on $0 \le t \le T$ by the conditional expectations

$$X_t = E(X \mid \mathcal{F}_t) \text{ and } Y_t = E(Y \mid \mathcal{F}_t).$$

PROOF. The processes X_t and Y_t satisfy

$$dX_t = \phi(\omega,t)\,dB_t \quad \text{and} \quad dY_t = \psi(\omega,t)\,dB_t,$$

so, by the product rule for standard processes, we have

$$d(X_tY_t) = X_t\,dY_t + Y_t\,dX_t + \phi(\omega,t)\psi(\omega,t)\,dt.$$

When we integrate over $[0,T]$ and use the hypothesis (12.24), we immediately find the desired representation. □

Now, when we combine the last proposition with our observation in equation (12.23) about the support of the stochastic integral representation of the complex exponential $\exp(i\theta(B_{t+s} - B_s))$, we obtain a very important corollary.

COROLLARY 12.1. *For any $0 \le t_0 < t_1 < \cdots < t_n = T$, the random variable*

$$Z = \prod_{j=1}^{n-1} \exp\left(i\theta_j (B_{t_j} - B_{t_{j-1}})\right) \tag{12.25}$$

has a representation of the form

$$Z = E(Z) + \int_0^T \phi(\omega, t)\, dB_t \quad \text{where } \phi \in \mathcal{H}^2[0,T].$$

At this point, an experienced analyst might say that the rest of the proof of Theorem 12.2 is "just a matter of technique." What the analyst has in mind is the fact that the set of random variables of the form (12.25) is "clearly" so rich that any random variable in $L^2(\Omega, \mathcal{F}_T, P)$ can be approximated by a linear combination of such Z's. This approximation and our ability to represent the Z's as stochastic integrals with $\mathcal{H}^2$ integrands should then give us the ability to represent any element of $L^2(\Omega, \mathcal{F}_T, P)$ as such an integral.

Naturally, the experienced analyst is right. The plan is a sound one, and eventually it can be considered as routine. Nevertheless, there is a good bit to be learned from the details that arise, and, in order to give a complete argument, we will need to make more than one addition to our toolkit.

THE APPROXIMATION ARGUMENT

Our first step will be to show that the set of random variables with an $\mathcal{H}^2$ representation forms a closed set in L^2. We can even be explicit about the representation that one finds for a limit point.

PROPOSITION 12.4. *Suppose that X_n, $n = 1, 2, \ldots$, is a sequence of random variables that may be represented in the form*

$$X_n = E(X_n) + \int_0^T \phi_n(\omega, t)\, dB_t \quad \text{with } \phi_n \in \mathcal{H}^2. \tag{12.26}$$

If $X_n \to X$ in $L^2(dP)$, then there is a $\phi \in \mathcal{H}^2$ such that $\phi_n \to \phi$ in $L^2(dP \times dt)$ and

$$X = E(X) + \int_0^T \phi(\omega, t)\, dB_t. \tag{12.27}$$

PROOF. Since $\{X_n\}$ converges in $L^2(dP)$, we also have $E(X_n) \to E(X)$. We then note that $X_n - E(X_n)$ is a Cauchy sequence in $L^2(dP)$, so the Itô isometry tells us that ϕ_n is a Cauchy sequence in $L^2(dP \times dt)$. By the completeness of $\mathcal{H}^2$, there is a $\phi \in \mathcal{H}$ such that $\phi_n \to \phi$ in $L^2(dP \times dt)$. By the fact that $E(X_n) \to E(X)$ and a second application of Itô's isometry, we see that we can take the $L^2(dP)$ limits on both sides of equation (12.26) to get equation (12.27). □

THE DENSITY ARGUMENT

The proof of the martingale representation theorem is almost complete. Because of Corollary 12.1 (on representation) and Proposition 12.4 (on approximation), the proof of Theorem 12.2 will be finished once we have established the following density lemma.

LEMMA 12.1. *If $\mathcal{S}$ is the linear span of the set of random variables of the form*

$$\prod_{j=1}^{n} \exp[iu_j(B_{t_j} - B_{t_{j-1}})]$$

over all $n \geq 1$, $0 = t_0 < t_1 < \cdots < t_n = T$, and $u_j \in \mathbb{R}$, then $\mathcal{S}$ is dense in the space of square integrable complex valued random variables

$$L^2(\Omega, \mathcal{F}_T, P) = \{X : X \in \mathcal{F}_T \text{ and } E(|X|^2) < \infty\}.$$

This result is more in the vein of analysis than probability, and its proof will force us to bore down into sediment that we have not touched before. We could proceed directly with that proof, but to do so might strain our patience, even though the proof is interesting and uses important methods. To keep our spirits up, we will return to the main trail and defer the proof of Lemma 12.1 until Section 12.6 on bedrock approximation techniques.

LOOKING BACK: "INDUCTION" AND ANALYSIS

This proof of the martingale representation theorem follows an important pattern that one finds in many parts of mathematics. The pattern is a kind of "induction," though naturally it is not the same as formal mathematical induction, where one proves a proposition, say $P(1)$, and then argues the general case by showing that proposition $P(n)$ implies proposition $P(n+1)$. Here, the process is more flexible, but the parallel with formal induction is still reasonably close.

One still begins with a special case where the conjectured theorem is true — a step that is analogous to proving $P(1)$. One then works with that special case to build up a larger set of cases where the conjecture is true — a step that is analogous to showing that $P(n)$ implies $P(n+1)$. Finally, after constructing a large reservoir of cases where the conjecture holds, one shows that an approximation argument suffices to imply the full conjecture. This last step is *loosely* analogous to the step of invoking the principle of mathematical induction.

12.3. Continuity of Conditional Expectations

Before proceeding to the discussion of the chapter's third representation theorem, we should take the opportunity to show how the martingale representation theorem can be used to solve an interesting technical problem. If $\{\mathcal{F}_t : 0 \leq t \leq T\}$ is the usual Brownian filtration and $X \in \mathcal{F}_T$ has a finite expectation, anyone would certainly expect the process defined by

$$X_t \stackrel{\text{def}}{=} E(X|\mathcal{F}_t) \quad \text{for } t \in [0, T]$$

to be continuous. True enough, X_t does indeed have a continuous version, but an honest barehanded proof is not so easy. To make things worse, there are even nice filtrations where the analogous conditional expectations do not have a continuous version.

In order to prove the continuity of the process $\{X_t\}$, one is forced to find some way to exploit the special properties of the standard Brownian filtration. The martingale representation theorem provides us with one way to make the critical link. Although this device may seem a bit extreme (and it is a bit sneaky), the continuity of $\{X_t\}$ ends up being a consequence of the continuity of the Itô integral.

PROPOSITION 12.5. *If* $\{ \mathcal{F}_t : 0 \le t \le T\}$ *is the standard Brownian filtration and* X *is an* $\mathcal{F}_T$*-measurable random variable with* $E(|X|) < \infty$*, then there is a continuous process* X_t *such that for each* $t \in [0,T]$ *we have*

$$X_t = E(X|\mathcal{F}_t) \text{ with probability one;}$$

that is, the process $E(X|\mathcal{F}_t)$ *has a continuous version.*

PROOF. Since $E(|X|1(|X| \ge N)) \to 0$ as $N \to \infty$, we can choose an increasing sequence $N_1 \le N_2 \le \cdots$ with $N_k \to \infty$ such that $E(|X|1(|X| \ge N_k)) \le 2^{-2k}$, so, in particular, the truncated variables defined by $X(0) = X1(|X| < N_1)$ and $X(k) = X1(N_k \le |X| < N_{k+1})$ for $k \ge 1$ also satisfy $||X(k)||_1 \le 2^{-2k}$ for $k \ge 1$.

Each of the variables $X(k)$ is bounded, so for each $k \ge 0$ we have a representation for $X(k)$ as

$$X(k) = \int_0^T \phi_k(\omega, s)\, dB_s \text{ with } \phi_k \in \mathcal{H}^2, \tag{12.28}$$

and if we take the conditional expectation, we have

$$E(X(k)|\mathcal{F}_t) = \int_0^t \phi_k(\omega, s)\, dB_s. \tag{12.29}$$

Our candidate for the continuous version of $E(X|\mathcal{F}_t)$ is then given by

$$X_t \overset{\text{def}}{=} \sum_{k=0}^{\infty} \int_0^t \phi_k(\omega, s)\, dB_s. \tag{12.30}$$

To check that X_t is well defined and continuous, we first note that by Doob's L^1 maximal inequality and equation (12.28) we have for all $k \ge 1$ that

$$P\left(\sup_{0 \le t \le T} \left| \int_0^t \phi_k(\omega, s)\, dB_s \right| \ge 2^{-k} \right) \le 2^{-k}.$$

The Borel–Cantelli lemma then tells us that with probability one the sum in equation (12.30) is uniformly convergent, so $\{X_t : 0 \le t \le T\}$ is indeed a continuous process.

To check that $X_t = E(X|\mathcal{F}_t)$ with probability one, we first note that by summing equation (12.29) we have

$$E[\, X1(|X| < N_{m+1}) \,|\, \mathcal{F}_t] = \sum_{k=0}^{m} \int_0^t \phi_k(\omega, s)\, dB_s. \tag{12.31}$$

Now, $X1(|X| < N_m)$ converges in L^1 to X, so the DCT tells us the left-hand side of equation (12.31) converges in L^1 to $E(X|\mathcal{F}_t)$, and, consequently, there is a subsequence m_n such that

$$E[X1(|X| < N_{m_n}) \,|\, \mathcal{F}_t] \to E(X|\mathcal{F}_t) \quad \text{a.s.}$$

We already know the right-hand side of equation (12.31) converges with probability one to X_t, so when we take limits in equation (12.31) along the subsequence m_n, we see that $X_t = E(X|\mathcal{F}_t)$ with probability one, just as we needed to complete the proof. □

12.4. Representation via Time Change

At last, we come to the most widely used of the chapter's three representation theorems. After learning how a stochastic integral may be written as a Brownian motion run by an altered clock, one's view of stochastic integrals is changed forever.

THEOREM 12.4 (Stochastic Integrals as a Time Change of Brownian Motion). *Suppose that* $\phi \in \mathcal{L}^2_{\text{LOC}}[0,T]$ *for all* $0 \le T < \infty$, *and suppose that we set*

$$X_t = \int_0^t \phi(\omega, s)\, dB_s, \quad t \ge 0. \tag{12.32}$$

If we have

$$\int_0^\infty \phi^2(\omega, s)\, ds = \infty \text{ with probability one} \tag{12.33}$$

and if we set

$$\tau_t = \min\{ u : \int_0^u \phi^2(\omega, s)\, ds \ge t \}, \tag{12.34}$$

then the process $\{X_{\tau_t} : 0 \le t < \infty\}$ *is a standard Brownian motion.*

PROOF. As in our proof of the martingale representation theorem, we begin by considering an exponential process, although this time we will use the more general process

$$\begin{aligned} Z_t &= \exp\left(i\theta \int_0^t \phi(\omega, s)\, dB_s + \frac{1}{2}\theta^2 \int_0^t \phi^2(\omega, s)\, ds \right) \\ &= \exp\left(i\theta \int_0^t \phi(\omega, s)\, dB_s \right) \cdot \exp\left(\frac{1}{2}\theta^2 \int_0^t \phi^2(\omega, s)\, ds \right). \end{aligned}$$

By analogy with our earlier work, we can guess that Z_t is a local martingale, and we can check this guess simply by calculating dZ_t. A pleasant way to organize this computation is to let X'_t and Y'_t denote the last two exponential factors and use Itô's formula to calculate the differentials

$$dX'_t = i\theta\phi(\omega, t)X'_t\, dB_t - \frac{1}{2}\theta^2\phi^2(\omega, t)X'_t\, dt \quad\text{and}\quad dY'_t = \frac{1}{2}\theta^2\phi^2(\omega, t)Y'_t\, dt,$$

so that now we can apply the product rule to obtain

$$\begin{aligned} dZ_t &= d(X'_tY'_t) = X'_t\, dY'_t + Y'_t\, dX'_t \\ &= \frac{1}{2}\theta^2\phi^2(\omega, t)X'_tY'_t\, dt + i\theta\phi(\omega, t)X'_tY'_t\, dB_t - \frac{1}{2}\theta^2\phi^2(\omega, t)X'_tY'_t\, dt \\ &= i\theta\phi(\omega, t)Z_t\, dB_t. \end{aligned}$$

This formula tells us that Z_t is a dB_t integral, and consequently Z_t is a local martingale.

One can easily check (say as in Exercise 12.4) that the process $\{Z_t\}$ is constant on all of the intervals $[\tau_{u-}(\omega), \tau_u(\omega)]$, so by Proposition 7.13 we know that $\{Z_{\tau_t}, \mathcal{F}_{\tau_t}\}$ is a continuous local martingale. Since we also have the deterministic bound

$$|Z_{\tau_t}| \le \exp(\theta^2 T/2) \text{ for all } t \in [0, T],$$

we then see that the local martingale $\{Z_{\tau_t}, \mathcal{F}_{\tau_t}\}$ is in fact an honest martingale on $[0, T]$, and, since T is arbitrary, it is even a martingale on all of $[0, \infty)$.

The importance of the last observation is that when we expand the martingale identity $E(Z_{\tau_t} \mid \mathcal{F}_{\tau_s}) = Z_{\tau_s}$, we find

$$E(\exp(i\theta(X_{\tau_t} - X_{\tau_s})) \mid \mathcal{F}_{\tau_s}) = \exp(-(t-s)\theta^2/2), \tag{12.35}$$

and the identity (12.35) tells us everything we want to know. In particular, Lemma 12.2 below will show that equation (12.35) implies that $X_{\tau_t} - X_{\tau_s}$ is independent of $\mathcal{F}_{\tau_s}$ and that $X_{\tau_t} - X_{\tau_s}$ has a Gaussian distribution with mean zero and variance $t - s$. Because we already know that $X_0 = 0$ and since the process $\{X_{\tau_t}\}$ is continuous (again by the argument of Exercise 12.4), nothing more is needed to conclude that $\{X_{\tau_t}\}$ is indeed a standard Brownian motion. □

LEMMA 12.2. *Suppose $\mathcal{F}'$ and $\mathcal{F}''$ are σ-fields such that $\mathcal{F}' \subset \mathcal{F}''$. If X is an $\mathcal{F}''$-measurable random variable such that*

$$E(e^{i\theta X} \mid \mathcal{F}') = e^{-\theta^2\sigma^2/2} \quad \text{for all } \theta \in \mathbb{R}, \tag{12.36}$$

then X is independent of the σ-field $\mathcal{F}'$ and X has the Gaussian distribution with mean zero and variance σ^2.

PROOF. In longhand equation (12.36) tells us that

$$E(e^{i\theta X} 1_A) = P(A)\, e^{-\theta^2\sigma^2/2} \tag{12.37}$$

for all θ and all $A \in \mathcal{F}'$. We therefore find that by taking $A = \Omega$ the condition (12.37) implies that X is Gaussian with mean zero and variance σ^2. Now, if A is any element of $\mathcal{F}'$ with $P(A) > 0$, we can also define a new probability measure (and expectation) by setting

$$P_A(B) = P(B \cap A)/P(A) \text{ and } E_A(Z) = E[Z1_A]/P(A)$$

for any $B \in \mathcal{F}''$ and any integrable Z that is $\mathcal{F}''$-measurable.

By equation (12.37), we then see that $E_A(e^{i\theta X}) = e^{-\theta^2\sigma^2/2}$, or, in other words, the P_A-characteristic function of X is again that of a Gaussian with mean zero and variance σ^2. By uniqueness of the characteristic function, we then have

$$P_A(X \le x) = \Phi(x/\sigma),$$

which is shorthand for

$$P(\{X \le x\} \cap A) = \Phi(x/\sigma)P(A) = P(X \le x)P(A).$$

This equation tells us that X is independent of any $A \in \mathcal{F}'$ with $P(A) > 0$, and, since X is trivially independent of any A with $P(A) = 0$, we see that X is independent of $\mathcal{F}'$. □

12.5. Lévy's Characterization of Brownian Motion

One immediate corollary of the time change representation theorem for stochastic integrals is the characterization it provides for continuous martingales with quadratic variation t. Broadly speaking, such a martingale must coincide with Brownian motion, and, curiously enough, this exceptionally useful fact was first obtained by P. Lévy without the use of stochastic integrals.

THEOREM 12.5 (Lévy's Characterization of Brownian Motion). *Suppose that the process $\{M_t : 0 \le t < \infty\}$ is a continuous martingale with respect to the standard Brownian filtration. If $E(M_t^2) < \infty$ and $\langle M \rangle_t = t$ for all $0 \le t < \infty$, then $\{M_t - M_0 : 0 \le t < \infty\}$ is a standard Brownian motion.*

To see how this result may be derived from Theorem 12.4, we first note that the L^2 martingale representation theorem tells us that there is a process β such that $\beta \in \mathcal{H}^2[0,T]$ for all $0 \le T < \infty$ such that

$$X_t = M_t - M_0 = \int_0^t \beta(\omega, s)\, dB_s \quad \text{for all } t \ge 0.$$

The assumption that $\langle M \rangle_t = t$ then tells us that

$$t = \int_0^t \beta^2(\omega, s)\, ds, \tag{12.38}$$

and the formula for the time change τ_t given by Theorem 12.4 tells us that $\tau_t = t$. The main assertion of Theorem 12.4 is that X_{τ_t} is a Brownian motion, but since $\tau_t = t$ this simply says $M_t - M_0$ is a Brownian motion.

A Second View of the Characterization

In fact, one can give a proof of the Brownian motion characterization theorem that is a bit more concrete than the appeal to Theorem 12.4 would suggest, and Theorem 12.5 is important enough to deserve an independent proof. All one needs to show is the identity

$$E[e^{i\theta(M_t - M_s)} \mid \mathcal{F}_s] = e^{-\theta^2(t-s)/2}, \tag{12.39}$$

and this turns out to be a nice exercise with Itô's formula.

The martingale representation theorem applied to M_t tells us there is a $\beta(\omega, t)$ in $\mathcal{H}^2[0,T]$ such that $dM_t = \beta(\omega, t)\, dB_t$, and our assumption on the quadratic variation of M_t further tells us $dM_t \cdot dM_t = dt$, so Itô's formula implies

$$\begin{aligned} de^{i\theta(M_t - M_s)} &= i\theta e^{i\theta(M_t - M_s)} dM_t - \frac{1}{2}\theta^2 e^{i\theta(M_t - M_s)} dM_t \cdot dM_t \\ &= i\theta e^{i\theta(M_t - M_s)} \beta(t, \omega)\, dB_t - \frac{1}{2}\theta^2 e^{i\theta(M_t - M_s)}\, dt. \end{aligned}$$

When we integrate this equation, we find

$$e^{i\theta(M_t - M_s)} - 1 = i\theta \int_s^t e^{i\theta(M_u - M_s)} \beta(u, \omega)\, dB_u - \frac{1}{2}\theta^2 \int_s^t e^{i\theta(M_u - M_s)}\, du,$$

and if we then take expectations, we find

$$E[e^{i\theta(M_t - M_s)} | \mathcal{F}_s] - 1 = -\frac{1}{2}\theta^2 \int_s^t E[e^{i\theta(M_u - M_s)} | \mathcal{F}_s]\, du. \tag{12.40}$$

The last equation tells us that

$$\phi(t) = E[e^{i\theta(M_t - M_s)} | \mathcal{F}_s]$$

satisfies

$$\phi(t) = 1 - \frac{1}{2}\theta^2 \int_s^t \phi(u)\, du, \tag{12.41}$$

or

$$\phi'(t) = -\frac{1}{2}\theta^2 \phi(t) \quad \text{and} \quad \phi(s) = 1.$$

The unique solution of this equation is just $\exp(-\theta^2(t-s)/2)$, so the proof of the theorem is complete.

Wholesale Translations from Brownian Motion to Martingales

Now, we must attend to a second important consequence of the time change representation provided by Theorem 12.4. In a nutshell, this theorem tells us that we can convert almost any fact that we know about Brownian motion into a corresponding fact for a continuous martingale. The recipe is simple, but sometimes the consequences can be striking.

To give just one example, suppose we first recall the law of the iterated logarithm for Brownian motion, which tells us that

$$\limsup_{t\to\infty} \frac{B_t}{\sqrt{2t\log\log t}} = 1$$

with probability one. Now, by Lévy's theorem, if $\{M_t, \mathcal{F}_t\}$ is a continuous martingale with the representation

$$M_t = \int_0^t \phi(\omega, s)\, dB_s$$

and with

$$P\left(\int_0^\infty \phi^2(\omega, s)\, ds = \infty\right) = 1,$$

then we also have

$$\limsup_{t\to\infty} \frac{M_{\tau_t}}{\sqrt{2t\log\log t}} = 1$$

with probability one when τ_t is given by

$$\tau_t = \inf\left\{u : \int_0^u \phi^2(\omega, s)\, ds \geq t\right\}.$$

Obviously, this story has more variations than could be imagined even by a modern Sheherazade — a thousand and one nights would never suffice to tell all the tales that spring from the marriage of each continuous martingale and each theorem for Brownian motion.

12.6. Bedrock Approximation Techniques

We have often used the simple idea that a continuous function on $[0, T]$ is uniquely determined by its values on the set of rational points in $[0, T]$. We will now develop a characterization of a probability measure that is analogous in that it shows that a probability measure may be uniquely determined by its values on a small subset of the measurable sets. The most immediate reason for making this excursion into the foundations of probability theory is that the tools that provide this characterization also serve to complete the density argument we needed to finish our proof the martingale representation theorem. As a bonus, these new tools will also prove useful in our development of Girsanov theory.

Useful Set Systems

Every probability measure has a σ-field as its natural home, but at times our work is made easier if we take advantage of some set systems that are simpler than σ-fields. Two of the handiest of these helpers are the π-systems and the λ-systems.

DEFINITION 12.1 (π-System). *A collection of sets $\mathcal{C}$ is called a* π-system *provided that it is closed under pairwise intersections; that is, $\mathcal{C}$ is a π-system provided that*

$$A \in \mathcal{C} \text{ and } B \in \mathcal{C} \Rightarrow A \cap B \in \mathcal{C}.$$

DEFINITION 12.2 (λ-System). *A collection $\mathcal{C}$ of subsets of a sample space Ω is called a* λ-system *provided that it has three simple properties:*

(i) $\Omega \in \mathcal{C}$,

(ii) $A, B \in \mathcal{C}$ *and* $B \subset A \;\Rightarrow\; A \setminus B \in \mathcal{C}$,

and

(iii) $A_n \in \mathcal{C},\; A_n \subset A_{n+1}, \;\Rightarrow \cup_{n=1}^{\infty} A_n \in \mathcal{C}$.

From these properties, one expects a close connection between λ-systems and σ-fields. In fact, a λ-system always satisfies the first two requirements of a σ-field, but it can fail to have the third. To check this, we first note that if $\mathcal{C}$ is a λ-system, then $\emptyset \in \mathcal{C}$ since $\emptyset = \Omega \setminus \Omega$, and if $A \in \mathcal{C}$, then $A^c \in \mathcal{C}$ since $A^c = \Omega \setminus A \in \mathcal{C}$. Finally, to see that a λ-system does not need to be closed under unions —even finite unions — we only need to consider the case of $\Omega = \{1, 2, 3, 4\}$ and

$$\mathcal{C} = \{A \subset \Omega : |A| = 0, |A| = 2, \text{ or } |A| = 4\}.$$

One can easily check that $\mathcal{C}$ is indeed a λ-system, but since $\{a, b\} \in \mathcal{C}$, $\{a, c\} \in \mathcal{C}$ and $\{a, b\} \cup \{a, c\} = \{a, b, c\} \notin \mathcal{C}$, we see that $\mathcal{C}$ fails to be a σ-field.

In this example, we may also note that $\mathcal{C}$ also fails to be closed under intersections, say by considering $\{a, b\} \cap \{a, c\} = \{a\} \notin \mathcal{C}$. Remarkably, if we plug this one small gap, then a λ-system must actually be a σ-field. This useful fact deserves to be set out as a proposition.

PROPOSITION 12.6. *If $\mathcal{C}$ is a λ-system and a π-system, then $\mathcal{C}$ is a σ-field.*

PROOF. We already know that any λ-system satisfies the first two properties of a σ-field, so we only need to check that $\mathcal{C}$ satisfies the third property as well. If we take any B_1 and B_2 in $\mathcal{C}$, then their complements are in $\mathcal{C}$, and since $\mathcal{C}$ is a π-system, the intersection of these is again in $\mathcal{C}$. Now, if we take complements again, we still stay in $\mathcal{C}$, so we find

$$B_1 \cup B_2 = (B_1^c \cap B_2^c)^c \in \mathcal{C}.$$

Closure under unions of pairs implies closure under any finite union, so if $B_1, B_2, \ldots$ is any countable sequence of elements of $\mathcal{C}$, we see $A_n = B_1 \cup B_2 \cup \cdots B_n$ is in $\mathcal{C}$, and since we have $A_n \subset A_{n+1}$, the third property of a λ-system tells us $\cup_n A_n \in \mathcal{C}$. Finally, since we have $\cup_n A_n = \cup_n B_n$, we see $\cup_n B_n \in \mathcal{C}$, so $\mathcal{C}$ does indeed have all of the properties that are required of a σ-field. □

There is one further property of λ-systems that will be useful for us, and it also provides a parallel with σ-fields. We will need to know that for any collection $\mathcal{C}$ of subsets of a sample space Ω there is a *smallest λ-system containing $\mathcal{C}$*. This λ-system is denoted by $\lambda(\mathcal{C})$, and one can easily check that such a smallest λ-system actually exists. We only need to note that $\sigma(\mathcal{C})$ is a λ-system that contains $\mathcal{C}$ and the intersection of the set of all λ-systems that contain $\mathcal{C}$ is also a λ-system. Trivially, this intersection is the unique minimal λ-system that contains $\mathcal{C}$, so $\sigma(\mathcal{C})$ is well defined.

The π-λ Theorem

We now come to an important result that accounts for our interest in π-systems and λ-systems. There are many times when one needs to show that some result holds for all the events in some σ-field, and the next theorem provides the most useful tool that we have for proving such results.

Theorem 12.6 (π-λ Theorem). *If $\mathcal{A}$ is a π-system and $\mathcal{B}$ is a λ-system, then*

$$\mathcal{A} \subset \mathcal{B} \quad \Rightarrow \quad \sigma(\mathcal{A}) \subset \mathcal{B}.$$

Proof. First, we note that $\mathcal{A} \subset \lambda(\mathcal{A}) \subset \mathcal{B}$ by the minimality of $\lambda(\mathcal{A})$, so to prove the theorem we only need to show that $\lambda(\mathcal{A})$ is also a π-system. Now, for any $C \in \lambda(\mathcal{A})$, we define a class of sets $\mathcal{G}_C$ by setting

$$\mathcal{G}_C = \{\, B \in \lambda(\mathcal{A}) : B \cap C \in \lambda(\mathcal{A}) \,\},$$

and we think of $\mathcal{G}_C$ as the set of elements of $\lambda(\mathcal{A})$ that are "good with respect to C" in the sense that the intersection of C and any B in $\mathcal{G}_C$ must again be in $\lambda(\mathcal{A})$. Now we have four simple observations about $\mathcal{G}_C$:

(i) For any $C \in \lambda(\mathcal{A})$, the set $\mathcal{G}_C$ is a λ-system (as one can easily check).

(ii) $A \in \mathcal{A} \Rightarrow \mathcal{A} \subset \mathcal{G}_A$ (since $\mathcal{A}$ is a π-system).

(iii) $A \in \mathcal{A} \Rightarrow \lambda(\mathcal{A}) \subset \mathcal{G}_A$ (since by (ii) $\mathcal{G}_A$ is a λ-system that contains $\mathcal{A}$).

(iv) $A \in \mathcal{A}$ and $B \in \lambda(\mathcal{A}) \Rightarrow A \cap B \in \lambda(\mathcal{A})$ (because of observation (iii)).

Now, the proof of the theorem depends on only one further observation. For any B in $\lambda(\mathcal{A})$, observation (iv) tells us that $\mathcal{A} \subset \mathcal{G}_B$, and since $\mathcal{G}_B$ is a λ-system, the minimality of $\lambda(\mathcal{A})$ tells us that $\lambda(\mathcal{A}) \subset \mathcal{G}_B$ for any $B \in \lambda(\mathcal{A})$. The last assertion is just another way of saying that $\lambda(\mathcal{A})$ is a π-system, so, by our first observation, the proof of the theorem is complete. □

A Uniqueness Theorem

One of the easiest consequences of the π-λ theorem is also one of the most important. This is the fact that the values of a probability measure on a π-system $\mathcal{A}$ uniquely determine the values of the measure on all of $\sigma(\mathcal{A})$. There are untold applications of this theorem, and its proof also helps explain why λ-systems crop up so naturally. Time after time, one finds that if $\mathcal{C}$ is the class of sets for which a given identity holds, then $\mathcal{C}$ is in fact a λ-system.

Theorem 12.7 (Uniqueness of Probability Measures). *Suppose that P and Q are two probability measures on the measurable space $(\Omega, \mathcal{F})$. If $\mathcal{A} \subset \mathcal{F}$ is a π-system and $P(A) = Q(A)$ for all $A \in \mathcal{A}$, then $P(A) = Q(A)$ for all $A \in \sigma(\mathcal{A})$. In particular, if $\mathcal{F} = \sigma(\mathcal{A})$, then P and Q are equal.*

Proof. As advertised, the key step is to introduce the class of sets $\mathcal{C}$ that satisfy the identity $P(B) = Q(B)$; that is, we define $\mathcal{C}$ by

$$\mathcal{C} = \{\, B : P(B) = Q(B) \,\}.$$

The fact that P and Q are probability measures lets us check that $\mathcal{C}$ is a λ-system. By hypothesis, we have that $\mathcal{A} \subset \mathcal{C}$ and that $\mathcal{A}$ is a π-system, so by the π-λ theorem we find that $\sigma(A) \subset \mathcal{C}$, just as we wanted to know. □

The Monotone Class Theorem

The monotone class theorem is another easy corollary of the π-λ theorem. In some applications, it has a modest advantage over the direct application of the π-λ theorem because it focuses on functions rather than sets. We will use this theorem in the proof of the density lemma that provoked this foray into the foundations of probability theory.

THEOREM 12.8 (Monotone Class Theorem). *Suppose the set $\mathcal{H}$ consists of bounded functions from a set Ω to $\mathbb{R}$ with the following three properties:*

(i) $\mathcal{H}$ is a vector space,

(ii) $\mathcal{H}$ contains the constant 1,

and

(iii) $f_n \in \mathcal{H}$, $f_n \uparrow f$, and f bounded $\Rightarrow f \in \mathcal{H}$.
If $\mathcal{H}$ contains the indicator function 1_A for each $A \in \mathcal{A}$, and if $\mathcal{A}$ is a π-system, then $\mathcal{H}$ contains all of the bounded functions that are $\sigma(\mathcal{A})$-measurable.

PROOF. If $\mathcal{C}$ is the class of all sets C such that the indicator function of C is in $\mathcal{H}$, then the properties (i),(ii), and (iii) permit us to check that $\mathcal{C}$ is a λ-system. Because $\mathcal{C}$ contains the π-system $\mathcal{A}$, the π-λ theorem then tells us that $\sigma(\mathcal{A}) \subset \mathcal{C}$. Now, if f is any bounded nonnegative $\sigma(\mathcal{A})$-measurable function, then f is the limit of an increasing sequence of $\sigma(\mathcal{A})$ simple functions. Such simple functions are in $\mathcal{H}$, so f is in $\mathcal{H}$. Finally, any bounded $\sigma(\mathcal{A})$-measurable function g is the difference of two nonnegative $\sigma(\mathcal{A})$-measurable functions, so g is in $\mathcal{H}$ because $\mathcal{H}$ is a vector space. The bottom line is that $\mathcal{H}$ must contain all of the bounded functions that are $\sigma(\mathcal{A})$-measurable. □

Finite-Dimensional Approximations

Our intuition about Brownian motion strongly suggests that we should be able to approximate any $\mathcal{F}_T$-measurable random variable as well as we like by a function that only depends on the values of Brownian motion at a finite set of times. The monotone class theorem turns out to be just the right tool for giving a rigorous interpretation of this intuitive idea.

LEMMA 12.3 (Finite Time Set Approximations). *If $\mathcal{D}$ denotes the set of random variables that can be written as*

$$f(B_{t_1}, B_{t_2} - B_{t_1}, \dots, B_{t_n} - B_{t_{n-1}})$$

for some $1 \leq n \leq \infty$, some $0 = t_0 < t_1 < \cdots < t_n = T$, and some bounded Borel function $f : \mathbb{R}^n \mapsto \mathbb{R}$, then $\mathcal{D}$ is dense in $L^2(\Omega, \mathcal{F}_T, P)$.

PROOF. We let $\mathcal{H}_0$ denote the set of all bounded random variables that can be written as the limit of a monotone increasing sequence of elements of $\mathcal{D}$ and then we take $\mathcal{H}$ to be the vector space generated by $\mathcal{H}_0$. One can easily check that $\mathcal{H}$ satisfies the hypotheses (i),(ii), and (iii) of the monotone class theorem.

Next, if we let $\mathcal{I}$ denote the collection of sets of the form

$$\{B_{t_1} < x_1, B_{t_2} < x_2, \dots, B_{t_n} < x_n)$$

then $\mathcal{I}$ is a π-system, and $\mathcal{H}$ certainly contains all of the indicator functions of the elements of $\mathcal{I}$. By the monotone class theorem, we therefore find that $\mathcal{H}$ contains all of the bounded $\sigma(\mathcal{I})$-measurable functions. Finally, by the definition of the standard Brownian filtration, the σ-field $\mathcal{F}_T$ is equal to the augmentation of $\sigma(\mathcal{I})$ by the class of null sets, so for any bounded simple $\mathcal{F}_T$-measurable function X there is a bounded simple $\sigma(\mathcal{I})$-measurable Y such that $P(X \neq Y) = 0$. Because $\mathcal{H}$ contains all of the bounded functions in $L^2(\Omega, \sigma(\mathcal{I}), P)$, it is certainly dense in $L^2(\Omega, \sigma(\mathcal{I}), P)$. Consequently, $\mathcal{D}$ is also dense in $L^2(\Omega, \mathcal{F}_T, P)$. □

A Fact from Hilbert Space

In a finite-dimensional vector space V, we know that if a linear subspace S does not span all of V, then we can find a nonzero $v \in V$ such that $v \cdot s = 0$ for all $s \in S$. An analogous fact is also true in the infinite-dimensional vector space $L^2(dP)$, and it is the last fact we need to develop before we complete our proof of the martingale representation theorem. For the statement of the next lemma, we recall that if $\mathcal{S}$ is a subspace of a Hilbert space $\mathcal{H}$, then $\mathcal{S}^\perp$ denotes the set of all $h \in \mathcal{H}$ such that h is orthogonal to s for all $s \in \mathcal{S}$.

LEMMA 12.4. *If $\mathcal{D}$ is a closed subspace of $L^2(dP)$, and $\mathcal{S} \subset \mathcal{D}$, then*

$$\mathcal{D} \cap \mathcal{S}^\perp = 0 \quad \Rightarrow \quad \bar{\mathcal{S}} = \mathcal{D}.$$

PROOF. Since $\mathcal{D}$ is a closed linear subspace of a Hilbert space, $\mathcal{D}$ is also a Hilbert space. The closed linear subspace $\bar{\mathcal{S}} \subset \mathcal{D}$ therefore has an orthogonal complement $\mathcal{S}^\perp$ such that any f in $\mathcal{D}$ can be written uniquely as $f = g + h$, where $g \in \bar{\mathcal{S}}$ and $h \in \mathcal{S}^\perp$, say by the Hilbert space projection theorem given in the Appendix of Mathematical Tools. Now, since $\mathcal{D}$ is closed and $\mathcal{S} \subset \mathcal{D}$, we have $f - g = h \in \mathcal{D}$ and $f - g = h \in \mathcal{S}^\perp$, so the hypothesis $\mathcal{D} \cap \mathcal{S}^\perp = 0$ tells us that $f - g = h = 0$. In other words, $f = g \in \bar{\mathcal{S}}$ and, since $f \in \mathcal{D}$ was arbitrary, we see $\mathcal{D} \subset \bar{\mathcal{S}}$, as we wanted to show. □

Density of the Z's

We have finally collected all of the required tools, and we are ready to address the problem that provided the stimulus for this section on bedrock approximation techniques. We want to complete the proof of the martingale representation theorem, and the only step that remains is to show that if $\mathcal{S}$ is the linear span of the set of random variables of the form

$$Z = \prod_{j=1}^{n} \exp[iu_j(B_{t_j} - B_{t_{j-1}})] \qquad (12.42)$$

over all $n \geq 1$, $0 = t_0 < t_1 < \cdots < t_n = T$, and all $u_j \in \mathbb{R}$, then $\mathcal{S}$ is dense in the space of square integrable complex-valued random variables,

$$L^2(\Omega, \mathcal{F}_T, P) = \{X : X \in \mathcal{F}_T \text{ and } E(|X|^2) < \infty\}.$$

To prove this fact, we first fix the set $\mathcal{T}$ of times $0 = t_0 < t_1 < \cdots < t_{n-1} < t_n = T$ and let $\mathcal{D}_\mathcal{T}$ denote the set of all $X \in L^2(\Omega, \mathcal{F}_T, P)$ that may be written in the form

$$X = f(B_{t_1}, B_{t_2} - B_{t_1}, \ldots, B_{t_n} - B_{t_{n-1}}),$$

where $f : \mathbb{R}^n \mapsto \mathbb{C}$ is a bounded Borel function. The set $\mathcal{D}_\mathcal{T}$ is a closed linear subspace of $L^2(\Omega, \mathcal{F}_T, P)$, and, if $\mathcal{S}_\mathcal{T}$ denotes the linear span of all Z of the form

(12.42), then the linear span is a subset of $\mathcal{D}_T$. We want to show that $\mathcal{S}_T$ is dense in $\mathcal{D}_T$.

Now, suppose that $X \in \mathcal{D}_T$ and $E(XZ) = 0$ for all Z of the form (12.42). If we let $\psi(x)$ denote the density of $(B_{t_1}, B_{t_2} - B_{t_1}, \dots, B_{t_n} - B_{t_{n-1}})$, then for all $(u_1, u_2, \dots, u_n)$, we have

$$\int_{\mathbb{R}^n} f(x_1, x_2, \dots, x_n) \prod_{j=1}^{n} \exp(iu_j x_j)\psi(x_1, x_2, \dots, x_n)\, dx_1 dx_2 \dots dx_n = 0. \tag{12.43}$$

This equation says that the Fourier transform of $f(\vec{x})\psi(\vec{x})$ is zero, so by the uniqueness of the Fourier transform $f(\vec{x})\psi(\vec{x})$ must be zero. Because ψ never vanishes, we can conclude that f is identically equal to zero.

What this means in terms of the spaces $\mathcal{S}_T$ and $\mathcal{D}_T$ is that $\mathcal{D}_T \cap \mathcal{S}_T^{\perp} = 0$, so by Lemma 12.4 we see that $\mathcal{S}_T$ is dense in $\mathcal{D}_T$. In turn, this tells us that

$$\mathcal{S} \stackrel{\text{def}}{=} \bigcup_T \mathcal{S}_T \text{ is dense in } \bigcup_T \mathcal{D}_T \stackrel{\text{def}}{=} \mathcal{D}.$$

We already know by Lemma 12.3 that $\mathcal{D}$ is dense in $L^2(\Omega, \mathcal{F}_T, P)$ so we see that $\mathcal{S}$ is also dense in $L^2(\Omega, \mathcal{F}_T, P)$. This, at long last, completes the proof of the Lemma 12.1, and, as a consequence, the proof of the martingale representation theorem is also complete.

12.7. Exercises

This chapter has itself been quite an exercise, and many readers may have found the need to invest some time supplementing their background knowledge. Here, the first two formal exercises offer practice with local martingales and localization arguments, while the third exercise offers a bare-handed view of the time change representation theorem. Finally, Exercise 12.4 fills in some details that were omitted in the proof of Theorem 12.4, and Exercise 12.5 shows how the hypotheses of Theorem 12.5 may be relaxed.

EXERCISE 12.1 (A $\mathcal{L}^2_{\text{LOC}}$ Counterexample to Itô's Isometry). Give an example of an $X \in \mathcal{F}_T$ and a $\phi \in \mathcal{L}^2_{\text{LOC}}$ such that we have the representation

$$X = \int_0^T \phi(\omega, s)\, dB_s \quad \text{and we have} \quad E(X^2) < \infty,$$

but nevertheless we still have

$$\int_0^T E[\phi^2(\omega, s)]\, ds = \infty.$$

This result shows rather dramatically that Itô's isometry does not hold on $\mathcal{L}^2_{\text{LOC}}$. Hint: Consider the example that was used to show nonuniqueness in Dudley's theorem.

EXERCISE 12.2 (A More General Martingale Representation Theorem). Suppose that $\{M_t, \mathcal{F}_t\}$ is a continuous martingale where $\{\mathcal{F}_t\}$ is the standard Brownian filtration. Use a localization argument and our L^2 representation theorem to show there is a $\phi(\omega, s) \in \mathcal{L}^2_{\text{LOC}}[0, T]$ such that

$$M_t = M_0 + \int_0^t \phi(\omega, s)\, dB_s \quad \text{for all } t \in [0, T].$$

EXERCISE 12.3. Consider the continuous martingale $M_t = B_t^2 - t$, and let τ_t be the time change guaranteed by Theorem 12.4. Give a direct proof that M_{τ_t} is a Brownian motion.

EXERCISE 12.4. The time change process defined by equation (12.34) will have jumps if the process $\phi(\omega, s)$ has intervals on which it vanishes, so the continuity properties of the process $\{X_{\tau_t} : t \geq 0\}$ defined in Theorem 12.4 deserve to be checked. As a first step, explain why one has the existence of the limits

$$\tau_{t+} \stackrel{\text{def}}{=} \lim_{s \searrow t} \tau_s \quad \text{and} \quad \tau_{t-} \stackrel{\text{def}}{=} \lim_{s \nearrow t} \tau_s,$$

and then explain why they satisfy

$$\int_{\tau_{t-}}^{\tau_{t+}} \phi^2(\omega, s)\, ds = 0 \quad \text{for all } \omega \text{ and all } t \geq 0.$$

Next, explain how this identity implies that the hypotheses of Proposition 7.13 are satisfied and that one can safely conclude that $\{X_{\tau_t}, \mathcal{F}_{\tau_t}\}$ is a continuous local martingale. Finally, explain why an analogous argument shows that the exponential process $\{Z_{\tau_t}, \mathcal{F}_{\tau_t}\}$ used in the proof of Theorem 12.4 is also a continuous local martingale.

EXERCISE 12.5. We smoothed the organization of the proof of Theorem 12.5 by assuming that $E[M_t^2] < \infty$ for all $0 \leq t < \infty$, but one can show that this hypothesis may be dropped. Prove this fact by showing that if $\{M_t, \mathcal{F}_t\}$ is a continuous martingale with $\langle M \rangle_t = t$ for all $0 \leq t < \infty$, then $E[M_t^2] < \infty$ for all $0 \leq t < \infty$.

Hint: Consider the process $X_t = M_t^2 - \langle M \rangle_t$ and consider stopping times τ_m, $m = 1, 2, \ldots$ that make $\{X_{t \wedge \tau_m}, \mathcal{F}_t\}$ a bounded martingale. Finish off by using the martingale property and Fatou's lemma to show $E[M_t^2] \leq t$.

CHAPTER 13

Girsanov Theory

Can a stochastic process *with drift* also be viewed as a process *without drift*? This modestly paradoxical question is no mere curiosity. It has many important consequences, the most immediate of which is the discovery that almost any question about Brownian motion with drift may be rephrased as a parallel (but slightly modified) question about standard Brownian motion.

The collection of theorems that tell us how to make drift disappear is commonly called Girsanov theory, although the important contributions of I.V. Girsanov were neither the first nor the last in this noble line. Today, Girsanov theory creates most of its value by providing us with a powerful tool for the construction of new martingales. In particular, we can apply Girsanov theory in the Black–Scholes model to find a probability measure that makes the present value of the stock price into a martingale, and, rather amazingly, the arbitrage price of any contingent claim can be expressed in terms of the conditional expectation of the claim with respect to this new probability measure.

Our study of Girsanov theory begins with the investigation of a simple (but crafty!) simulation technique called *importance sampling.* The idea of importance sampling is then extended in a natural way to random processes, and in short order this extension yields our first Girsanov theorem. To illustrate the effectiveness of even this simple theorem, we derive the elegant Lévy–Bachelier formula for the density of the first time that Brownian motion hits a sloping line.

Eventually, we will find Girsanov theorems of several flavors, and we will also find that they can be established by several different methods. One side benefit of our development of Girsanov theory is that it leads us to develop a new perspective on continuous random processes. Here, we will find that a continuous stochastic process is often best viewed as a random variable with a value that we regard as a *point* chosen from the *path space* $C[0,T]$.

13.1. Importance Sampling

We begin our investigation by considering a practical problem that people all over the planet mess up every day. In the simplest instance, we consider the calculation of $E[f(X)]$, where f is a known function and X is a normal random variable with mean zero and variance one. A natural (but naive) way to calculate $E[f(X)]$ is to use direct simulation; specifically, one can try to estimate $E[f(X)]$ by taking an approximation of the form

$$E[f(X)] \approx \frac{1}{n}\sum_{i=1}^{n} f(X_i), \tag{13.1}$$

where n is a large (but tractable) integer and the X_i are computer-generated pseudorandom numbers that are advertised as being independent and having the standard normal distribution.

One can quibble about the validity of the pseudorandom sequence or argue about the rule for choosing n, but, in some situations, the defects of direct simulation go much deeper. Sadly, there are very simple f for which the naive approximation (13.1) fails to produce even one significant digit of $E[f(X)]$. This failure can take place even if the pseudorandom numbers are perfect and if n is taken to be as large as the number of electrons in the universe.

To see the problem in the simplest case, we may as well take $f(x)$ to be the indicator function $1(x \geq 30)$. For this particular choice, we know several ways to calculate $E[f(X)]$ to many significant digits, but we will ignore these alternatives for the moment in order to explore what we can expect from equation (13.1). The first observation is that if we let N denote the number of terms in equation (13.1) until the first nonzero summand, then the rapid decay of the tail probability of a normal random variable tells us that $E[N] > 10^{100}$. Furthermore, to have any real confidence that we have correctly calculated the first significant digit of $E[f(X)]$, we probably should not stop our simulation before we have calculated a hundred or so nonzero summands. These two observations reveal that it is infeasible to calculate the first significant digit of $E[f(X)]$ by direct simulation (13.1).

This may seem to argue that for $f(x) = 1(x \geq 30)$ one cannot calculate $E[f(X)]$ by simulation, but nothing could be farther from the truth. In fact, a properly designed simulation will permit us to use a reasonably sized n and readily available pseudorandom numbers to compute three or more significant digits of $E[f(X)]$ in just a few seconds. All we need to do is to find a way to focus our efforts on what is really important.

The problem with naive direct simulation is that one may draw samples $\{X_i\}$ that are terribly far away from any point where important information is to be found about f. If our problem had been to estimate $E[g(X)]$ where $g(x) = 1(x \leq 0)$, then both the standard normal samples and the most informative behavior of g would have been centered at zero. In such a case, direct simulation would have done a thoroughly satisfactory job. These observations suggest that we should find some way to transform our original problem to one where the samples that we draw are focused more directly on the *important* behavior of f.

Shift the Focus to Improve a Monte Carlo

The idea is to look for some way to gain a bit of flexibility over the place where we draw our samples, and, fortunately, there is a natural computation that supports the feasibility of this idea. If we let E_μ denote expectation under the model that $X \sim N(\mu, 1)$, then we can always rewrite $E_0[f(X)]$ as an E_μ expectation:

$$
\begin{aligned}
E_0[f(X)] &= \frac{1}{\sqrt{2\pi}} \int_{-\infty}^{\infty} f(x) e^{-x^2/2}\, dx \\
&= \frac{1}{\sqrt{2\pi}} \int_{-\infty}^{\infty} f(x) e^{-(x-\mu)^2/2} e^{\mu^2/2} e^{-\mu x}\, dx \qquad (13.2) \\
&= E_\mu\left[f(X) e^{-\mu X + \mu^2/2} \right].
\end{aligned}
$$

The punch line is that one has a whole parametric family of alternatives to the naive simulation given by equation (13.1). By the *mean shifting identity* (13.2), we could just as well use any of the estimations

$$E_0[f(X)] \approx \frac{1}{n}\sum_{i=1}^{n} g(Y_i), \tag{13.3}$$

where

$$g(y) = f(y)e^{-\mu y + \mu^2/2} \text{ and } Y_i \sim N(\mu, 1).$$

Now, with μ completely at our disposal, we can surely improve on the estimation efficiency of the naive simulation (13.1). For example, if we could evaluate the coefficient of variation $c(\mu) = |E[g(Y_1)]|/(\text{Var}[g(Y_1)])^{\frac{1}{2}}$ as a function of μ, then the ideal choice for μ would be the value that minimizes $c(\mu)$. Sadly, such optimization schemes are not realistic. After all, we are trying to use simulation to find $E[f(X)]$ — which is just $E[g(Y_1)]$ for $\mu = 0$ — so even the boldest optimist must admit that we cannot expect to know $c(\mu)$.

Nevertheless, we are far from lost. We do not need to make an optimal choice of μ in order to make a good choice. In practical problems, we often have some intuition about the region where important information is to be discovered about f, and we almost always make a major improvement over naive simulation if we choose a value of μ that shifts the sampling nearer to that location. For example, in our original problem with $f(x) = 1(x > 30)$, we can just take $\mu = 30$. We will then find that the importance sampling method (13.3) is literally billions of times more efficient than naive simulation (13.1).

13.2. Tilting a Process

Importance sampling is not limited to functions of a single variable, or even to functions of a finite number of variables. There are direct analogs for processes. To begin with the simplest case, we take Brownian motion with drift

$$X_t = B_t + \mu t \qquad t \geq 0,$$

and we consider the calculation of the expectation

$$E[f(X_{t_1}, X_{t_2}, \dots, X_{t_n})],$$

where $0 = t_0 < t_1 < t_2 < \cdots < t_n \leq T$ and where $f : \mathbb{R}^n \to \mathbb{R}$ is a bounded Borel function.

In order to exploit the independent increment property of X_t, we first note that there is a $g : \mathbb{R}^n \mapsto \mathbb{R}$ such that $f(x_1, x_2, \dots, x_n) = g(x_1, x_2 - x_1, \dots, x_n - x_{n-1})$, so we can focus attention on the density of the vector $(X_{t_1}, X_{t_2} - X_{t_1}, \dots, X_{t_n} - X_{t_{n-1}})$. We may write this density as the product of the normalizing constant

$$C = (2\pi)^{-n/2} t_1^{-1/2}(t_2 - t_1)^{-1/2} \cdots (t_n - t_{n-1})^{-1/2}$$

and an exponential function of the x_i's,

$$\prod_{i=1}^{n} \exp\left(-\{(x_i - x_{i-1}) - \mu(t_i - t_{i-1})\}^2/2(t_i - t_{i-1})\right),$$

where we have taken $x_0 \equiv 0$ to keep the formulas tidy. If we expand the quadratic exponents in the last expression and then collect terms, we can obtain an alternative product where terms that contain μ are aggregated into the second factor:

$$\prod_{i=1}^{n} \exp\Big(-(x_i - x_{i-1})^2/2(t_i - t_{i-1})\Big) \cdot \prod_{i=1}^{n} \exp\left(\mu(x_i - x_{i-1}) - \frac{1}{2}\mu^2(t_i - t_{i-1})\right).$$

Now, we see just how lucky we are. When we collect the exponents in the second factor, the sums of the linear terms telescope to leave us with the simple product

$$\left\{\prod_{i=1}^{n} \exp\left(-(x_i - x_{i-1})^2/2(t_i - t_{i-1})\right)\right\} \cdot \exp\left(\mu x_n - \frac{1}{2}\mu^2 t_n\right).$$

Comparing this product to the density of the vector $(B_{t_1}, B_{t_2} - B_{t_1}, \ldots, B_{t_n} - B_{t_{n-1}})$, we see our original expectation of a function of the process $X_t = B_t + \mu t$ can be expressed as a modified expectation of a function of the Brownian motion B_t:

$$E[f(X_{t_1}, X_{t_2}, \ldots, X_{t_n})] = E\left[f(B_{t_1}, B_{t_2}, \ldots, B_{t_n}) \exp\left(\mu B_{t_n} - \frac{1}{2}\mu^2 t_n\right)\right]. \tag{13.4}$$

This identity is a direct analog to the importance sampling formula (13.2), although this time we see a couple of new twists.

Perhaps the most striking feature of equation (13.4) is the unheralded appearance of the familiar martingale

$$M_t = \exp\left(\mu B_t - \frac{1}{2}\mu^2 t\right),$$

and several bonuses come with the emergence of M_t. The most immediate benefit of the martingale property of M_t is that it lets us rewrite equation (13.4) in a slightly more symmetrical way. We only need to notice that the random variable $f(X_{t_1}, X_{t_2}, \ldots, X_{t_n})$ is $\mathcal{F}_{t_n}$-measurable and $t_n \leq T$, we can also write equation (13.4) as

$$E[f(X_{t_1}, X_{t_2}, \ldots, X_{t_n})] = E[f(B_{t_1}, B_{t_2}, \ldots, B_{t_n}) M_T]. \tag{13.5}$$

This important identity is often called the *tilting formula*, and we can think of M_T as a kind of correction factor that reweights (or tilts) the probability of Brownian motion paths so that in the end they have the probabilities one would expect from the paths of a Brownian motion with drift μ. One should note that the tilting factor M_T in formula (13.5) does not depend on the sampling times $0 < t_1 < t_2 < \cdots < t_n \leq T$. This feature of M_T turns out to be critical.

Later, we will develop several useful consequences of the quantitative properties of the tilting factor M_T, but, before we dig into any new abstractions, we should first test our new tool on an honest problem. Encouragingly enough, we will find that even the modest tilting formula (13.5) gives us a nearly automatic way to calculate interesting quantities such as the density of the first time that Brownian motion hits a sloping line.

Functions of a Brownian Path

If we let $S(n) = \{iT/n : 0 \leq i \leq n\}$, then the continuity of Brownian motion tells us that for all ω we have

$$\lim_{n\to\infty} \max_{t\in S(n)} (B_t + \mu t) = \max_{t\in[0,T]} (B_t + \mu t),$$

and consequently we also have for all x that

$$\lim_{n\to\infty} 1\left(\max_{t\in S(n)}(B_t+\mu t)<x\right) = 1\left(\max_{t\in[0,T]}(B_t+\mu t)<x\right).$$

Now, the tilting formula (13.5) also tells us that for all n we have

$$E\big[1\big(\max_{t\in S(n)}(B_t+\mu t)<x\big)\big] = E\left[1\left(\max_{t\in S(n)} B_t<x\right)\exp(\mu B_T-\frac{1}{2}\mu^2T)\right], \tag{13.6}$$

so to get a general property of Brownian motion, we only need to check what happens when $n\to\infty$. The integrand on the left-hand side of equation (13.6) is bounded, and the integrand on the right-hand side is dominated by the integrable function $\exp(\mu B_T-\frac{1}{2}\mu^2T)$, so we can take the limits in equation (13.6) to find the interesting identity

$$E\left[1\left(\max_{t\in[0,T]}(B_t+\mu t)<x\right)\right] = E\left[1\left(\max_{t\in[0,T]} B_t<x\right)\exp\left(\mu B_T-\frac{1}{2}\mu^2T\right)\right]. \tag{13.7}$$

This formula tells us that to calculate the distribution of the maximum of Brownian motion with drift, all we need to do is work out the right-hand side of equation (13.7).

Hitting Time of a Sloping Line: Direct Approach

We already know that for any $a>0$ the hitting time $\tau_a=\min\{t\colon B_t=a\}$ has a density that can be written compactly as

$$f_{\tau_a}(t)=\frac{a}{t^{3/2}}\phi\left(\frac{a}{\sqrt{t}}\right)\quad \text{for } t\geq 0. \tag{13.8}$$

One of the remarkable early successes of Bachelier was his discovery that the density of the hitting time of a *sloping line* has a formula that is strikingly similar to that of formula (13.8). In particular, if the line L has the equation $y=a+bt$ with $a>0$, then the first hitting time of the line L,

$$\tau_L=\inf\{\,t\colon B_t=a+bt\,\},$$

has a density that is given by

$$f_{\tau_L}(t)=\frac{a}{t^{3/2}}\phi\left(\frac{a+bt}{\sqrt{t}}\right)\quad \text{for } t\geq 0. \tag{13.9}$$

This formula creates an unforgettable parallel between the problem of hitting a line $y=a+bt$ and that of hitting a level $y=a$. Even more, it gives an elegant quantitative expression to several important qualitative distinctions, especially those that depend on the sign of b.

For example, when $b\leq 0$, the function $f_{\tau_L}(t)$ integrates to 1, reflecting the fact that Brownian motion will hit such a line with probability one. On the other hand, if $b>0$, we find that $f_{\tau_L}(t)$ integrates to a number $p_L<1$, and this reflects the fact that there is exactly a probability of $1-p_L>0$ that a standard Brownian motion will never reach the line L.

Now, to prove the formula (13.9), we first note that $\tau_L>t$ if and only if $B_s-bs<a$ for all $0\leq s<t$, and as a consequence the distribution of τ_L is easily

addressed by the process tilting formula (13.5), or, rather, the corollary (13.7) from which we find

$$P(\tau_L > t) = P\left(\max_{0\le s\le t}(B_s - sb) < a\right)$$

$$(13.10) \qquad = E\left[1\left(\max_{0\le s\le t} B_s < a\right)\exp\left(-bB_t - \frac{1}{2}b^2 t\right)\right].$$

What makes this formula practical is that long ago, in equation (5.14), we found that the joint density of B_t and $B_t^* = \max_{0\le s\le t} B_s$ is given by

$$f_{(B_t,B_t^*)}(u,v) = \frac{2(2v-u)}{\sqrt{2\pi t^3}}\exp\left(-\frac{(2v-u)^2}{2t}\right) \quad \text{for all } (u,v)\in D,$$

where $D = \{(u,v) : \max(0,u) \le v\}$. When we substitute this formula into equation (13.10) we find that

$$P(\tau_L \ge t) = \iint_{D\cap\{v\le a\}} \frac{2(2v-u)}{\sqrt{2\pi t^3}}\exp\left(-\frac{(2v-u)^2}{2t}\right)\exp\left(-bu - \frac{1}{2}b^2 t\right) du\, dv.$$

There is nothing between this double integral and the lovely formula (13.9) for the density of the hitting time of a line except a little calculus, and, even though the integral is messy, one can always deal with mess.

The part of our derivation that most deserves sincere appreciation is that equation (13.7) gives us an *automatic* way to convert almost any question about Brownian motion with drift into a question that only depends on standard (driftless) Brownian motion. In the particular case of the derivation of equation (13.9), we will soon find that we can even avoid the pesky double integral. As it happens, a slightly more abstract view of the tilting formula (13.5) will lead us quite naturally to a martingale method that replaces the double integral for $P(\tau_L > t)$ by a one-dimensional integral that is no trouble at all.

13.3. Simplest Girsanov Theorem

Our view of Brownian motion never focused too closely on the underlying measure space, and, by and large, we have profited from keeping a respectful distance. Nevertheless, the time has come to take a second look at our basic probability model, and, in particular, we need to develop the idea that the sample path of a continuous process on $[0,T]$ may be viewed simply as a *point* in the space $C[0,T]$ that has been chosen at random according to a probability measure.

When we view $C[0,T]$ as a metric space under the usual supremum norm, then the metric determines the class of open sets in $C[0,T]$, and the Borel σ-field $\mathcal{B}$ of $C[0,T]$ is defined to be the smallest σ-field that contains all of these open sets. Our new measurable space $(\Omega,\mathcal{F})$ takes $\Omega = C[0,T]$ to be the sample space and takes $\mathcal{F} = \mathcal{B}$ to be the base σ-field.

Now, all we need to construct useful probability measures on the space $(\Omega,\mathcal{F})$ is to observe that any continuous stochastic process $\{X_t : 0 \le t \le T\}$ that is defined on *any* probability space $(\tilde\Omega,\tilde{\mathcal{F}},\tilde P)$ suggests a way to define a measure on $(\Omega,\mathcal{F})$. Given such a process, we first define a mapping $X : \tilde\Omega \mapsto C[0,T]$ by taking $X(\tilde\omega)$ to be the continuous function on $[0,T]$ that is specified by

$$t \mapsto X_t(\tilde\omega) \text{ for } t \in [0,T],$$

and then we define a new probability measure Q on $(C[0,T],\mathcal{B})$ by taking

$$Q(A) = \tilde{P}(X^{-1}(A)). \tag{13.11}$$

We often say that a probability measure Q obtained in this way is the measure on $(C[0,T],\mathcal{B})$ that is *induced* by the continuous stochastic process $\{X_t\}$, or say that Q is the probability measure on $(C[0,T],\mathcal{B})$ that *corresponds* to the process $\{X_t\}$.

We now simply let P denote the measure on $C[0,T]$ that corresponds to standard Brownian motion and let Q denote the measure on $C[0,T]$ that corresponds to Brownian motion with constant drift μ.

THEOREM 13.1 (Simplest Girsanov Theorem: Brownian Motion with Drift). *If the process $\{B_t\}$ is a P-Brownian motion and Q is the measure on $C[0,T]$ induced by the process $X_t = B_t + \mu t$, then every bounded Borel measurable function W on the space $C[0,T]$ satisfies*

$$E_Q(W) = E_P(WM_T), \tag{13.12}$$

where M_t is the P-martingale defined by

$$M_t = \exp(\mu B_t - \mu^2 t/2). \tag{13.13}$$

PROOF. Since any bounded Borel function W is a limit of bounded simple functions, the proof of the theorem only requires that we prove the identity (13.12) for all $W = 1_A$, where A is a Borel set in $C[0,T]$. Next let $\mathcal{I}$ denote the class of all sets that are of the form

$$A = \{\omega : \omega(t_i) \in [a_i, b_i] \text{ for all } 1 \le i \le n\}$$

for some integer n and real a_i, b_i. The tilting formula (13.5) tells us that equation (13.12) holds for all $W = 1_A$ with $A \in \mathcal{I}$. The collection $\mathcal{I}$ is also a π-system, and, if we let $\mathcal{C}$ denote the class of all A for which $W = 1_A$ satisfies the identity (13.12), then we can easily check that $\mathcal{C}$ is a λ-system. Since $\mathcal{I} \subset \mathcal{C}$, the π-λ theorem then tells us that equation (13.12) holds for all $A \in \sigma(\mathcal{I})$. Finally, since $\sigma(\mathcal{I})$ is precisely the class of Borel sets of $C[0,T]$, the proof of the theorem is complete. □

Before we move on to the extension of the simplest Girsanov theorem, we should take another quick look at the Lévy–Bachelier formula. This brief detour gives a nice illustration of one of the ways that stopping times may be used to simplify calculations.

HITTING A SLOPING LINE: THE MARTINGALE VIEW

The fact that the drift correction process $\{M_t\}$ is a martingale turns out to be important in many applications of Girsanov theory, and we should always remain alert to the possibility of exploiting this connection. The problem of calculating the density of the first hitting time of a sloping line provides a classic case. If we use the martingale property of M_t from the outset, we can completely avoid the messy double integral that we found before, and, with the right perspective, we will quickly find a simple one-dimensional integral that yields the desired density with minimal calculation.

We noted earlier that the first time τ_L that B_t hits the line $y = a+bt$ is precisely the first time τ_a that the process $X_t = B_t - bt$ hits the level $y = a$, so now if we let

Q denote the measure on $C[0,T]$ induced by the process $\{X_t\}$, then the simplest Girsanov theorem tells us

$$P(\tau_L \le t) = Q(\tau_a \le t) = E_Q(1(\tau_a \le t)) = E_P(1(\tau_a \le t)M_T),$$

where M_t is the P-martingale defined by equation (13.13).

Now, to exploit the martingale property of M_t, we first note that $\{\tau_a \le t\}$ is an $\mathcal{F}_{t\wedge\tau_a}$-measurable event, so if we condition on the σ-field $\mathcal{F}_{t\wedge\tau_a}$, we find

$$\begin{aligned} E_P(1(\tau_a \le t)M_T) &= E_P(1(\tau_a \le t)M_{t\wedge\tau_a}) \\ &= E_P(1(\tau_a \le t)\exp\left(-ab - b^2\tau_a/2\right)) \\ &= \int_0^t \exp\left(-ab - b^2 s/2\right) \frac{a}{s^{3/2}}\phi\left(\frac{a}{\sqrt{s}}\right)\, ds. \end{aligned}$$

Differentiation of the last integral is a piece of cake, and, after completing the square in the exponent, we are again pleased to find the Lévy–Bachelier formula for the density of the first hitting time of a sloping line:

$$f_{\tau_L}(t) = \frac{a}{t^{3/2}}\phi\left(\frac{a+bt}{\sqrt{t}}\right) \quad \text{for } t \ge 0. \tag{13.14}$$

EQUIVALENCE, SINGULARITY, AND QUADRATIC VARIATION

Before we take on more elaborate results, we should reflect for a moment on some simple qualitative questions that are answered by Girsanov's theorem. For example, suppose we observe two independent processes

$$X_t = \mu t + \sigma B_t \quad \text{and} \quad \tilde{X}_t = \tilde{\mu} t + \tilde{\sigma} B_t$$

during the time period $[0,T]$. Can we decide *with certainty* that the two processes are genuinely different, rather than just independent realizations of equivalent processes?

The answer to this question may be surprising. If $\sigma \ne \tilde{\sigma}$, then the fact is that we can tell with certainty that the two processes are different. The values are taken by μ and $\tilde{\mu}$, do not matter a bit, and neither does it matter how small one might take $T > 0$. When $\sigma \ne \tilde{\sigma}$, a billionth of a second is more than enough time to tell the two processes are different. On the other hand, if $\sigma = \tilde{\sigma}$, then we can *never* be certain on the basis of observation during a finite time interval whether the processes X_t and $\tilde{X}_t$ are not equivalent.

The reason for this is simple. If we take $t_i = iT/2^n$ for $0 \le i \le 2^n$, then we have

$$\lim_{n\to\infty} \sum_{i=0}^{2^n-1} (X_{t_i} - X_{t_{i-1}})^2 = \sigma^2 T,$$

where the convergence takes place with probability one. The bottom line is that with probability one we can use our observed path to compute σ^2 exactly. If $\sigma \ne \tilde{\sigma}$ then even the tiniest interval of time suffices for us to be able to tell with certainty that we are watching two different processes.

Two probability measures P_1 and P_2 on a measurable space $(\Omega, \mathcal{F})$ are said to be *equivalent* if

$$P_1(A) = 0 \Leftrightarrow P_2(A) = 0,$$

and are said to be *singular* if there exists an $A \in \mathcal{F}$ such that

$$P_1(A) = 1 \text{ and } P_2(A) = 0.$$

Now, thanks to Girsanov's theorem, we know that for any $T < \infty$, if $P_{\mu,\sigma}$ is the probability measure induced on $C[0,T]$ by the continuous process $X_t = \mu t + \sigma B_t$, then $P_{\mu,\sigma}$ is equivalent to $P_{0,\sigma}$ for all μ. Also, by our observations concerning quadratic variation, we know that P_{μ,σ_1} and P_{μ,σ_2} are singular if and only if $\sigma_1 \neq \sigma_2$.

13.4. Creation of Martingales

If we consider the process $X_t = B_t + \mu t$ together with the measure space $(\Omega, \mathcal{F}, P)$, where B_t is a standard Brownian motion, then X_t is certainly not a martingale. Nevertheless, if we define a new probability measure Q on $(\Omega, \mathcal{F})$ by taking

$$Q(A) = E_P[1_A \exp(-\mu B_T - \mu^2 T/2)],$$

then the simplest Girsanov theorem tells us precisely that on the probability space $(\Omega, \mathcal{F}, Q)$ the process X_t is a standard Brownian motion and a fortiori X_t is a Q-martingale.

This miracle applies to processes that go far beyond drifting Brownian motion, and, subject to a few modest restrictions, a great number of continuous processes that are adapted to the filtration of Brownian motion may be transformed into a martingale by an appropriate change of measure. For our first example of such a transformation, we will consider the present value of the stock price in the Black–Scholes model.

Discounted Stock Price

If we go back to the Black–Scholes model, we had a stock process that we took to be

$$dS_t = \mu S_t \, dt + \sigma S_t \, dB_t,$$

and we had a bond model given by

$$d\beta_t = r\beta_t \, dt \quad \beta_0 = 1,$$

so, if we solved these equations, we found

$$S_t = S_0 \exp\left(t(\mu - \sigma^2/2) + \sigma B_t\right) \text{ and } \beta_t = e^{rt}.$$

The present value of the time t stock is defined to be the stock price discounted by the bond price, so, in symbols, the discounted stock price is given by

$$D_t = S_t/\beta_t = S_0 \exp\left(t\left(\mu - r - \frac{1}{2}\sigma^2\right) + \sigma B_t\right). \tag{13.15}$$

If we apply Itô's formula, we find the SDE for D_t to be

$$dD_t = (\mu - r)D_t \, dt + \sigma D_t dB_t, \tag{13.16}$$

and we can see from either equation (13.16) or equation (13.15) that D_t is not a martingale with respect to P except in the special case when $r = \mu$. Nevertheless, a similar story could have been told for the process $X_t = B_t + \mu t$, and in that case we found a measure that makes X_t a martingale. We must ask whether there might also be a measure Q under which D_t is a martingale.

THE QUEST FOR A NEW MEASURE

In order to make the connection between D_t and drifting Brownian motion more explicit, we first note that we can rewrite the SDE for D_t in the form

$$dD_t = \sigma D_t \, d\left\{ t\frac{(\mu - r)}{\sigma} + B_t \right\} = \sigma D_t \, dX_t,$$

where $X_t = B_t + \hat{\mu} t$, and the drift parameter $\hat{\mu}$ is taken to be $(\mu - r)/\sigma$. In longhand, this equation tells us that

$$D_t - D_0 = \int_0^t \sigma D_s \, dX_s, \tag{13.17}$$

and now our job is to interpret the previous equation.

Naturally, there are two interpretations — actually infinitely many — but only two really matter. We can interpret equation (13.17) under the measure P that makes B_t a standard Brownian motion, or we can interpret it under the measure Q defined by

$$Q(A) = E_P[1_A \exp(-\hat{\mu} B_T - \hat{\mu}^2 T/2)].$$

Now, since X_t is a standard Brownian motion with respect to Q and the integrand D_t is a nice continuous process, we know that the stochastic integral given by equation (13.17) is again a well-defined Itô integral.

The idea that we may have multiple interpretations of a single stochastic integral may seem a little odd at first, but, in fact, the only reason we might feel ill at ease with this multiplicity of interpretations is that when we did our earlier work with stochastic integrals we kept only one probability measure in mind. At the time, we had no reason to be alert to the richer possibility that any one of our integrals could be viewed as a member of many different probability spaces.

13.5. Shifting the General Drift

Thus far, we have only seen how one can remove a constant drift, but the drift-removal idea works much more generally. The next theorem shows how a Brownian motion plus a general drift process may still be viewed as a standard Brownian motion.

THEOREM 13.2 (Removing Drift). *Suppose that $\mu(\omega, t)$ is a bounded, adapted process on $[0,T]$, B_t is a P-Brownian motion, and the process X_t is given by*

$$X_t = B_t + \int_0^t \mu(\omega, s) \, ds. \tag{13.18}$$

The process M_t defined by

$$M_t = \exp\left(-\int_0^t \mu(\omega, s) \, dB_s - \frac{1}{2} \int_0^t \mu^2(\omega, s) \, ds \right)$$

is a P-martingale and the product $X_t M_t$ is also a P-martingale. Finally, if Q denotes the measure on $C[0,T]$ defined by $Q(A) = E_P[1_A M_T]$, then X_t is a Q-Brownian motion on $[0,T]$.

PROOF. By Itô's formula we have $dM_t = -\mu(\omega, s)M_t\,dB_t$, so M_t is certainly a P-local martingale. The boundedness of $\mu(\omega, s)$ then lets us check without difficulty that $\sup_{0\le t\le T} M_t$ has a finite expectation, so (after recalling Exercise 7.3) we see that M_t is indeed an honest P-martingale.

Thus, the main task is to show that under Q the process $\{X_t\}$ has the same joint distributions as Brownian motion. Remarkably, this follows almost immediately from the elegant fact that for all bounded *deterministic* $f : [0,T] \to \mathbb{C}$ one has the key formula

$$E_Q\left[\exp\left(\int_0^T f(s)\,dX_s\right)\right] = \exp\left(\frac{1}{2}\int_0^T f^2(s)\,ds\right).$$

To exploit this formula, we simply take any $\theta_j \in \mathbb{R}$ for $j = 1, 2, ..., N$ and take any $0 = t_0 < t_1 < \cdots < t_{N-1} < t_N = T$ so that the deterministic function

$$f(s) = \sum_{j=1}^{N} i\theta_j\, 1(t_{j-1} \le s < t_j) \quad \text{for } 0 \le s \le T,$$

and the key formula may be combined to tell us

$$E_Q\left[\exp\left(i\sum_{j=1}^{N} \theta_j(X_{t_j} - X_{t_{j-1}})\right)\right] = \exp\left(-\frac{1}{2}\sum_{j=1}^{N} \theta_j^2(t_j - t_{j-1})\right).$$

This identity says that the difference vector $(X_{t_1}, X_{t_2}-X_{t_1}, ..., X_{t_N}-X_{t_{N-1}})$ has the same characteristic function under Q as the difference vector of Brownian motion, so $\{X_t\}$ is in fact a Q-Brownian motion.

All that remains is to prove the key formula, and, as a first step, we simply apply the definition of X_t and Q to rewrite the left-hand side of the formula as

$$E_P\left[\exp\left(\int_0^T (f(s) - \mu(\omega, s))\,dB_s + \int_0^T f(s)\mu(\omega, s)\,ds - \frac{1}{2}\int_0^T \mu^2(\omega, s)\,ds\right)\right].$$

Now, since f is deterministic, this expectation may also be written as

$$\exp\left(\frac{1}{2}\int_0^T f^2(s)ds\right) E_P\left[\exp\left(\int_0^T (f(s)-\mu(\omega, s))dB_s - \frac{1}{2}\int_0^T (f(s)-\mu(\omega, s))^2 ds\right)\right]$$

and this brings us to a critical observation. The second factor here is the expectation of an exponential martingale that is associated with $\alpha(\omega, s) = \mu(\omega, s) - f(s)$ in the same way that $\mu(\omega, s)$ was associated with M_t. Consequently, this expectation must equal one, and the proof of the key formula is suddenly complete. □

Two Drifts and One Volatility

If one can remove drift by a change of measure, one can surely add drift by a similar change. When the two processes are put together, we only need good bookkeeping to keep track of the transformations. If we are careful, we can even incorporate a general volatility into our drift-shifting processes.

THEOREM 13.3 (Swapping the Drift). *Suppose that X_t is a standard process that we can write as*

$$X_t = x + \int_0^t \mu(\omega, s)\, ds + \int_0^t \sigma(\omega, s)\, dB_s. \tag{13.19}$$

If the ratio $\theta(\omega, t) = (\mu(\omega, t) - \nu(\omega, t))/\sigma(\omega, t)$ is bounded, then the process

$$M_t = \exp\left(-\int_0^t \theta(\omega, s)\, dB_s - \frac{1}{2}\int_0^t \theta^2(\omega, s)\, ds\right)$$

is a P-martingale and the product $X_t M_t$ is a P-martingale. Finally, if we define a measure Q on $C[0,T]$ by

$$Q(A) = E_P(1_A M_T), \tag{13.20}$$

then the process defined by

$$\tilde{B}_t = B_t + \int_0^t \theta(\omega, s)\, ds$$

is a Q-Brownian motion, and the process X_t has the representation

$$X_t = x + \int_0^t \nu(\omega, s)\, ds + \int_0^t \sigma(\omega, s)\, d\tilde{B}_s. \tag{13.21}$$

PROOF. We have already done most of the work during the proof of the preceding theorem. The fact that M_t is a P-martingale follows the now standard calculation, and the fact that $\tilde{B}_t$ is a Q-Brownian motion is exactly the *content* of Theorem 13.2. Finally, to see that equation (13.21) holds for X_t, we first note $X_0 = x$ and simplify equation (13.21) to find

$$\begin{aligned} dX_t &= \nu(\omega, t)\, dt + \sigma(\omega, t)\, d\tilde{B}_t \\ &= \nu(\omega, t)\, dt + \sigma(\omega, t)\{dB_t + \theta(\omega, t)\, dt\} \\ &= \nu(\omega, t)\, dt + \sigma(\omega, t)\, dB_t + \{\mu(\omega, t) - \nu(\omega, t)\}\, dt \\ &= \mu(\omega, t)\, dt + \sigma(\omega, t)\, dB_t. \end{aligned}$$

The last expression matches up with our target equation (13.19), so the proof of the theorem is complete. □

DID WE NEED BOUNDEDNESS?

Theorem 13.2 imposed a boundedness condition on μ and Theorem 13.3 imposed boundedness on θ, but one should check that the proofs of these theorems used these hypotheses only to obtain the martingale property for the associated exponential local martingales. In both of these theorems, one can replace the boundedness hypothesis by the more general (but more evasive) assumption that the exponential process M_t is a martingale. Such a substitution may seem odd, but there are times when the added flexibility is useful. Such a hypothesis shifts the responsibility to the user to find more powerful criteria for M_t to be a martingale. The next section shows how to rise to this challenge.

13.6. Exponential Martingales and Novikov's Condition

One of the key issues in the use of Girsanov theory is the articulation of circumstances under which an exponential local martingale is an honest martingale. Sometimes, we can be content with a simple sufficient condition such as boundedness, but at other times we need serious help. The next theorem provides a sufficient condition that is among the best that theory has to offer.

THEOREM 13.4 (The Novikov Sufficient Condition). *For any* $\mu \in \mathcal{L}^2_{\text{LOC}}[0,T]$, *the process defined by*

$$M_t(\mu) = \exp\left(\int_0^t \mu(\omega,s)\,dB_s - \frac{1}{2}\int_0^t \mu^2(\omega,s)\,ds\right) \tag{13.22}$$

is a martingale, provided that μ *satisfies the* Novikov condition

$$E\left[\exp\left(\frac{1}{2}\int_0^T \mu^2(\omega,s)\,ds\right)\right] < \infty. \tag{13.23}$$

Understanding the Condition

One of Pólya's bits of advice in *How to Solve It* is to "understand the condition." Like many of the other pieces of Pólya's problem-solving advice, this seems like such basic common sense that we may not take the suggestion as seriously as perhaps we should. Here the suggestion is particularly wise.

When we look at the condition (13.23) and angle for a deeper understanding, one of the observations that may occur to us is that if μ satisfies the condition then so does $\lambda\mu$ for any $|\lambda| \le 1$. At first, there may not seem like there is much force to this added flexibility, but it does set us onto a promising path.

A Plan Suggested by Power Series

From our work in Chapter 7, we already know that the nonnegative local martingale $M_t(\mu)$ is a supermartingale, and by Proposition 7.11 we also know that $M_t(\mu)$ will be an honest martingale on $[0,T]$ if $E[M_T(\mu)] = 1$. This modest observation suggests a marvelous plan.

If we introduce the function $H(\lambda) = E[M_T(\lambda\mu)]$, where λ is a real parameter, then the proof is complete if we show $H(1) = 1$, but, if the theorem is true (as we strongly suspect!), we should actually have $H(\lambda) = 1$ for all $|\lambda| \le 1$. It is trivial that $H(0) = 1$, and from the definition of M_t we might suspect that we would have an easier time proving $H(\lambda) = 1$ for $\lambda \in (-1,0]$ than for positive $\lambda > 0$. At this point, some experience with power series suggests that if we can prove $H(\lambda) = 1$ for all λ in an interval such as $(-1,0]$, then we should have great prospects of proving that $H(\lambda) = 1$ for all $\lambda \le 1$. Even without such experience, the plan should be at least modestly plausible, and, in any event, we will need to make some small modifications along the way.

First a Localization

As usual when working with local processes, we do well to slip in a localization that makes our life as easy as possible. Here, we want to study $M_t(\lambda\mu)$ for negative λ, so we want to make sure that the exponent in $M(\mu)$ is not too small. For this purpose, we will use the related process

$$Y_t = \int_0^t \mu(\omega, s)\, dB_s - \int_0^t \mu^2(\omega, s)\, ds$$

and introduce the stopping time

$$\tau_a = \inf\{ t : Y_t = -a \text{ or } t \geq T \}.$$

The next proposition gives us some concrete evidence that our plan is on track.

PROPOSITION 13.1. *For all $\lambda \leq 0$, we have the identity*

$$E[M_{\tau_a}(\lambda\mu)] = 1. \tag{13.24}$$

PROOF. As we have seen several times before, Itô's formula tells us that the process $M_t(\lambda\mu)$ satisfies $dM_t(\lambda\mu) = \lambda\mu(\omega, t)M_t(\lambda\mu)dB_t$ and as a consequence we have the integral representation

$$M_{\tau_a}(\lambda\mu) = 1 + \int_0^{\tau_a} \lambda\mu(\omega, s)M_s(\lambda\mu)\, dB_s. \tag{13.25}$$

Now, to prove (13.24), we only need to show that the integrand in equation (13.25) is in $\mathcal{H}^2$, or, in other words, we must show

$$E\left[\int_0^{\tau_a} \mu^2(\omega, s)M_s^2(\lambda\mu)\, ds\right] < \infty. \tag{13.26}$$

Here we first note that for $s \leq \tau_a$ we have

$$\begin{aligned} M_s(\lambda\mu) &= \exp\left(\lambda \int_0^s \mu(\omega, s)\, dB_s - \frac{\lambda^2}{2} \int_0^s \mu^2(\omega, s)\, ds\right) \\ &= \exp(\lambda Y_s) \exp\left((\lambda - \lambda^2/2) \int_0^s \mu^2(\omega, s)\, ds\right) \\ &\leq \exp(a|\lambda|), \end{aligned} \tag{13.27}$$

where in the last step we use the definition of τ_a and the fact that $\lambda - \lambda^2/2 \leq 0$ for $\lambda \leq 0$. Next, we note that the simple bound $x^2 \leq 2\exp(x^2/2)$ and Novikov's condition combine to tell us that

$$E\left(\int_0^T \mu^2(\omega, s)\, ds\right) \leq 2E\left[\exp\left(\frac{1}{2}\int_0^T \mu^2(\omega, s)\, ds\right)\right] < \infty. \tag{13.28}$$

Finally, in view of the bounds (13.27) and (13.28), we see that equation (13.26) holds, so the proof of the proposition is complete. □

POWER SERIES AND POSITIVE COEFFICIENTS

At this point, one might be tempted to expand $E(M_{\tau_a}(\lambda\mu))$ as a power series in λ in order to exploit the identity (13.24), but this frontal assault runs into technical problems. Fortunately, these problems can be avoided if we can manage to work with power series with nonnegative coefficients. The next lemma reminds us how pleasantly such series behave. To help anticipate how the lemma will be applied, we should note that the inequality (13.29) points toward the supermartingale property of $M_{\tau_a}(\lambda\mu)$ whereas the equality (13.30) connects with the identity that we just proved in Proposition 13.1.

LEMMA 13.1. *If $\{c_k(\omega)\}$ is a sequence of nonnegative random variables and $\{a_k\}$ is a sequence of real numbers such that the two power series*

$$f(x,\omega) = \sum_{k=0}^{\infty} c_k(\omega)\, x^k \text{ and } g(x) = \sum_{k=0}^{\infty} a_k x^k$$

satisfy

$$E[f(x,\omega)] \le g(x) < \infty \quad \text{for all } x \in (-1,1] \tag{13.29}$$

and

$$E[f(x,\omega)] = g(x) \quad \text{for all } x \in (-1,0], \tag{13.30}$$

then

$$E[f(1,\omega)] = g(1). \tag{13.31}$$

PROOF. Since $c_k(\omega) \ge 0$, we can apply Fubini's theorem and the bound (13.29) to get

$$E[f(x,\omega)] = \sum_{k=0}^{\infty} E(c_k)\, x^k \le g(x) < \infty \quad \text{for all } x \in (-1,1], \tag{13.32}$$

whereas Fubini's theorem and the identity (13.30) give us

$$E[f(x,\omega)] = \sum_{k=0}^{\infty} E(c_k)\, x^k = \sum_{k=0}^{\infty} a_k x^k \quad \text{for all } x \in (-1,0]. \tag{13.33}$$

Now, by the uniqueness of power series, the last identity tells us $E(c_k) = a_k$ for all $k \ge 0$, and this implies that the identity (13.33) actually holds for all $x \in (-1,1)$. Finally, since $f(x,\omega)$ is a monotone function of x on $[0,1)$, we can take the limit $x \uparrow 1$ in the identity $g(x) = E(f(x,\omega))$ on $[0,1)$ to conclude that $g(1) = E(f(1,\omega))$. □

EXTENDING THE IDENTITY

By Proposition 13.1, we know that $E[M_{\tau_a}(\lambda\mu)] = 1$ for all $\lambda \le 0$, and we simply need to extend this identity to $\lambda \le 1$. When we write $M_t(\lambda\mu)$ in terms of Y_t, we find

$$M_t(\lambda\mu) = \exp\left(\lambda Y_t + (\lambda - \lambda^2/2)\int_0^t \mu^2(\omega,s)\,ds\right),$$

and the relationship of Y_t to the level a can be made more explicit if we consider

$$e^{\lambda a} M_t(\lambda\mu) = \exp\left(\lambda(Y_t + a) + (\lambda - \lambda^2/2)\int_0^t \mu^2(\omega,s)\,ds\right). \tag{13.34}$$

Now, if we reparameterize the preceding expression just a bit, we will be able to obtain a power series representation for $e^{\lambda a} M_{\tau_a}(\lambda\mu)$ with nonnegative coefficients.

Specifically, we first choose z so that $\lambda - \lambda^2/2 = z/2$, and we then solve the quadratic equation to find two candidates for λ. Only the root $\lambda = 1 - \sqrt{1-z}$ will satisfy $\lambda \le 1$ when $|z| \le 1$, so we will use the substitutions

$$\lambda - \lambda^2/2 = z/2 \text{ and } \lambda = 1 - \sqrt{1-z}$$

to replace the λ's by the z in the identity (13.34).

In these new variables, the power series for $e^{\lambda a}M_{\tau_a}(\lambda\mu)$ is given by

$$f(\omega,z)=\exp\left((1-\sqrt{1-z})(Y_{\tau_a}+a)+\frac{z}{2}\int_0^{\tau_a}\mu^2(\omega,s)\,ds\right)\\=\sum_{k=0}^{\infty}c_k(\omega)z^k, \tag{13.35}$$

and, because the power series for e^z and $1-\sqrt{1-z}$ have only positive coefficients, we see that $c_k(\omega)\geq 0$ for all $k\geq 0$.

Now, because $e^{(1-\sqrt{1-z})a}M_t((1-\sqrt{1-z})\mu)$ is a supermartingale for any $z\leq 1$, we can also take the expectation in equation (13.35) to find

$$E[f(\omega,z)]\leq \exp\big(a(1-\sqrt{1-z}\,)\big)\stackrel{\text{def}}{=}g(z)\stackrel{\text{def}}{=}\sum_{k=0}^{\infty}a_kz^k. \tag{13.36}$$

The identity of Proposition 13.1 tells us that for all $z\in(-1,0]$ we have

$$E[f(\omega,z)]=\exp\big(a(1-\sqrt{1-z}\,)\big)=g(z), \tag{13.37}$$

so all of the conditions of Lemma 13.1 are in place, and we can apply the lemma to conclude that

$$E[f(\omega,1)]=e^a,$$

so when we unwrap the definition of f, we find

$$E[M_{\tau_a}(\mu)]=1.$$

All that remains to complete the proof of Theorem 13.4 is to show that τ_a can be replaced by T in the previous identity.

FINAL STEP: DELOCALIZATION

The natural plan is to let $a\to\infty$ in $E[M_{\tau_a}(\mu)]=1$ so that we may conclude $E[M_T(\mu)]=1$. This plan is easily followed. The first step is to note that the identity $E[M_{\tau_a}(\mu)]=1$ gives us

$$\begin{aligned}1&=E[M_{\tau_a}(\mu)1(\tau_a<T)]+E[M_{\tau_a}(\mu)1(\tau_a=T)]\\&=E[M_{\tau_a}(\mu)1(\tau_a<T)]+E[M_T(\mu)1(\tau_a=T)],\end{aligned}$$

and trivially we have

$$E[M_T(\mu)]=E[M_T(\mu)1(\tau_a=T)]+E[M_T(\mu)1(\tau_a<T)],$$

so we have

$$E[M_T(\mu)]=1-E[M_{\tau_a}(\mu)1(\tau_a<T)]+E[M_T(\mu)1(\tau_a<T)]. \tag{13.38}$$

Now, on the set $\{\tau_a<T\}$ we have $Y_{\tau_a}=-a$ so

$$\begin{aligned}M_{\tau_a}(\mu)1(\tau_a<T)&=1(\tau_a<T)\exp\left(Y_{\tau_a}+\frac{1}{2}\int_0^{\tau_a}\mu^2(\omega,s)\,ds\right)\\&\leq e^{-a}\exp\left(\frac{1}{2}\int_0^{T}\mu^2(\omega,s)\,ds\right),\end{aligned}$$

and the Novikov condition tells us the exponential has a finite expectation so as $a \to \infty$ we find

$$E[M_{\tau_a}(\mu)1(\tau_a < T)] \le e^{-a} E\left[\exp\left(\frac{1}{2}\int_0^T \mu^2(\omega, s)\,ds\right)\right] \to 0. \tag{13.39}$$

The continuity of Y_t implies that $1(\tau_a < T) \to 0$ for all ω, and the supermartingale property gave us $E[M_T(\mu)] \le 1$, so now by the dominated convergence theorem, we find

$$E[M_T(\mu)1(\tau_a < T)] \to 0 \quad \text{as } a \to \infty. \tag{13.40}$$

Finally, if we apply the limit results (13.39) and (13.40) in the identity (13.38), then we see at last that $E[M_T(\mu)] = 1$ and we have confirmed that $\{M_t : 0 \le t \le T\}$ is an honest martingale.

Looking Back: The Nature of the Pattern

In our development of the martingale representation theorem we found an analogy between mathematical induction and the way we worked our way up from a special case to the general theorem. Here, the analogy is more strained, but perhaps it still merits consideration. We began with the trivial observation that $H(0) = 1$ (analogous to the proposition $P(1)$ in mathematical induction), and this observation motivated us to study the more general case $H(\lambda) = 1$ for $\lambda \le 0$ (analogous to showing $P(n) \Rightarrow P(n+1)$). Finally, function theoretic facts were used to show that $H(\lambda) = 1$ for $\lambda \le 1$, and this last step was (very loosely!) analogous to invoking the principle of mathematical induction.

13.7. Exercises

The first exercise is just a warm-up based on the tilting formula and the simplest Girsanov theorem, but the second exercise is both lovely and sustained. It outlines a proof that Brownian motion will write your name with probability one during any time interval when you care to watch.

The next two exercises aim to provide insight into Girsanov's theorem. The first of these explores one of the ways in which Girsanov's theorem is sharp, and the last exercise shows that except for a tiny ϵ one can prove Girsanov's theorem using little more than Hölder's inequality.

Exercise 13.1 (Warm-ups).

(a) Show that the tilting formula (13.5) implies the elementary mean shifting formula (13.2). Use this opportunity to sort out when one has a plus μ or a minus μ.

(b) Use the simplest Girsanov theorem to show for $\mu > 0$ we have

$$E\left[e^{-\mu B_T} \max_{0\le t\le T} B_t\right] \simeq \frac{1}{2\mu} e^{\mu^2 T/2} \quad \text{and} \quad E\left[e^{\mu B_T} \max_{0\le t\le T} B_t\right] \simeq \mu T e^{\mu^2 T/2}$$

as $T \to \infty$.

EXERCISE 13.2 (Brownian Motion Writes Your Name). Prove that Brownian motion in $\mathbb{R}^2$ will write your name (in cursive script, without dotted i's or crossed t's).

To get the pen rolling, first take B_t to be two-dimensional Brownian motion on $[0, 1]$, and note that for any $[a, b] \subset [0, 1]$ that the process

$$X_t^{(a,b)} = (b-a)^{1/2}(B_{a+t/(b-a)} - B_a)$$

is again a Brownian motion on $[0, 1]$. Now, take $g : [0, 1] \mapsto \mathbb{R}^2$ to be a parameterization of your name, and note that Brownian motion spells your name (to precision ϵ) on the interval (a, b) if

$$\sup_{0 \le t \le 1} |X_t^{(a,b)} - g(t)| \le \epsilon. \tag{13.41}$$

(a) Let A_k denote the event that inequality (13.41) holds for $a = 2^{-k-1}$ and $b = 2^{-k}$. Check that the A_k are independent events and that one has $P(A_k) = P(A_1)$ for all k. Next, use the Borel–Cantelli lemma to show that if $P(A_1) > 0$ then infinitely many of the A_k will occur with probability one.

(b) Consider an extremely dull individual whose signature is maximally undistinguished so that $g(t) = (0, 0)$ for all $t \in [0, 1]$. This poor soul does not even make an X; his signature is just a dot. Show that

$$P\left(\sup_{0 \le t \le 1} |B_t| \le \epsilon\right) > 0. \tag{13.42}$$

(c) Finally, complete the solution of the problem by using (13.42) and an appropriate Girsanov theorem to show that $P(A_1) > 0$; that is, prove

$$P\left(\sup_{0 \le t \le 1} |B_t - g(t)| \le \epsilon\right) > 0. \tag{13.43}$$

EXERCISE 13.3 (Sharpness of Novikov's Condition). We begin with a warm-up example with $T = \infty$. We have not established any Girsanov theorems on the infinite interval $[0, \infty)$, but we can still learn something from exploring what can go wrong there. First, we choose an $a > 0$ and an $0 < \epsilon < \frac{1}{2}$ so that we may introduce the hitting time

$$\tau_L = \inf\{t : B_t = -a + (1-\epsilon)t\}.$$

(a) Use the Lévy–Bachelier formula to show that

$$E\left(\exp\left(\left(\frac{1}{2} - \epsilon\right)\tau_L\right)\right) = e^{a(1-2\epsilon)}. \tag{13.44}$$

(b) Let $\mu(\omega, s) = 1(s \le \tau_L)$ and show that

$$E\left(\exp\left(\left(\frac{1}{2} - \epsilon\right)\int_0^\infty \mu^2(\omega, s)\, ds\right)\right) = e^{a(1-2\epsilon)} < \infty \tag{13.45}$$

but the exponential process

$$M_t(\mu) = \exp\left(\int_0^t \mu(\omega, s)\, dB_s - \frac{1}{2}\int_0^t \mu^2(\omega, s)\, ds\right) \tag{13.46}$$

satisfies

$$E(M_\infty) = E(M_{\tau_L}) = e^{-2\epsilon a} < 1. \tag{13.47}$$

(c) Now, use the previous observations to give an example of a $\mu(\omega, s)$ with $0 \leq s \leq 1$ such that

$$E\left(\exp\left(\left(\frac{1}{2}-\epsilon\right)\int_0^1 \mu^2(\omega,s)\,ds\right)\right) < \infty, \tag{13.48}$$

but for which we have

$$E(M_1(\mu)) < 1. \tag{13.49}$$

Such a μ demonstrates that one cannot relax the Novikov condition by replacing $\frac{1}{2}$ by any smaller number. In order to construct μ, one may want to recall that the process

$$X_t = \int_0^t \frac{1}{1-s}\,dB_s \quad 0 \leq t < 1$$

is equivalent to the process $\{B_{z(t)}\}$, where $z(t) = t/(1-t)$ for $0 \leq t < 1$.

EXERCISE 13.4 (Lazy Man's Novikov). If $\epsilon > 0$, the condition

$$E\left[\exp\left(\left(\frac{1}{2}+\epsilon\right)\int_0^T \mu^2(\omega,s)\,ds\right)\right] < \infty \tag{13.50}$$

is stronger than the Novikov condition, but it still holds under many of the cases where one would apply Theorem 13.4. The purpose of this exercise is to outline a proof of Theorem 13.4 where the condition (13.50) replaces the Novikov condition. The proof is quite straightforward, and it reveals that all the subtlety in Theorem 13.4 comes from squeezing out that last ϵ.

(a) Let $\tau_n = \min\{s : |M_s\sigma(\omega,s)| \geq t, \text{ or } s \geq T\}$, and let $\alpha > 1$ and $\beta > 0$ be parameters to be chosen later. Let

$$X(t) = \int_0^t \mu(\omega,s)\,dB_s \text{ and } Y(t) = \int_0^t \mu^2(\omega,s)\,ds,$$

so that we can write

$$M^{\alpha}_{t\wedge\tau_n} = \exp\left(\alpha X(t\wedge\tau_n) - \frac{1}{2}(\alpha+\beta)Y(t\wedge\tau_n) + \frac{1}{2}\beta Y(t\wedge\tau_n)\right). \tag{13.51}$$

Now, show that there are suitable choices for $\alpha > 1$, $\beta > 0$ together with suitable choices for p and q so that Hölder's inequality and (13.51) imply

$$E(M^{\alpha}_{t\wedge\tau_n}) \leq E\left[\exp\left((\frac{1}{2}+\epsilon)\int_0^T \mu^2(\omega,s)\,ds\right)\right]. \tag{13.52}$$

(b) By the condition (13.50), the right-hand side of equation (13.52) is finite, and by inspection the resulting bound is *independent* of n. Use this fact and the notion of uniform integrability to prove that $E(M_T) = 1$, so M_t is a martingale on $[0, T]$, exactly as we hoped.

CHAPTER 14

Arbitrage and Martingales

The martingale theory of arbitrage pricing is one of the greatest triumphs of probability theory. It is of immense practical importance, and it has a direct bearing on financial transactions that add up to billions of dollars per day. It is also of great intellectual appeal because it unites our economic understanding of the world with genuinely refined insights from the theory of stochastic integration.

Strangely, this high-profile theory also has a rough-and-ready side. Its central insight is in some ways embarrassingly crude, and, ironically, the whole theory rests on a bold guess that is prefectly natural to a streetwise gambler—yet much less natural to ivory tower economists or to portfolio-optimizing arbitrageurs. Still, the gambler's guess has a rigorous justification, and, in the end, the gambler's unfettered insight provides the world with scientific progress on an impressive scale.

14.1. Reexamination of the Binomial Arbitrage

Several chapters ago, we began our discussion of arbitrage pricing with the investigation of a simple one-period model where the stock price could take on only two possible values. Specifically, we considered an idealized world with one stock, one bond, and two times — time 0 and time 1. We assumed that the stock had a price of \$2 at time 0 and that its price at time 1 would either equal \$1 or \$4. We also assumed an interest rate of zero so that a bond with a price of \$1 at time 0 would also be worth \$1 at time 1.

The derivative security we studied was a contract that promised to pay \$3 when the stock moves to \$4 and to pay nothing if the stock moves to \$1. We looked for a way to replicate this payout (see Table 14.1) with a portfolio consisting of α units of stock and β units of bond, and by examination of the combined payout table we found that a portfolio that is long one unit of stock and short one unit of bond would exactly replicate the payout of the derivative X. The net out-of-pocket cost to create this portfolio was \$1, and the portfolio neither consumes nor creates any cash flow between period 0 and period 1, so by the arbitrage argument the unique arbitrage-free price for X is also precisely \$1.

TABLE 14.1. REPLICATION OF A DERIVATIVE SECURITY

	Portfolio	Derivative Security
Original cost	$\alpha S + \beta B$	X
Payout if stock goes up	$4\alpha + \beta$	3
Payout if stock goes down	$\alpha + \beta$	0

One of the points of this arbitrage pricing argument that seems to elude some people is that the arbitrage price of the derivative was determined without any regard for the probability p_{up} that the stock goes up or the corresponding probability $p_{down} = 1 - p_{up}$ that the stock goes down. The irrelevance of such probabilities may seem paradoxical at first, but on reflection we see that it expresses a naked mathematical truth. Moreover, the irrelevance of transition probabilities also provides us with clear answers to several nagging questions. For instance, we may have asked ourselves why utility theory did not enter into the valuation of the derivative security. Now, we can see that there was no need for utility theory because all of the uncertainty is driven out of our model by the arbitrage argument and the replicating portfolio. Probabilities do not enter the game at all, and, consequently, neither does utility theory.

A Clean Slate and Streetwise Inferences

For the moment, let us suppose that we forget this clear understanding of arbitrage valuation. Instead, let us share the untainted state of mind of a street-smart friend who has never heard about arbitrage but who does have extensive experience in the practical art of making sound bets. How would such a worldly individual price the derivative X?

For such a person, an all but inescapable first step is to try to *infer* the values of the transition probabilities for the price movements of the stock. Given such a task, about the only means at our disposal for making such an inference is to assume that the stock price offers a precisely fair gamble. This assumption is tantamount to saying that

$$2 = 4 \cdot p_{up} + 1 \cdot (1 - p_{up}),$$

and from this assumption we can infer that the transition probabilities must be

$$p_{up} = \frac{1}{3} \quad \text{and} \quad p_{down} = \frac{2}{3}.$$

Thus far, this inference looks feasible enough, even if a bit bold, but the proof of the pudding is the pricing of the derivative.

Streetwise Valuations

Here, the street-smart individual may squirm a bit. If a bet is to be made, the considerations of utility and risk preference cannot be ignored. Still, if the bet is small enough so that the funds at risk are within the (approximately) linear range of the gambler's utility, then our streetwise gambler will be content to estimate the fair price of the derivative contract X by its expected value under the inferred probability distribution $\pi = (p_{up}, p_{down})$. That is to say, our gambler's candidate for the fair price of the derivative contract is given by

$$E_\pi(X) = 3 \cdot p_{up} + 0 \cdot p_{down} = 3 \cdot \frac{1}{3} + 0 \cdot \frac{2}{3} = 1.$$

What is going on here? This ill-informed guttersnipe gambler has hit upon the same price for the derivative that we found with our careful economic reasoning. Worse yet, this was not just a lucky example. For any binomial model and any derivative contract, this apparently harebrained method of inferred probabilities turns out to yield the same value for the derivative contract that one finds by the economically rigorous arbitrage argument. Something serious is afoot.

14.2. The Valuation Formula in Continuous Time

The same street-smart reasoning can also be used in continuous-time models with nonzero interest rates. For us, such models provide the most natural domain for option pricing theory, so we will not tarry any longer with the analysis of discrete time models. The key step in continuous time is to find a proper analog to the inferred probability distribution that worked so well in the two-time model. Once we have found an appropriate candidate for the inferred probability distribution in continuous time, we can turn ourselves over to the machinery of stochastic calculus. Thereafter, simple calculation can be our guide.

To provide a concrete context for our analysis, we will take the general stock model

$$dS_t = \mu_t S_t \, dt + \sigma_t S_t \, dB_t, \tag{14.1}$$

where the only a priori constraints on the coefficients $\mu_t(\omega)$ and $\sigma_t(\omega)$ are those that we need to make $\{S_t\}$ a standard stochastic process in the sense of Chapter 8. For our bond model, we will also take a more general view and assume that

$$d\beta_t = r_t \beta_t \, dt, \tag{14.2}$$

or, equivalently,

$$\beta_t = \beta_0 \exp\left(\int_0^t r_s \, ds\right),$$

where the nonnegative process $\{r_t\}$ has the interpretation as an instantaneous risk-free interest rate. In passing, we also note that the last formula gives us a quick way to see that

$$d(\beta_t^{-1}) = -r_t \beta_t^{-1} \, dt, \tag{14.3}$$

a small fact that we will use shortly.

Present Value and Discounting

The street-smart gambler knows that a dollar today is better than a dollar tomorrow, and the search for the inferred probability measure must take account of this fact. Here, the street-smart gambler looks at the stock process S_t and notes that in a world with genuine interest rates one should try to think of the *discounted stock price* $D_t = S_t/\beta_t$ as a martingale. In other words, one should consider a probability measure Q on the path space $C[0,T]$ such that D_t is a martingale with respect to Q. Here, Q is the direct analog of the inferred transition probabilities $\pi = (p_{up}, p_{down})$ that we found so useful in the binomial model.

The derivative contract pays X dollars at time T, and the *present value* of such a payout is just X/β_T. In parallel with the gambler's analysis for derivative pricing in the binomial model, he would then regard the natural price for the derivative contract at time 0 to be the expectation of the present value of the contract under the inferred probability measure Q. Thus, at time 0 the gambler's formula for the fair price of the contingent claim would be

$$V_0 = E_Q(X/\beta_T).$$

For this formula to make sense, we must at least assume that X is integrable, but, to minimize technicalities, we will only consider contingent claims $X \in \mathcal{F}_T$ that satisfy $X \geq 0$ and $E_Q(X^2) < \infty$. This requirement does not impinge in any practical way on our ability to analyze real-world claims.

At a general time $t \in [0,T]$, the streetwise pricing formula is given by the conditional expectation $E_Q(X/(\beta_T/\beta_t) \mid \mathcal{F}_t)$, and because the time t bond price β_t is $\mathcal{F}_t$-measurable, the pricing formula can also be written as

$$V_t = \beta_t E_Q(X/\beta_T \mid \mathcal{F}_t) \quad \text{for } 0 \le t \le T. \tag{14.4}$$

As it sits, the gambler's candidate for the fair value of the contingent claim X is not much more than a bold guess. Still, there is some immediate wisdom to this guess. At the very least, V_t replicates the correct terminal value of the option because if we let $t = T$ in equation (14.4) then the two interest rate adjustments exactly cancel to give $V_T = X$.

A Further Word About Q

We have tried to follow the lead of the discrete-time problem in our design of the pricing formula (14.4), but there remains an important restriction on Q that must be brought out. If the only property that we require of Q is that the process $\{S_t/\beta_t\}$ be a martingale under Q, then we are forced to accept some pretty dumb choices for Q. In particular, one could take Q to be any measure on $C[0,T]$ that makes S_t/β_t a constant. Such a silly choice was not what the streetwise gambler had in mind.

When we look back at the discrete problem, we find that the probability measure $\pi = (p_{up}, p_{down})$ has one further property besides making the stock price a martingale. Hidden in the simple notation for π is the fact that π puts its mass on the same paths that are *possible* in the original scenario, and vice versa. To impose an analogous condition on Q in our continuous-time model, we need to introduce the notion of equivalent measures.

If P and Q are two probability measures on a measurable space $(\Omega, \mathcal{F})$, we say that P and Q are *equivalent* provided that for any A in $\mathcal{F}$ we have

$$P(A) = 0 \quad \text{if and only if} \quad Q(A) = 0.$$

Intuitively, this condition tells us that events under Q are possible (or impossible) if and only if they are possible (or impossible) under P. One should note that the equivalence of measures is quite a different notion from the equivalence of processes. Two measures can be quite different yet be equivalent, whereas two equivalent processes must be — well — equivalent.

The bottom line here is that there are two requirements that one must impose on the probability measure Q that we use in the martingale pricing formula (14.4). The first requirement is that Q be equivalent to P in the sense just defined, and the second requirement is the fundamental one that under Q the process S_t/β_t be a martingale. Logically enough, Q is often called the *equivalent martingale measure*, although the term is neither as euphonious nor as precise as one might hope.

Naturally, there are questions of existence and uniqueness that come to us hand-in-hand with our introduction of Q. These important issues will be dealt with in due course, but, before tackling any technical issues, we should take a harder look at some basic structural features of the pricing formula.

A Critical Question

Thus far, we have only modest heuristic support for the reasonability of the pricing formula (14.4). To be truly excited, we at least need to show that V_t is equal to the value of a self-financing portfolio in the underlying stock and bond; that is, we would need to show that V_t can also be written as

$$V_t = a_t S_t + b_t \beta_t \quad \text{for } 0 \le t \le T, \tag{14.5}$$

where the process V_t also satisfies the self-financing condition

$$dV_t = a_t dS_t + b_t d\beta_t \quad \text{for } 0 \le t \le T. \tag{14.6}$$

Remarkably enough, the determination of coefficients a_t and b_t that satisfy equations (14.5) and (14.6) is almost mechanical.

The Inventory

First, we should recall that there are two natural Q-martingales of immediate importance to us, and both of these deserve to be written in their representations as stochastic integrals with respect to the Q-Brownian motion $\tilde{B}_t$. The first of these martingales is the ***unadjusted conditional expectation***

$$U_t = E_Q(X/\beta_T \mid \mathcal{F}_t) = E_Q(X/\beta_T) + \int_0^t u(\omega, s)\, d\tilde{B}_s, \tag{14.7}$$

and the second is the ***discounted stock price*** that motivated us to introduce the inferred probability measure Q,

$$D_t = S_t/\beta_t = S_0/\beta_0 + \int_0^t d(\omega, s)\, d\tilde{B}_s. \tag{14.8}$$

The integrands $u(\omega, t)$ and $d(\omega, t)$ that represent the Q-martingales U_t and D_t will provide the building blocks for our portfolio weights a_t and b_t.

Finally, we need to keep in mind the simple bookkeeping relationships between U_t and D_t and their (adjusted and unadjusted) counterparts V_t and S_t. Specifically, we have

$$V_t = \beta_t U_t \quad \text{and} \quad S_t = \beta_t D_t. \tag{14.9}$$

For the moment, all we need from these equations is that the second one tells us how to express the differential of the stock price in terms of the differential of the discounted stock price, or vice versa. In particular, we will find that as we work toward formulas for the portfolio weights a_t and b_t we will need to use the fact that

$$dS_t = \beta_t\, dD_t + D_t\, d\beta_t \quad \text{and} \quad dD_t = \beta_t^{-1}\{dS_t - D_t\, d\beta_t\}. \tag{14.10}$$

The Portfolio Weights

Now, we come to a simple calculation that has some claim to being the most important calculation in this book. We will find that the direct expansion of the stochastic differential of V_t will give us explicit candidates for the portfolio weights a_t and b_t. We simply begin with $V_t = \beta_t U_t$ and work our way toward an equation that contains dS_t and $d\beta_t$ as the only differentials:

$$\begin{aligned}
dV_t &= d(\beta_t U_t) = \beta_t\, dU_t + U_t\, d\beta_t \\
&= \beta_t u(\omega, t)\, d\tilde{B}_t + U_t\, d\beta_t \\
&= \beta_t \frac{u(\omega,t)}{d(\omega,t)}\, dD_t + U_t\, d\beta_t \\
&= \beta_t \frac{u(\omega,t)}{d(\omega,t)} \beta_t^{-1}\, \{dS_t - D_t d\beta_t\} + U_t\, d\beta_t \\
&= \frac{u(\omega,t)}{d(\omega,t)}\, dS_t + \left\{ U_t - \frac{u(\omega,t)}{d(\omega,t)} D_t \right\} d\beta_t.
\end{aligned}$$

This calculation gives us the required candidates for the portfolio weights:

$$a_t = \frac{u(\omega,t)}{d(\omega,t)} \quad \text{and} \quad b_t = U_t - \frac{u(\omega,t)}{d(\omega,t)} D_t. \tag{14.11}$$

These formulas are very important for the martingale theory of pricing, and we will make repeated use of them in this chapter. Here, we should note that these portfolio allocations may be infeasible if the ratio $u(\omega,t)/d(\omega,t)$ is poorly behaved, and before a_t and b_t are put to any serious use one needs to check that they satisfy the integrability conditions that are required by the SDE for dV_t.

Exercise 14.2 illustrates that pathological behavior is possible even in a fairly reasonable model, but such examples should not make us paranoid. Pathologies are the exception rather than the rule, and, unless otherwise specified, we will always assume that the stock and bond models lead us to values for $u(\omega,t)$ and $d(\omega,t)$ that provide well-defined portfolio weights a_t and b_t that are suitably integrable for any contingent claim X that satisfies the standing conditions that $X \geq 0$ and $E_Q(X^2) < \infty$.

The Self-Financing Check

By our construction of the portfolio weights a_t and b_t, we know that

$$dV_t = a_t\, dS_t + b_t\, d\beta_t \quad \text{for } t \in [0,T],$$

so if we want to show that the portfolio determined by (a_t, b_t) is self-financing we only need to show that we also have

$$V_t = a_t S_t + b_t \beta_t. \tag{14.12}$$

When we evaluate the right-hand side of equation (14.12) by inserting the values of a_t and b_t given by equation (14.11), we find

$$\begin{aligned}
a_t S_t + b_t \beta_t &= \frac{u(\omega,t)}{d(\omega,t)} S_t + \left\{ U_t - \frac{u(\omega,t)}{d(\omega,t)} D_t \right\} \beta_t \\
&= U_t \beta_t + \frac{u(\omega,t)}{d(\omega,t)} S_t - \frac{u(\omega,t)}{d(\omega,t)} D_t \beta_t,
\end{aligned}$$

and since $U_t \beta_t = V_t$ and $D_t \beta_t = S_t$ the preceding formula simplifies to just V_t. The bottom line is that V_t does represent the value of a self-financing portfolio, and explicit formulas for the portfolio weights are given by the equations of (14.11).

THE EXISTENCE OF Q AND TWO TECHNICAL CONDITIONS

At this point, the pricing formula is starting to look rather compelling, but several issues still require attention before we go too much further. At a minimum, we must be sure that we really do have a probability measure Q that makes the discounted stock price $D_t = S_t/\beta_t$ into a Q-martingale. After all, the martingale pricing formula is simply a typographical fantasy if Q does not exist.

From our experience with Girsanov theory, we know that the first step toward the determination of such a measure is to work out the SDE for D_t. An easy way to find this SDE is simply to apply Itô's formula to D_t and turn the crank:

$$\begin{aligned} dD_t &= d(S_t/\beta_t) = \beta_t^{-1}dS_t + S_t d\beta_t^{-1} \\ &= \beta_t^{-1}S_t(\mu_t\,dt + \sigma_t dB_t) - r_t\beta_t^{-1}S_t\,dt \\ &= D_t\{(\mu_t - r_t)\,dt + \sigma_t dB_t\}. \end{aligned} \tag{14.13}$$

Now, from this SDE, we see that D_t would be a local martingale if we could only remove the drift term $\mu_t - r_t$. Fortunately, we also know from Girsanov theory that this is easily done.

If we define the measure Q by taking $Q(A) = E_P(1_A M_T)$, where M_t is the exponential process

$$M_t = \exp\left(-\int_0^t m_s\,dB_s - \frac{1}{2}\int_0^t m_s^2\,ds\right) \text{ with } m_t = \{\mu_t - r_t\}/\sigma_t, \tag{14.14}$$

then we know by Theorem 13.2 that if $m_t = (\mu_t - r_t)/\sigma_t$ is bounded (or even if m_t just satisfies the Novikov condition), then M_t is a martingale, and the process defined by

$$d\tilde{B}_t = dB_t + m_t\,dt$$

is a Q-Brownian motion. Finally, the SDE for D_t given by equation (14.13) can be written in terms of $\tilde{B}_t$ as

$$dD_t = D_t\sigma_t\,d\tilde{B}_t = S_t\sigma_t/\beta_t\,d\tilde{B}_t, \tag{14.15}$$

so we see that D_t is a Q-local martingale. One way to check that D_t is actually an honest martingale is by showing that $E_Q(\sup_{0\le t\le T} D_t) < \infty$; that is, one checks the sufficient condition given in Exercise 7.3. An alternative criterion that is a bit more powerful can be obtained by noting that the SDE (14.15) tells us that D_t will be a Q-martingale provided that σ_t satisfies the Novikov condition under Q. In particular, if the process σ_t is bounded, then equation (14.15) tells us that D_t is a Q-martingale without further ado.

THE MARKET PRICE OF RISK

The remarkable quantity $m_t = (\mu_t - r_t)/\sigma_t$ that appears in the exponential martingale (14.14) has an economic interpretation in addition to its technical significance. The ratio $(\mu_t - r_t)/\sigma_t$ measures (in units of σ_t) the excess of the rate of return of the risky security S_t over the riskless security β_t. For this reason, m_t is often called the *market price of risk*. Models for which $m_t = 0$ are called *risk neutral models*, and by the form of the Girsanov transform (14.14), we see that such models have the amusing property that $P = Q$.

THE UNIQUENESS OF Q

If S_t/β_t is a constant process, then any Q makes S_t/β_t into a martingale, so one does not have uniqueness of the equivalent martingale measure unless some conditions are imposed on the stock and bond processes. While one might be able to concoct some situation where the pricing formula is well-defined even when the choice of Q is ambiguous, one should consider such a situation to be a fluke; in any reasonable application, Q must be unique for the value of $\beta_t E_Q(X/\beta_T \mid \mathcal{F}_t)$ to be unique. The following proposition spells out a simple but natural situation where one is sure to have the required uniqueness.

PROPOSITION 14.1 (On the Uniqueness of Q). *Suppose that $\{B_t : 0 \le t \le T\}$ is a Q-Brownian motion and $\mathcal{F}_t = \sigma\{B_s : 0 \le s \le t\}$ for $0 \le t \le T$. Further, suppose that $\{D_t, \mathcal{F}_t\}_{0\le t\le T}$ is a Q-martingale with an integral representation that is given by*

$$D_t = D_0 + \int_0^t d(s,\omega)\, dB_s \qquad 0 \le t \le T \tag{14.16}$$

where $d(s,\omega) \neq 0$, except possibly on a set with $dt \times dQ$ measure zero. If Q_0 is a probability measure on $\mathcal{F}_T$ that is equivalent to Q and if $\{D_t, \mathcal{F}_t\}_{0\le t\le T}$ is also a Q_0-martingale, then $Q(A) = Q_0(A)$ for all $A \in \mathcal{F}_T$.

PROOF. If $A \in \mathcal{F}_T$ and $Q(A) = 0$, then the equivalence of Q and Q_0 tells us that $Q_0(A) = 0$, so by the Radon-Nikodym theorem, such as one may find in Dudley (1989, pp. 134–138) or in Williams (1991, pp. 145–149), there exists an $\mathcal{F}_T$-measurable random variable in $L^1(dQ)$ such that $Q_0(A) = E_Q[1_A Y]$ for all A in $\mathcal{F}_T$. Moreover, we can use Y to build a continuous Q-martingale $\{M_t, \mathcal{F}_t\}$ by setting $M_t = E_Q(Y|\mathcal{F}_t)$ for $0 \le t \le T$, and, since $M_0 = E_Q[Y] = Q_0(\Omega) = 1$, the L^1-martingale representation theorem then tells us that we may write

$$M_t = 1 + \int_0^t \phi(\omega, s)\, dB_s \qquad 0 \le t \le T. \tag{14.17}$$

This formula now suggests an elegant plan. If one shows that $\phi(\omega, s) = 0$ almost surely, then we have $M_T = 1$, and, since M_T also equals Y, we further find that $Q_0(A) = E_Q(Y 1_A) = E_Q(M_T 1_A) = Q(A)$ for all $A \in \mathcal{F}_T$.

The desired uniqueness thus depends on showing $\phi(\omega, s) \equiv 0$, and this depends in turn on two basic observations. First, one can easily check that $D_t M_t$ is indeed a Q-local martingale simply by working with the definition or by following the hints in Exercise 14.5. Second, the product rule gives us

$$\begin{aligned} d(D_t M_t) &= D_t dM_t + M_t dD_t + dD_t \cdot dM_t \\ &= D_t \phi(\omega, t)\, dB_t + M_t d(\omega, t)\, dB_t + \phi(\omega, t) d(\omega, t)\, dt, \end{aligned}$$

so, for $D_t M_t$ to be a Q-local martingale, the drift term $\phi(\omega, t) d(\omega, t)\, dt$ must vanish. By hypothesis $d(\omega, t) \neq 0$ except on a set with $dt \times dQ$ measure zero, so we see $\phi(\omega, t)$ must vanish almost everywhere, and the proof is therefore complete. □

The existence and uniqueness of the equivalent martingale measure Q is a *sine qua non* of the martingale method of contingent claim valuation, and the results of the last two subsections deserve to be summarized in a proposition.

PROPOSITION 14.2 (Existence and Uniqueness of Q). *If the market price for risk $m_t = (\mu_t - r_t)/\sigma_t$ satisfies the P-Novikov condition*

$$E_P\left[\exp\left(\frac{1}{2}\int_0^T m_t^2\,dt\right)\right] < \infty,$$

then there is a probability measure Q that is equivalent to P such that S_t/β_t is a Q-local martingale. Moreover, if the volatility coefficient σ_t satisfies the Q-Novikov condition

$$E_Q\left[\exp\left(\frac{1}{2}\int_0^T \sigma_t^2\,dt\right)\right] < \infty,$$

then S_t/β_t is an honest Q-martingale. Finally, if the integrand that represents S_t/β_t as an Itô integral does not vanish except on a set with $dt \times dQ$ measure zero, then there is only one Q equivalent to P such that S_t/β_t is a Q-martingale.

14.3. The Black–Scholes Formula via Martingales

Among the cases where the martingale valuation formula $\beta_t E_Q(X/\beta_T|\mathcal{F}_t)$ may be worked out explicitly, there is no more instructive example than that provided by the European call option under the classical Black–Scholes model. Here, of course, the stock and bond price processes are given by the SDEs

$$dS_t = \mu S_t\,dt + \sigma S_t\,dB_t \quad \text{and} \quad d\beta_t = r\beta_t\,dt, \tag{14.18}$$

and the basic task is to calculate the time-zero value

$$V_0 = e^{-rT}E_Q[(S_T - K)_+]. \tag{14.19}$$

There are several ways to organize this calculation, but two basic steps seem inevitable. First, we must understand the measure Q that makes $D_t = S_t/\beta_t$ a martingale, and then we must exploit that understanding to calculate the expectation in equation (14.19). We always have the option of writing Q explicitly in terms of P and an exponential martingale, but one often does well to avoid this direct translation. In most cases, we can get what we need more quickly just by using the fact that we know the distribution of $\{S_t\}$ under Q.

FINDING Q TO MAKE A MARTINGALE

One natural way to find a probability measure Q that turns $D_t = S_t/\beta_t$ into a martingale is to take advantage of the SDE for D_t. In particular, we look for a way to rewrite that SDE so that Girsanov theory will tell us how to construct a Q that makes D_t into a local martingale.

In the case of the classical economy specified by the model (14.18), we can simply solve the stock and bond SDE to find an explicit representation of the process $D_t = S_t/\beta_t$ as

$$S_t/\beta_t = S_0\exp\left((\mu - r - \sigma^2/2)t + \sigma B_t\right), \tag{14.20}$$

and we can then use this equation to calculate the SDE of $D_t = S_t/\beta_t$. Specifically, we can apply Itô's formula to the representation (14.20) to find

$$\begin{aligned} d(S_t/\beta_t) &= (\mu - r)(S_t/\beta_t)\,dt + \sigma(S_t/\beta_t)\,dB_t \\ &= \sigma(S_t/\beta_t)\left\{d\left(t(\mu - r)/\sigma + B_t\right)\right\}. \end{aligned} \tag{14.21}$$

This SDE tells us that $D_t = S_t/\beta_t$ will be a Q-local martingale provided that Q is chosen so that the process defined by

$$\tilde{B}_t \stackrel{\text{def}}{=} \frac{\mu - r}{\sigma} t + B_t$$

is a Q-Brownian motion. In terms of the process $\{\tilde{B}_t\}$, the SDE for D_t given by equation (14.21) can be written as $dD_t = \sigma D_t d\tilde{B}_t$, and, since σ is just a constant here, this SDE tells us that D_t is in fact an honest martingale under the measure Q. Finally, since $\tilde{B}_t$ (and B_t) can be written as a monotone function of S_t/β_t, we also see that

$$\sigma\{S_t/\beta_t : 0 \le t \le T\} = \mathcal{F}_T,$$

so Proposition 14.1 confirms that Q is in fact the unique measure equivalent to P such that S_t/β_t is a Q-martingale.

How to Work Out a Q-Expectation

Now that Q has been defined, the next task is to make an explicit calculation of the valuation formula

$$V_0 = e^{-rT} E_Q[(S_T - K)_+]. \tag{14.22}$$

The defining property of Q is that $\tilde{B}_t = t(\mu - r)/\sigma + B_t$ is Q-Brownian motion, so the most logical way to compute the expectation in equation (14.22) is to express S_t in terms of $\tilde{B}_t$. This reexpression will let us throw the whole calculation back on our understanding of Brownian motion.

We can write S_T as a function of B_T, and we can write B_T as a function of $\tilde{B}_T$, so if we put the two steps together, we find

$$S_T = S_0 \exp(rT - \sigma^2 T/2 + \sigma \tilde{B}_T).$$

The time-zero valuation formula for a European call option with strike price K can then be written as

$$\begin{aligned} V_0 &= e^{-rT} E_Q[(S_T - K)_+] \\ &= e^{-rT} E_Q[(S_0 \exp(rT - \sigma^2 T/2 + \sigma \tilde{B}_T) - K)_+], \end{aligned}$$

and at this point only calculus remains. If we note that $Y = -\sigma^2 T/2 + \sigma \tilde{B}_T$ is a Gaussian variable with mean $-\sigma^2 T/2$ and variance $\sigma^2 T$, then V_0 equals the expectation of a function of Y that we may write explicitly as an integral:

$$e^{-rT} \int_{\log(K/S_0) - rT}^{\infty} \{S_0 e^{rT + y} - K\} e^{-(y + \sigma^2 T/2)^2 / 2\sigma^2 T} \frac{dy}{\sigma\sqrt{2\pi T}}. \tag{14.23}$$

This integral can be written more informatively as two integrals

$$\begin{aligned} S_0 \int_{\log(K/S_0) - rT}^{\infty} & e^{y} e^{-(y + \sigma^2 T/2)^2 / 2\sigma^2 T} \frac{dy}{\sigma\sqrt{2\pi T}} \\ & - e^{-rT} K \int_{\log(K/S_0) - rT}^{\infty} e^{-(y + \sigma^2 T/2)^2 / 2\sigma^2 T} \frac{dy}{\sigma\sqrt{2\pi T}}, \end{aligned}$$

and this formula already begins to take the shape of the Black–Scholes formula. All that remains is the computation of two integrals that are precisely of the type that we found in our first solution of the Black–Scholes PDE. With just a bit of diligence, the sum of these integrals may be reduced to $f(0, S_0)$, where $f(t, x)$ is the solution of the Black–Scholes PDE that is given by

$$x\Phi\left(\frac{\log(x/K) + (r + \frac{1}{2}\sigma^2)\tau}{\sigma\sqrt{\tau}}\right) - Ke^{-r\tau}\Phi\left(\frac{\log(x/K) + (r - \frac{1}{2}\sigma^2)\tau}{\sigma\sqrt{\tau}}\right), \tag{14.24}$$

where we use τ as the usual shorthand for $T - t$.

The bottom line here is that the martingale valuation formula (14.22) for the call option is not so abstract after all. In fact, even with a bit less work than was needed to solve the Black–Scholes PDE, we were able to recapture the classical Black–Scholes formula for the arbitrage price of the European call option. This is certainly a worthwhile achievement which can only add to our enthusiasm for the martingale arbitrage pricing formula. Still, before we become too devoted to the martingale method, we should check how well it measures up to the other big achievement of the PDE method — the explicit determination of the asset allocations for the replication portfolio.

The Portfolio Weights

When we followed the PDE approach to the evaluation of the European call option via the Black–Scholes PDE, we found delightfully explicit formulas for the asset allocation of the replicating portfolio. Specifically, if $f(x, t)$ is the solution to the Black–Scholes PDE given by equation (14.24), then the portfolio weights

$$a_t = f_x(t, S_t) \quad \text{and} \quad b_t = \frac{1}{r\beta_t}\left\{f_t(t, S_t) + \frac{1}{2}f_{xx}(t, S_t)\sigma^2 S_t^2\right\} \tag{14.25}$$

yield a self-financing portfolio $a_tS_t + b_t\beta_t$ that replicates the cash flow of the call option. Can we obtain such nice formulas by the martingale method?

We have already found that when we apply the martingale method to the problem of valuing the general claim X, we find formulas for a_t and b_t that can be written as

$$a_t = \frac{u(\omega, t)}{d(\omega, t)} \quad \text{and} \quad b_t = U_t - \frac{u(\omega, t)}{d(\omega, t)}D_t. \tag{14.26}$$

For the purposes of theoretical analysis these formulas are quite useful, but at least at first glance they are undeniably more abstract than the concrete formulas (14.25). Given values of t, S_t, and the model parameter σ, the PDE formulas (14.25) tell us exactly how to construct our portfolio, but if we hope to rely on (14.26) we have more work to do. Nevertheless, we will find in Exercise 14.4 that the abstract formulas (14.26) can be used to deduce the concrete formulas (14.25), even though the latter are powerless to return the favor in the general case.

Two Portfolio Properties

Even though a bit of work is needed to obtain useable portfolio weights from the martingale pricing formula, there are many other questions that it answers very quickly. In particular, for the European call option, the martingale pricing formula

$$V_t = \beta_t E_Q[(S_T - K)_+/\beta_T \mid \mathcal{F}_t] \quad \text{for } 0 \le t \le T \tag{14.27}$$

automatically tells us that

$$V_t > 0 \quad \text{for all } 0 \le t < T, \tag{14.28}$$

since for any value of S_t with $t < T$ the probability law Q assigns positive probability to the event $S_T > K$. This strict positivity of V_t for $t < T$ may seem like a modest property, but we will find shortly that it has some meaningful consequences.

A second nice property of the portfolio value V_t is that the discounted value V_t/β_t is a Q-martingale. We designed Q so that S_t/β_t would be a martingale, but the martingale property of V_t/β_t comes to us as a bonus from the representation

$$V_t/\beta_t = E_Q[(S_T - K)_+/\beta_T \mid \mathcal{F}_t]. \tag{14.29}$$

The tower property implies that all such conditional expectations are martingales, and at first glance the martingale property of V_t/β_t may seem a little toothless. Still, in due course, we will find that it also has important consequences. For the moment, we will pursue a different corollary of the martingale pricing formula — one that is undeniably interesting because it answers a question that has been floating in the background since we first began discussing options.

14.4. American Options

An especially pleasing consequence of the martingale pricing formula is the insight that it provides into the pricing of American options, a topic we have neglected for too long. The most important of these options is the call option with strike price K, and in this case if we exercise the option at time $0 \leq t \leq T$ our payout is $h(S_t)$, where $h(x) = (x - K)_+$.

We can now use the martingale theory of arbitrage pricing to establish the important fact that there is never any benefit to exercising an American call option at any time t prior to the terminal time T. In order to establish such a result, we need to show that there is no *strategy* for determining the time to exercise the option that is superior to the simple strategy that executes the option at time T.

In this context, the idea of a strategy turns out to be rather simple. It is just a rule that tells us when to exercise the option, and, since any rule can only be based on the information that is available up to the time of exercising the call, we see that strategies correspond precisely to stopping times.

Given any exercise strategy τ, the basic problem is to determine the value of following that strategy starting from time zero. If we focus on contingent claims such as the American option that has a time t payout of $h(S_t)$, then the time-zero present value of that payout is just $h(S_t)/\beta_t$. Similarly, if we consider the exercise strategy τ, then the time-zero present value of that payout is given by $h(S_\tau)/\beta_\tau$. The streetwise valuation of this payout is its expected value under Q, and no exercise strategy τ can yield a value of $E_Q(h(S_\tau)/\beta_\tau)$ that is larger than

$$V_0^A \stackrel{\text{def}}{=} \sup_\tau E_Q[h(S_\tau)/\beta_\tau]. \tag{14.30}$$

In particular, the constant $\tau \equiv T$ is also an exercise strategy, and the time-zero martingale valuation of a European option with terminal time T has the pricing formula

$$V_0^E \stackrel{\text{def}}{=} E_Q[h(S_T)/\beta_T|\mathcal{F}_0] = E_Q[h(S_T)/\beta_T],$$

so we always have the trivial inequality

$$V_0^E \leq V_0^A. \tag{14.31}$$

Without imposing some constraints on h, we can have strict inequality in the bound (14.31), but, fortunately, the class of h for which we have *equality* is reasonably large. In particular, this lucky class is large enough to contain $h(x) = (x-K)_+$, so for the American call options there is no strategy τ that does better than simply exercising the option at the terminal time T.

PROPOSITION 14.3 (Condition for No Early Exercise). *Suppose that the function $h : \mathbb{R} \to \mathbb{R}$ is convex and $h(0) = 0$. Suppose also that the processes S_t and β_t are $\{\mathcal{F}_t\}$ adapted, $\beta_t > 0$, β_t is nondecreasing, and the ratio S_t/β_t is a Q-martingale. If $M_t = h(S_t)/\beta_t$ is integrable for $0 \le t \le T$, then $\{M_t, \mathcal{F}_t\}$ is a Q-submartingale for $0 \le t \le T$.*

PROOF. We first observe that if $0 \le p \le 1$, then for any $x \ge 0$ we have the trivial identity $p \cdot x = p \cdot x + (1-p) \cdot 0$, so the convexity of h tells us that

$$h(px) \le p \cdot h(x) + (1-p) \cdot h(0) = p \cdot h(x). \tag{14.32}$$

Now, for any $t \ge 0$ and $s \ge 0$ we have

$$\begin{aligned} E_Q(M_{t+s} \,|\, \mathcal{F}_t) &= E_Q[h(S_{t+s})/\beta_{t+s} \,|\, \mathcal{F}_t] \\ &= \frac{1}{\beta_t} E_Q\left[h(S_{t+s}) \frac{\beta_t}{\beta_{t+s}} \,\Big|\, \mathcal{F}_t\right] \\ &\ge \frac{1}{\beta_t} E_Q\left[h\Big(\frac{\beta_t}{\beta_{t+s}} S_{t+s}\Big) \,\Big|\, \mathcal{F}_t\right], \end{aligned} \tag{14.33}$$

where in the last inequality we used the elementary bound (14.32). Now, if we first apply Jensen's inequality and then use the martingale property of $\{S_t/\beta_t\}$, we also have the bound

$$\begin{aligned} E_Q\left[h\left(\frac{\beta_t}{\beta_{t+s}} S_{t+s}\right) \,\Big|\, \mathcal{F}_t\right] &\ge h\left(E_Q\left[\frac{\beta_t}{\beta_{t+s}} S_{t+s} \,\Big|\, \mathcal{F}_t\right]\right) \\ &= h\left(\beta_t E_Q\left[\frac{S_{t+s}}{\beta_{t+s}} \,\Big|\, \mathcal{F}_t\right]\right) = h(S_t). \end{aligned}$$

When we combine the last estimate with the bound (14.33), we come precisely to the submartingale property of $M_t = h(S_t)/\beta_t$. □

The preceding proposition is of great reassurance to us. While there do exist some isolated circumstances of European style option contracts in the real world, they are almost vanishingly rare compared to American style contracts. Now that we know that there is no economic benefit to the early exercise of an American call, we see that our valuation of the European call carries over to that of the American call. Without the modest Proposition 14.3, the valuation of the European call would almost have been a hollow exercise.

PUTS ARE NOT COVERED

The payout function at time t for a put option is given by $h(S_t) = (S_t - K)_-$, and in this case $h(x)$ is again convex. Unfortunately, $h(0) \ne 0$ so Proposition 14.3 does not permit us to reduce the valuation of American puts to the case of European puts. Moreover, this fact is not just a coincidence; one can easily find situations where the *immediate exercise* of an American put option will produce a value that is larger than the value of the corresponding European option.

This places the problem of valuing American put options in a very difficult spot. The requisite mathematical tools have much more to do with the theory of free boundary problems for partial differential equations than with stochastic calculus, and the whole topic is best left for another time and another place. In any case, several important foundational questions must be faced before we are at liberty to worry about the special problems presented by American puts.

14.5. Self-Financing and Self-Doubt

Pólya once wisely advised that anyone who is lucky enough to have two things to say should first say one and then the other, not both at the same time. Ever since we began our discussion of self-financing strategies in continuous time, there have been some important observations that have gone unsaid. The time has come for them to take their place in the conversation.

The sad fact is that the self-financing condition is considerably more subtle in continuous time than it is in discrete time. Many statements that are honestly self-evident in discrete time are neither so self-evident nor so honest in continuous time. We could have engaged these subtleties before we saw how the martingale pricing formula recaptured the Black–Scholes formula, or before we saw how it implied that one should not exercise an American call option before the terminal time. We naturally chose first to build a store of energy and courage before engaging those model nuances that may cause us to doubt our intuition or ourselves.

The crux of the problem is that in continuous time the set of self-financing portfolios is so large that some pathological portfolios manage to creep in. Our task now is to study some of those portfolios in order to assess and repair any damage that they might do to martingale pricing theory. Fortunately, by adding two natural supplements to the basic self-financing condition, we will be able to rule out the pathological portfolios. The resulting theory will then support all of the work that we have done earlier.

A Huge Class of Self-Financing Portfolios

The notion of a self-financing portfolio is a pillar of arbitrage pricing theory, yet so far we have not taken time to examine this important notion on its own merits. As we suggested a bit earlier, the self-financing condition is trickier in continuous time than in discrete time, and we will surely need some concrete examples to help us sort out the subtleties.

To begin, we can take a general stock model

$$dS_t = \mu_t S_t \, dt + \sigma_t S_t \, dB_t \quad \text{with} \quad S_0 = 1, \tag{14.34}$$

where the only restrictions on the coefficients μ_t and σ_t are those that are needed in order for S_t to be a standard process. Actually, the generality provided by the SDE (14.34) will only be used to illustrate how little is required of S_t in the construction of the self-financing strategies. In the examples that are important to us, we will simply take S_t to be the familiar geometric Brownian motion.

Our choice for the bond model is much more constrained. We will take a zero interest rate model so that $\beta_t \equiv 1$ for all t. The benefit of this restriction is that the trivial SDE $d\beta_t = 0$ causes the self-financing condition to be radically simplified. Remarkably enough, this strong restriction on β_t is not as draconian as one might think. In fact, Exercise 14.3 outlines an argument that shows how almost all problems involving self-financing portfolios can be reduced to the case where the interest rate is zero.

By definition, the portfolio $V_t = a_t S_t + b_t \beta_t$ is self-financing whenever we have $dV_t = a_t dS_t + b_t d\beta_t$, and by our choice of the bond model we have $d\beta_t = 0$ so the self-financing condition boils down to just the simple relation

$$dV_t = a_t dS_t. \tag{14.35}$$

If we integrate this SDE, we obtain a representation for V_t that we may equate to $a_t S_t + b_t \beta_t = a_t S_t + b_t$ to get an equation for b_t, which we may write as

$$b_t = V_0 + \int_0^t a_u \, dS_u - a_t S_t. \tag{14.36}$$

The bottom line here is that for *essentially any* choice of a_t one can take the b_t defined by equation (14.36) to obtain a self-financing pair (a_t, b_t). The only constraint on this construction is the modest requirement that the processes a_t and S_t be nice enough to guarantee the existence of the integral in equation (14.36). In the situations where one applies this construction, this constraint is usually trivial to verify, so, in the end, we have a very powerful machine for the construction of self-financing portfolios.

A Wild Self-Financing Portfolio

Now, we want to exploit the general construction to give examples of wild behavior in some simple self-financing portfolios. Our first step is to specialize the stock model (14.34) to one that facilitates computation, and the natural choice is simply to take

$$dS_t = \sigma S_t \, dB_t \qquad S_0 = 1, \tag{14.37}$$

so that S_t is just a geometric Brownian motion with geometric drift parameter $\mu = 0$.

Now, from the general recipe for the construction of self-financing portfolios with a constant bond price $\beta_t \equiv 1$, we know that we have almost complete flexibility in our choice of the portfolio weight a_t. If we keep an eye toward our past experience with Gaussian processes, then an interesting choice of a_t that presents itself is simply

$$a_t = \frac{1}{\sigma S_t (T-t)} \qquad 0 \le t < T. \tag{14.38}$$

Since $d\beta_t = 0$, we then have

$$dV_t = a_t \, dS_t = \frac{1}{\sigma S_t (T-t)} \sigma S_t \, dB_t = \frac{1}{T-t} \, dB_t, \tag{14.39}$$

and, consequently, V_t has the representation

$$V_t = V_0 + \int_0^t \frac{1}{T-s} \, dB_s. \tag{14.40}$$

We are already well acquainted with this process, and in particular we know from our first work with Gaussian processes that if we let

$$\tau_t = \int_0^t \frac{1}{(T-s)^2} \, ds,$$

then we may write $V_t - V_0 = B_{\tau_t}$, where B_t is a standard Brownian motion. When one pairs this fact with the general observation that the portfolio with weights given by equations (14.38) and (14.36) is self-financing, one has the raw material for many interesting examples (or counterexamples!) for problems in mathematical finance.

Unbounded Wealth versus Banking Standards

The portfolio $V_t = a_t S_t + b_t \beta_t$ constructed above is self-financing, and since V_0 was left arbitrary we may as well take $V_0 = 1$. Now, by the representation $V_t - V_0 = B_{\tau_t}$ and the fact that $\tau_t \to \infty$ as $t \to T$, we see that for any M the stopping time $\nu = \min\{t : V_t = M\}$ satisfies $P(\nu < T) = 1$ and $V_\nu = M$. In other words, the self-financing portfolio determined by the weights (a_t, b_t) is guaranteed to make an arbitrarily large amount of money during the time period $[0, T]$. This simple two security portfolio has all the essential features of the long-sought philosopher's stone; it is a tool for generating untold riches.

Our intuition tells us something has gone wrong, yet when we look back we find there is no room to criticize the stock model or the bond model. In fact, with only a little more calculation, we could have produced a similarly pathological portfolio under the completely standard Black–Scholes model that we have been using all along.

Nothing is wrong with the model or with the verification of the self-financing property, yet something *has* gone wrong. The problem is that our model diverges from the real world in a way that any decent banker would spot in an instant. Just consider what happens as the time gets nearer to T if the investor in charge of the portfolio V_t has not yet reached his goal. In that case, the investor borrows more and more heavily, and he pours the borrowed funds into the risky asset. Any banker worth his salt who observes such investment behavior will pull the investor's credit line immediately. The simplest rules of lending practice are enough to prohibit the management of a portfolio by any strategy (a_t, b_t) like that defined by equations (14.36) and (14.38).

Consequences of a Credit Constraint

The modeler's problem is not to improve on banking practice. The challenge is rather to see how banking standards translate into appropriate constraints on the class of allowable self-financing strategies. Fortunately, this translation is simple. If we restrict our attention to just those self-financing portfolios that satisfy the *credit constraint* $V_t \geq 0$ for all $t \in [0, T]$, then we can show that all sorts of get-rich-quick schemes will be ruled out automatically. The mathematical basis for this observation is given by the next proposition.

Proposition 14.4 (Credit Constraint Supermartingale). *Consider the general stock and bond model*

$$dS_t = \mu_t S_t \, dt + \sigma_t S_t \, dB_t \quad \text{and} \quad d\beta_t = r_t \beta_t \, dt. \tag{14.41}$$

If there is a unique probability measure Q that is equivalent to P such that S_t/β_t is a square integrable Q-martingale, and if the portfolio $V_t = a_t S_t + b_t \beta_t$ is self-financing, then the discounted portfolio value V_t/β_t is a Q-local martingale. Moreover, if $V_t \geq 0$ for all $0 \leq t \leq T$, then V_t/β_t is also a Q-supermartingale.

PROOF. The first step is to relate the SDE for V_t/β_t to that of S_t/β_t. This only requires a product rule calculation:

$$\begin{aligned} d(V_t/\beta_t) &= \beta_t^{-1} dV_t - V_t r_t \beta_t^{-1}\, dt \\ &= \beta_t^{-1}\{a_t dS_t + b_t r_t \beta_t\, dt\} - \{a_t S_t + b_t \beta_t\} r_t \beta_t^{-1}\, dt \\ &= a_t\{\beta_t^{-1} dS_t - S_t r_t \beta_t^{-1}\, dt\} = a_t d(S_t/\beta_t). \end{aligned} \tag{14.42}$$

Now, since S_t/β_t is a Q-martingale, the martingale representation theorem and localization tell us there is an α_t such that

$$d(S_t/\beta_t) = \alpha_t\, d\tilde{B}_t, \tag{14.43}$$

where $\tilde{B}_t$ is a standard Brownian motion under Q; moreover, if we thumb back to equation (14.15) to recall that $dD_t = \sigma_t D_t d\tilde{B}_t$, then we even find that $\alpha_t = \sigma_t S_t/\beta_t$, although such an explicit formula is not really necessary.

Simply by combining equations (14.42) and (14.43), we have

$$d(V_t/\beta_t) = a_t \alpha_t\, d\tilde{B}_t,$$

and this SDE is enough to certify that V_t/β_t is a local martingale. Finally, we proved long ago (in the course of establishing Proposition 7.11) that any nonnegative local martingale is a supermartingale, so the proof of the proposition is complete. □

The credit constraint $V_t \geq 0$ applied in Proposition 14.4 is more liberal than the conditions that are universally imposed by lenders and regulatory agencies, so we will not miss out on any realistic trading strategies if we restrict our attention to just those self-financing pairs (a_t, b_t) that satisfy $V_t \geq 0$ for all $t \in [0, T]$. This class of strategies is called SF^+, and it is an important one for financial theory.

SF^+ AND RISK-FREE BONDS

The lending rules of banks and regulators give us the best reason for limiting our consideration to the trading strategies in SF^+, but there are also beneficial modeling consequences to this restriction. In order to make one of these precise, we will need the notion of a *risk-free bond arbitrage.*

DEFINITION 14.1 (Risk-Free Bond Arbitrage). *A portfolio strategy* (a_t, b_t) *is called a* risk-free bond arbitrage *on* $[0, T]$ *if there is a* $\lambda \in \mathbb{R}$ *such that*

$$P(V_0/\beta_0 \leq \lambda) = 1, \quad P(V_T/\beta_T \geq \lambda) = 1, \tag{14.44}$$

and

$$P(V_T/\beta_T > \lambda) > 0. \tag{14.45}$$

Any portfolio V_t that satisfies (14.44) is *guaranteed* to provide a yield during the time period $[0, T]$ that is at least as great as the yield of the risk-free bond, and by equation (14.45) it has a chance of providing a better yield. Such a deal is simply too good. A trading universe that supports the existence of such portfolios would not have any demand for risk-free bonds, yet the real world shows there is a substantial demand for such bonds. One of the reassuring features of the class of SF^+ strategies is that it rules out the possibility of a risk-free bond arbitrage.

PROPOSITION 14.5 (SF^+ and Risk-free Bond Arbitrage). *If (a_t, b_t) is a risk-free bond arbitrage, then $(a_t, b_t) \notin SF^+$.*

PROOF. If (a_t, b_t) satisfies (14.44) and (14.45), then the fact that P and Q are equivalent measures implies that the inequalities (14.44) and (14.45) remain true when P is replaced by Q. This in turn implies that

$$E_Q(V_0/\beta_0) \leq \lambda \quad \text{and} \quad E_Q(V_T/\beta_T) > \lambda. \tag{14.46}$$

On the other hand, if $(a_t, b_t) \in SF^+$, then Proposition 14.4 implies $\{V_t/\beta_t\}$ is a Q-supermartingale. Since the equations of (14.46) are incompatible with the fact that supermartingales have nonincreasing expectations, we see $(a_t, b_t) \notin SF^+$ as claimed. □

PROBLEMS REMAIN EVEN WITHIN SF^+

Remarkably, there are still some pathological strategies even within the nice class SF^+. These odd strategies are easy to overlook because they call on the investor to behave in a way that is stupid, or even self-destructive. For this reason, these are called ***suicide strategies***, and they will lead us to another important restriction on the set of candidate strategies. Formally, we say that the portfolio weight pair (a_t, b_t) is a suicide strategy provided that $V_0 > 0$ and $V_T = 0$. Such portfolios are just the opposite of those that generate unrealistic wealth. These portfolios are born to lose.

To see why suicide strategies deserve our attention, suppose we are interested in the value of a time T contingent claim X and that we have a strategy (a_t, b_t) from SF^+ such that $V_T = a_T S_T + b_T \beta_T = X$. Someone who has not worried about the subtle twists and turns of continuous-time processes might confidently argue that the only arbitrage-free price for X at time zero must be $V_0 = a_0 S_0 + b_0 \beta_0$, but our experience with continuous-time portfolios may make us cautious about such a claim. Such self-doubts would be well placed.

To see the problem, suppose that (a'_t, b'_t) is a suicide strategy in SF^+ and that V'_t is the associated portfolio. If we consider the new portfolio strategy defined by taking the sum $(a_t + a'_t, b_t + b'_t)$, then again we have a strategy from SF^+, but in this case the time-zero value of the portfolio is $V_0 + V'_0 > V_0$. The arbitrage argument for SF^+ trading strategies may be applied equally well to the two prices $V_0 + V'_0$ and V_0, so who is to say which value is the true arbitrage value of X? The existence of even one suicide strategy in SF^+ is enough to put us on the horns of a dilemma.

EXISTENCE OF SUICIDE STRATEGIES

Life would be beautiful if suicide strategies did not exist in SF^+, but, alas, they do exist. We can even build an example out of the same raw material we used to construct our first pathological self-financing strategy, the one that generated such great wealth in the blink of an eye.

To construct our new example, we begin with the same simple stock and bond model

$$dS_t = \sigma S_t\, dB_t \text{ where } S_0 = 1 \quad \text{and } \beta_t \equiv 1 \quad \text{for all } t \geq 0, \tag{14.47}$$

and we recall from our earlier analysis that (a_t, b_t) is a self-financing strategy when a_t and b_t are given by

$$a_t = \frac{1}{\sigma S_t(T-t)} \quad 0 \leq t < T, \quad \text{and} \quad b_t = V_0 + \int_0^t a_u\, dS_u - a_t S_t. \tag{14.48}$$

Furthermore, we recall that this choice of (a_t, b_t) yields a portfolio value V_t with the property that $V_t = V_0 + B_{\tau_t}$, where $\{B_t\}$ is a Brownian motion and $\tau_t \to \infty$ as $t \to T$.

Now, to get the desired suicide strategy, we modify (a_t, b_t) just a bit. To begin, we let ν denote the first time t such that $V_t = 0$, and we note by the representation of V_t as a time change of Brownian motion that we have $P(\nu < T) = 1$. We then define new portfolio weights (a'_t, b'_t) by setting

$$(a'_t, b'_t) = \begin{cases} (a_t, b_t) & \text{for } t \in [0, \nu] \\ (0, 0) & \text{for } t \in (\nu, T]. \end{cases} \tag{14.49}$$

To check that (a'_t, b'_t) is self-financing, we only note that we have

$$dV'_t = a'_t\, dS_t + b'_t\, d\beta_t$$

under each of the two separate cases of equation (14.49). Also, by the definition of ν as the *first* hitting time of the level 0, we have $V'_t \geq 0$ for all $t \in [0, T]$, and since $P(\nu < T) = 1$ we also have $P(V'_T = 0) = 1$. Finally, the choice of $V_0 = V'_0$ can be made arbitrarily in (14.48), so by taking $V_0 = 1$ we see that (a'_t, b'_t) is an element of SF^+ with all of the properties required by a suicide portfolio.

Elimination of Suicide Strategies

Suicide strategies exist in SF^+, and they wreak havoc on naive applications of the arbitrage argument, so — by hook or by crook — they must be banished. One of the easiest ways to eliminate suicide strategies is to restrict our attention only to those strategies in SF^+ that have the property that V_t/β_t is a martingale under the equivalent martingale measure Q. At first glance, this may seem like an odd requirement, but after a bit of checking one can confirm that it is both effective and appropriate.

To see why this restriction works, we just note that any suicidal strategy in SF^+ has the property that

$$P(V_0/\beta_0 > 0) = 1 \text{ and } P(V_T/\beta_T = 0) = 1,$$

so, by the equivalence of Q and P, we also have

$$Q(V_0/\beta_0 > 0) = 1 \text{ and } Q(V_T/\beta_T = 0) = 1.$$

These identities imply that

$$E_Q(V_0/\beta_0) > 0 \text{ and } E_Q(V_T/\beta_T) = 0, \tag{14.50}$$

and the inequalities of (14.50) are incompatible with the assumption that V_t/β_t is a Q-martingale.

A small point worth recording here is that there is nothing about equation (14.50) that contradicts the *supermartingale* property of V_t/β_t, which we get for free when we assume that $(a_t, b_t) \in SF^+$. To rule out the suicide strategies, one must impose additional restrictions on (a_t, b_t), and, once one decides to rule out (14.50), the idea of requiring V_t/β_t to be a Q-martingale is natural enough.

14.6. Admissible Strategies and Completeness

Our discussion of self-financing strategies has disclosed the uncomfortable fact that continuous-time models can strain our intuition. To be sure, the difficulties that we found were met with sensible antidotes, but one cannot help remaining a bit apprehensive. How can we be sure that there are not further difficulties that remain undiscovered? The challenge now is to show that such undiscovered difficulties do not exist.

To meet this challenge, we must formalize the ideas that have been discovered in the preceding sections, and we must formulate theorems that address our worries. We will continue to work with the general stock and bond model in the form

$$dS_t = \mu_t S_t\,dt + \sigma_t S_t\,dB_t \quad \text{and} \quad d\beta_t = r_t \beta_t\,dt. \tag{14.51}$$

To begin, we will assume nothing about the coefficients in the SDEs (14.51) except that the processes r_t and σ_t are both nonnegative and that $\{S_t\}$ and $\{\beta_t\}$ are both standard stochastic processes in the sense of Chapter 8.

A Trio of Fundamental Definitions

The first step of the formalization is to lay out three basic definitions. The first two formalize ideas that have been introduced earlier, and the third introduces the important idea of *completeness* for a financial model.

DEFINITION 14.2 (Contingent Claim). *If there is a unique Q equivalent to P such that $\{S_t/\beta_t\}$ is a Q-martingale on $[0,T]$, then a nonnegative random variable $X \in \mathcal{F}_T$ is called a* contingent claim *provided that*

$$E_Q(X^2) < \infty. \tag{14.52}$$

DEFINITION 14.3 (Admissibility and Replication). *If $(a_t, b_t) \in SF^+$, then the pair (a_t, b_t) is called an* admissible strategy *provided that $\{V_t/\beta_t\}$ is a Q-martingale, where Q is the unique measure equivalent to P for which $\{S_t/\beta_t\}$ is a Q-martingale. The class of all admissible strategies is denoted by $\mathcal{A}$, and if $(a_t, b_t) \in \mathcal{A}$ satisfies*

$$a_T S_T + b_T \beta_T = X$$

for a contingent claim X, we say that (a_t, b_t) replicates *X.*

DEFINITION 14.4 (Completeness). *The model $\{(S_t, \beta_t) : t \in [0,T]\}$ is said to be* complete *provided that (1) there is a unique probability measure Q equivalent to P such that $\{S_t/\beta_t\}$ is a Q-martingale on $[0,T]$ and that (2) every time T contingent claim X can be replicated by a trading strategy in $\mathcal{A}$.*

Admissibility and Two Useful Consequences

The motivation for the class of admissible strategies differs considerably from the motivation for the class SF^+. We came to SF^+ because we had no alternative. We were driven by the rules of agencies and lenders to restrict our attention only to the strategies that satisfy the credit constraint that defines SF^+. To be sure, we were happy to find several theoretical benefits of working within SF^+, but these benefits were essentially forced on us.

In complete contrast, the decision to restrict our attention to the smaller class of admissible strategies is essentially *voluntary*. As a practical matter, investors are not institutionally constrained from following suicide strategies, even though few investors will wince at the loss of liberty that would come from proscribing such strategies. Nevertheless, a reasonable investor might worry that by restricting our attention only to the admissible strategies there might be some useful strategies thrown out along with the suicide strategies. We will soon allay this concern by showing that $\mathcal{A}$ contains all of the trading strategies we will ever need.

Before we prove this important result, we will first establish two other attractive properties of $\mathcal{A}$. The first of these tells us that if the contingent claim is replicated by a strategy (a_t, b_t) from SF^+ and a strategy (a'_t, b'_t) from $\mathcal{A}$, then the strategy from $\mathcal{A}$ is guaranteed to have an initial cost that is no greater than the initial cost of the strategy from SF^+. The bottom line is that when you can do the job in $\mathcal{A}$ and in SF^+, it cannot cost you a dime to use $\mathcal{A}$ — and it may save you a buck.

PROPOSITION 14.6 ($\mathcal{A}$ Is Optimal within SF^+). *For any pair of trading strategies such that*

$$(a'_t, b'_t) \in \mathcal{A} \quad \textit{and} \quad (a_t, b_t) \in SF^+ \tag{14.53}$$

which satisfy the terminal conditions

$$a'_T S_T + b'_T \beta_T = X \quad \textit{and} \quad a_T S_T + b_T \beta_T = X, \tag{14.54}$$

where X is a contingent claim, we have the inequality

$$a'_0 S_0 + b'_0 \beta_0 \le a_0 S_0 + b_0 \beta_0. \tag{14.55}$$

PROOF. Since we assume the strategy (a_t, b_t) is an element of SF^+, we have

$$V_t = a_t S_t + b_t \beta_t \ge 0 \quad \text{for all } 0 \le t \le T, \tag{14.56}$$

and consequently by Proposition 14.4 we see that V_t/β_t is a Q-supermartingale. The supermartingale property tells us that

$$a_0 S_0 + b_0 \beta_0 = E_Q(V_0/\beta_0) \ge E_Q(V_T/\beta_T) = E_Q(X/\beta_T), \tag{14.57}$$

while, on the other hand, for $V'_t = a'_t S_t + b'_t \beta_t$ the assumption $(a'_t, b'_t) \in \mathcal{A}$ and the definition of $\mathcal{A}$ tell us t V'_t/β_t is a Q-martingale. Thus, we also have

$$a'_0 S_0 + b'_0 \beta_0 = E_Q(V'_0/\beta_0) = E_Q(V'_T/\beta_T) = E_Q(X/\beta_T), \tag{14.58}$$

and, by the equality of the right-hand sides of equations (14.57) and (14.58), the proof of the inequality (14.55) is complete. □

The next proposition has less direct financial impact, but it provides a property of $\mathcal{A}$ that is quite useful computationally. It tells us that under very modest conditions on the model (S_t, β_t) there is at most one strategy in $\mathcal{A}$ that can be used to replicate a given contingent claim.

PROPOSITION 14.7 (Uniqueness within $\mathcal{A}$). *Suppose the two trading strategies (a_t, b_t) and (a'_t, b'_t) are both elements of $\mathcal{A}$. If both strategies replicate the contingent claim X at time T, so*

$$a'_T S_T + b'_T \beta_T = X \quad \text{and} \quad a_T S_T + b_T \beta_T = X, \tag{14.59}$$

then we have

$$P(a_t = a'_t \text{ and } b_t = b'_t) = 1 \quad \text{for almost all } t \in [0, T], \tag{14.60}$$

provided that

$$P(\sigma_t > 0 \text{ for all } t \in [0, T]) = 1.$$

PROOF. Since $\{V_t/\beta_t\}$ and $\{V'_t/\beta'_t\}$ are both Q-martingales, we have

$$V_t/\beta_t = E_Q(V_T/\beta_T|\mathcal{F}_t) \quad \text{and} \quad V'_t/\beta_t = E_Q(V'_T/\beta_T|\mathcal{F}_t),$$

and by equations (14.58) the right-hand side of each of these equals $E_Q(X/\beta_T|\mathcal{F}_t)$. Consequently, we have $V_t/\beta_t = V'_t/\beta_t$, and since we always have $0 < \beta_t < \infty$ we also have $V_t = V'_t$.

The self-financing property applied to V_t and V'_t then gives us the basic equality

$$a_t\, dS_t + b_t\, d\beta_t = a'_t\, dS_t + b'_t\, d\beta_t. \tag{14.61}$$

At this point, one could just assert the equality of the coefficients on the basis of the equality of the differentials, and, even though this direct assertion would be valid, there is some humor to arguing the uniqueness from first principles.

In particular, if we let $\Delta_t = V_t - V'_t$, then Δ_t is identically equal to zero, but amusingly enough we can still apply Itô's formula to Δ_t^2 to find

$$0 = d\Delta_t^2 = 2\Delta_t d\Delta_t + d\Delta_t \cdot d\Delta_t = (a_t - a'_t)^2 \sigma_t^2\, dt. \tag{14.62}$$

This equation tells us that $(a_t - a'_t)^2\sigma_t^2$ is equal to zero with probability one for almost every t, and since $P(\sigma_t > 0) = 1$ we see that $P(a_t = a'_t) = 1$ for almost every t. When we subtract $a_t S_t = a'_t S_t$ from the equality $V_t = V'_t$, we also get $b_t\beta_t = b'_t\beta_t$ with probability one for almost every t. Finally, since $r_t \geq 0$ we always have $\beta_t > 0$, and therefore we also find that $P(b_t = b'_t) = 1$ for almost every $t \in [0, T]$. □

COMPLETENESS OF A MARKET MODEL

The optimality and uniqueness properties expressed by Proposition 14.6 and Proposition 14.7 demonstrate desirable features of the class of admissible trading strategies, but these are faint virtues if admissible strategies do not exist for the problems that interest us. Fortunately, the next theorem confirms that admissible strategies exist in abundance under almost any reasonable stock and bond model.

THEOREM 14.1 (Completeness Theorem).
The stock and bond model $\{(S_t, \beta_t)\}$ given by the SDEs of (14.51) is complete, provided that four conditions are satisfied:

$$E_P\left[\exp\left(\frac{1}{2}\int_0^T m_t^2\,dt\right)\right] < \infty \text{ where } m_t = (\mu_t - r_t)/\sigma_t, \tag{14.63}$$

$$E_Q\left[\exp\left(\frac{1}{2}\int_0^T \sigma_t^2\,dt\right)\right] < \infty, \tag{14.64}$$

$$\sigma\{S_t/\beta_t : 0 \le t \le T\} = \mathcal{F}_T, \quad \text{and} \tag{14.65}$$

$$P\left(\int_0^T \mu_t^2/\sigma_t^2\,dt < \infty\right) = 1. \tag{14.66}$$

PROOF. The first three conditions are already familiar to us, and we know from Proposition 14.2 that they guarantee the existence of a unique probability measure Q that is equivalent to P for which the process $\{S_t/\beta_t\}$ is a Q-martingale. The critical task is to show that for any contingent claim X there is an admissible trading strategy (a_t, b_t) that replicates X at time T.

If we define the process $\{V_t\}$ by the familiar valuation formula,

$$V_t = \beta_t E_Q(X/\beta_T \mid \mathcal{F}_t) \quad \text{for all } t \in [0, T], \tag{14.67}$$

then the measurability property of a contingent claim tells us $X \in \mathcal{F}_T$, so by (14.67) we have $V_T = X$. Also, by (14.67) and the claim's positivity property $X \ge 0$, we have $V_t \ge 0$, and finally (14.67) teams up with the basic properties of conditional expectation to imply that the process $\{V_t/\beta_t\}$ is a Q-martingale. These trivial checks tell us that V_t has every property that is required of a portfolio in $\mathcal{A}$, *provided* that V_t corresponds to the value of a portfolio that is given by a self-financing strategy (a_t, b_t).

Fortunately, we already know from the fundamental calculation of Section 14.2 that there are portfolio weights (a_t, b_t) such that $V_t = \beta_t E_Q(X/\beta_T \mid \mathcal{F}_t)$ may be written as $a_t S_t + b_t \beta_t$. We even know that the strategy (a_t, b_t) has the *formal* self-financing property

$$dV_t = a_t\,dS_t + b_t\,d\beta_t. \tag{14.68}$$

All we need to do now is to check that the coefficients in the SDE (14.68) satisfy the integrability conditions that are needed for the SDE to be well-defined.

When we use the model SDEs to expand (14.68), we find

$$dV_t = a_t\,dS_t + b_t\,d\beta_t = a_t\mu_t S_t\,dt + a_t\sigma_t S_t\,dB_t + b_t\beta_t r_t\,dt, \tag{14.69}$$

and to verify the necessary integrability properties of the coefficients of dt and dB_t, we need to use the formulas for a_t and b_t that were derived earlier in (14.11). Specifically, we have

$$a_t = \frac{u(\omega,t)}{d(\omega,t)} \quad \text{and} \quad b_t = U_t - \frac{u(\omega,t)}{d(\omega,t)} D_t \tag{14.70}$$

where $D_t = S_t/\beta_t$, $U_t = E_Q(X/\beta_T|\mathcal{F}_t)$, and the processes $d(\omega,t)$ and $u(\omega,t)$ are related to D_t, U_t, and the Q-Brownian motion $\{\tilde{B}_t\}$ by the SDEs

$$dD_t = d(\omega,t)\,d\tilde{B}_t \quad \text{and} \quad dU_t = u(\omega,t)\,d\tilde{B}_t.$$

The only fact that we will need about $d(\omega, t)$ is that it has the representation

$$d(\omega, t) = \sigma_t(\omega)S_t(\omega)/\beta_t(\omega) \tag{14.71}$$

that was derived in equation (14.15). Finally, the only fact that we will need about $u(\omega, s)$ is that it is in $L^2(dt)$ for almost every ω. This basic property comes from the fact that the representation theorem for the $L^2(dQ)$ martingale $U_t = E_Q(X/\beta_T|\mathcal{F}_t)$ provides a unique representing integrand $u(\omega, s)$ that is an element of $\mathcal{H}^2(dQ)$.

The pieces are all assembled, and now we only need to complete the calculations. These may seem a bit routine, but the end result is important, and some aspects of the calculations are informative. What adds spice to these computations is that models that fail to be complete often owe that fact to the failure of one of these routine checks.

To begin, we note by the formula for a_t given in equation (14.70) and by the formula for $d(\omega, t)$ given by equation (14.71) that we have

$$a_t(\omega)\mu_t(\omega)S_t(\omega) = \frac{u(\omega,t)}{d(\omega,t)}\mu_t(\omega)S_t(\omega) = \frac{\mu_t(\omega)}{\sigma_t(\omega)}u(\omega,t)\beta_t(\omega), \tag{14.72}$$

so by Cauchy's inequality we have

$$\int_0^T |a_t(\omega)\mu_t(\omega)S_t(\omega)|\,dt \le \left(\int_0^T \frac{\mu_t^2(\omega)}{\sigma_t^2(\omega)}\,dt\right)^{\frac{1}{2}} \left(\int_0^T u^2(\omega,t)\beta_t^2(\omega)\,dt\right)^{\frac{1}{2}}. \tag{14.73}$$

By the fourth condition of the theorem, the first integral of the inequality (14.73) is finite with probability one. To confirm the almost sure finiteness of the second integral, we only need to note that $\beta_t(\omega)$ is continuous on $[0, T]$ so it is bounded and to recall that we already know that $u(\omega, t) \in L^2(dt)$ almost all ω. The first coefficient in the SDE (14.69) therefore passes its test, and we have just two more to go.

The coefficient of dB_t in the SDE (14.69) is given by

$$a_t(\omega)\sigma_t(\omega)S_t(\omega) = \frac{u(\omega,t)\beta_t(\omega)}{\sigma_t(\omega)S_t(\omega)}\sigma_t(\omega)S_t(\omega) = u(\omega,t)\beta_t(\omega), \tag{14.74}$$

and for this to be a valid dB_t coefficient we need to check that it is an element of $L^2(dt)$ for almost every ω. We already noted that $\beta_t(\omega)$ is bounded for all ω and that $u(\omega, t) \in L^2(dt)$ for almost all ω, so the product $u(\omega, t)\beta_t(\omega)$ checks out just as it should.

The third coefficient in the SDE (14.69) is

$$\begin{aligned} b_t(\omega)\beta_t(\omega)r_t(\omega) &= \left(U_t(\omega) - \frac{u(\omega,t)}{d(\omega,t)}D_t(\omega)\right)\beta_t(\omega)r_t(\omega) \\ &= U_t(\omega)\beta_t(\omega)r_t(\omega) - \beta_t(\omega)\frac{r_t(\omega)}{\sigma_t(\omega)}u(\omega,t). \end{aligned} \tag{14.75}$$

Because $U_t(\omega)\beta_t(\omega)$ is continuous on $[0, T]$ it is bounded, and because β_t is a standard process we know that r_t is integrable for almost every ω. Consequently, the first summand of equation (14.75) is integrable for almost every ω.

The second summand of equation (14.75) is more interesting. We know now for almost every ω that $\beta_t(\omega)$ is bounded and $u(\omega, t)$ is in $L^2(dt)$, so the integrability of the second summand (14.75) will follow from Cauchy's inequality if we show that r_t/σ_t is in $L^2(dt)$ for almost every ω. This turns out to be easy to see if we note that the Novikov condition for $m_t = (\mu_t - r_t)/\sigma_t$ certainly implies that m_t is in

$L^2(dt)$ for almost every ω, and condition four of the theorem tells us μ_t/σ_t is in $L^2(dt)$ for almost every ω. As a consequence, the difference

$$\frac{r_t}{\sigma_t} = \frac{\mu_t}{\sigma_t} - m_t$$

is in $L^2(dt)$ for almost every ω. This completes the last of the coefficient checks and therefore completes the proof of the theorem. □

14.7. Perspective on Theory and Practice

In retrospect, this long chapter has proceeded through three distinct phases. In the first phase, we saw how a streetwise gambler could guess the pricing formula

$$V_t = \beta_t E_Q(X/\beta_T \mid \mathcal{F}_t). \tag{14.76}$$

We then saw what this representation could do for us, and, in particular, we found that the simple formula (14.76) could recapture the Black–Scholes formula for a call option. We also saw how the representation (14.76) could be used to argue that an American call option should not be exercised before the terminal date.

In the second phase, we made some critical observations about self-financing strategies, and we took a deeper look at the arbitrage argument in continuous time. We observed that the class SF contains strategies that are tantamount to get-rich-quick schemes that will not be allowed in any mature financial system. We then argued from basic banking principles that the only strategies that one should consider are those that satisfy a credit constraint. This led us to consider the class SF^+, and we found several properties of SF^+ that support its viability as a domain for financial theory. Specifically, we found that SF^+ has the economically important feature of excluding the possibility of any risk-free bond arbitrage.

Nevertheless, we continued our self-doubting exploration of self-financing strategies, and we discovered that even SF^+ has serious problems. In particular, we were forced to face the fact that suicide strategies present serious difficulties to continuous-time arbitrage theory. This was our darkest hour, and in near desperation we turned to the class $\mathcal{A}$ of admissible strategies. At first, this class seemed terribly special, and arbitrage theory seemed to be at risk. The nagging worry was that the alternating pattern of pathological example and quick-fix might keep repeating *ad nauseam.*

We then came to the third and most formal phase of our development. We took up the cause of the admissible strategies in earnest, and we proved the important completeness theorem. This result confirmed that under rather general circumstances any contingent claim may be replicated by an admissible strategy. This fact combined with the simpler uniqueness and minimality properties of admissibility to show that the class of admissible strategies is actually a natural domain for the continuous-time theory of arbitrage pricing. Our fears of the alternating pattern then could be safely set aside.

We now come to a fourth and final phase — a celebration. At the heart of the celebration, we have to place the acknowledgment of what is genuinely new in the martingale theory of arbitrage pricing. If the theory only provided a new perspective and gave us a sharper look at American call options, these would be fine accomplishments, but the contributions of martingale pricing theory go much further.

New Possibilities

To put martingale pricing theory in perspective, we must think back to the SDE valuation method and to our development of the Black–Scholes PDE. The PDE method focuses entirely on derivative securities that may be expressed as a function of the *stock price at the terminal time*; that is, the PDE method only permits us to study contingent claims of the form $X = g(S_T)$. The martingale pricing formula (14.76) is not nearly so restrictive. It only requires that X be suitably measurable and suitably integrable. This increased generality opens up a world of practical possibilities.

For example, if the contingent claim X is taken to be the average stock price over the time period $[0, T]$, then we see that

$$X = \frac{1}{T}\int_0^T S_t\,dt \tag{14.77}$$

depends on the *whole path of the stock price* $\{S_t : 0 \le t \le T\}$. Such a contingent claim cannot be expressed as a function of S_T alone, yet the new pricing formula (14.76) is perfectly applicable. Similarly, if we want to price the contingent claim given by

$$X = \max_{t\in[0,T]} S_t, \tag{14.78}$$

then the PDE method is powerless, even though the martingale method is again in its element.

The options defined by equations (14.77) and (14.78) are simple examples of what are respectively called *Asian* and *Look-Back* options. There are untold variations on these options that financial engineers design on a daily basis to serve the risk-sharing needs of their clients. Essentially all of these new options call on the martingale theory of arbitrage pricing in order to value the options and to construct portfolios that hedge the risk of writing such options.

New Computations

A second important contribution of the martingale pricing formula (14.76) is the window that it opens on new computational techniques. As one might expect, the explicit calculation of the martingale pricing formula is a reasonably rare event, even though we have seen that the pricing formula (14.76) provides a path to the Black–Scholes formula that is a bit easier than solving the Black–Scholes PDE. Still, once a computation is possible there is not much point in worrying about which method may be a little easier or a little harder. The much more important issue is the range of problems that a new method *renders feasible*.

One of the major benefits of the valuation formula (14.76) is that it is almost always possible to use simulation methods to estimate an expectation such as $E_Q(X/\beta_T)$. To be sure, such simulations are not easy to do well, and they will almost always be less accurate than the numerical solution of a PDE when a PDE method is applicable. Nevertheless, skillful simulations may be coupled with other numerical methods to provide practical solutions to many valuation problems that fall completely outside the domain where PDE methods can be applied. For this reason alone, the valuation formula (14.76) has telling practical importance.

New Models

The third and final virtue of the martingale approach to arbitrage pricing is that it extends the class of stock and bond models that one can consider. In order to obtain a PDE by the coefficient matching method, one typically assumes that the coefficients of the stock model may be written as functions of the current stock price. One can weaken this requirement a bit — say to include coefficients that also depend on time and the bond price — but the PDE method can never deal with model coefficients that depend on the whole history of the stock price.

The martingale method imposes far less stringent constraints on the coefficients of the stock and bond models. All one ever requires is that the coefficients be nonanticipating processes that are not too large. This added flexibility permits one to consider a wide class of non-Markovian stock and bond models, and, although this possibility has not been exploited very much so far, non-Markovian models can be expected to gain importance. The traditional concerns over computational cost become much less significant every day.

Probabilists (and others) have long been warned by historians that one cannot wisely ignore the past, and streetwise gamblers are justifiably hesitant to assume that a stock's history is completely irrelevant given the current price. One can hardly doubt that non-Markovian models will have their day, and we should be pleased to know that the martingale theory of arbitrage pricing stands ready for use upon their arrival.

14.8. Exercises

The first problem asks for the arbitrage price of a contingent claim that depends on the whole sample path. The problem provides an example of a claim that cannot be priced by the PDE method, and it has the added virtue of having been used for several years as an interview question at a well-known investment bank. The second problem provides an instructive example of an incomplete model. It also shows how some basic computations become simpler under risk-neutral models, and it shows how the coefficient checks such as those in the proof of the completeness theorem can have real bite.

The third problem suggests a simple computation that supports the important observation that many portfolio problems may be reduced to the consideration of models that have zero interest rates. The fourth problem then indicates how the abstract portfolio weights given by martingale theory can be used to recapture the concrete portfolio weights obtained by PDE methods. Any one of these problems it well worth the investment of some part of a nice fall afternoon.

Exercise 14.1 (An Interview Question). Consider the classic Black–Scholes stock and bond model with constant μ, σ, and r. Find the arbitrage price of a contingent claim that pays M if the stock price gets to a level K or higher during the time period $[0,T]$ and that pays zero otherwise. In other words, find a formula for the arbitrage price of the claim

$$X = M1\Big(\sup_{t\in[0,T]} S_t \geq K \Big).$$

EXERCISE 14.2 (Pozdnyakov's Incompleteness Example). The stock and bond models given by

$$dS_t = S_t\,dt + tS_t\,dB_t \quad \text{and} \quad d\beta_t = \beta_t\,dt \tag{14.79}$$

are like those we have considered before *except* that the coefficient tS_t of dB_t vanishes at time $t = 0$. This one defect is enough to make the model incomplete.

(a) In the martingale method, the first step in computing the arbitrage price of the claim X is to find a measure Q that is equivalent to P such that $D_t = S_t/\beta_t$ is a Q-martingale. Calculate dD_t and argue that D_t is already an honest P-martingale.

(b) This means that the equivalent martingale measure Q is equal to P, and this interesting invariance property suggests that some claims may yield very simple formulas for the time t valuation V_t. For example, take $X = \exp(B_T + T/2)$ and calculate an explicit representation for

$$V_t = \beta_t E_Q(X/\beta_T \mid \mathcal{F}_t).$$

(c) Now work out the values of the portfolio weights a_t and b_t by first deriving formulas for the integrands of the stochastic integrals (14.7) and (14.8) that represent the U_t and D_t processes as stochastic integrals. In particular, show that

$$u(\omega, t) = \exp(B_t - t/2) \quad \text{and} \quad d(\omega, t) = t \cdot S_t/\beta_t, \tag{14.80}$$

and use these formulas to obtain

$$a_t = \exp(B_t - t/2)/(tS_t e^{-t}) \text{ and } b_t = \exp(B_t - t/2)\left\{1 - \frac{1}{t}\right\}. \tag{14.81}$$

(d) Finally, observe that for all $t \in (0, T]$ we have

$$\int_0^t b_s \beta_s\,ds \equiv \infty \quad \text{with probability one,}$$

and argue that this implies that the model defined by (14.79) is not complete.

EXERCISE 14.3 (Generality of Zero Interest Rates).
In the examination of the binomial arbitrage, we found considerable convenience in the assumption that the interest rate was zero. Very pleasantly, the nature of self-financing strategies is such that we can almost always restrict attention to this case.

Suppose that the processes $\{S_t\}$ and $\{\beta_t\}$ satisfy the equations of the SDEs

$$dS_t = \mu(\omega, t)\,dt + \sigma(\omega, t)\,dB_t \quad \text{and} \quad d\beta_t = r\beta_t\,dt. \tag{14.82}$$

If the trading strategy (a_t, b_t) is self-financing for the processes S_t and β_t in the sense that it satisfies

$$a_t S_t + b_t \beta_t = a_0 S_0 + b_0 \beta_0 + \int_0^t a_t\,dS_t + \int_0^t b_t\,d\beta_t, \tag{14.83}$$

then for any measurable adapted $\gamma_t > 0$, we also have

(14.84)

$$a_t(\gamma_t S_t) + b_t(\gamma_t \beta_t) = a_0(\gamma_0 S_0) + b_0(\gamma_0 \beta_0) + \int_0^t a_t\,d(\gamma_t S_t) + \int_0^t b_t\,d(\gamma_t \beta_t),$$

provided that all the indicated integrals are well defined.

EXERCISE 14.4 (Abstract to Concrete Portfolio Weights). In general, the portfolio weights (a_t, b_t) of martingale pricing theory are given by the familiar formulas

$$a_t = \frac{u(\omega,t)}{d(\omega,t)} \quad \text{and} \quad b_t = U_t - \frac{u(\omega,t)}{d(\omega,t)} D_t. \tag{14.85}$$

On the other hand, the PDE method can be applied under the classic Black–Scholes model with constant μ, σ, and r to show that for the call option with strike price K the replicating portfolio has the concrete weights

$$a_t = f_x(t, S_t) \quad \text{and} \quad b_t = \frac{1}{r\beta_t}\left\{ f_t(t,S_t) + \frac{1}{2} f_{xx}(t,S_t)\sigma^2 S_t^2 \right\}, \tag{14.86}$$

where $f(t,x)$ is given by the Black–Scholes formula (14.24).

Show that the concrete weights (14.86) may be obtained from the more abstract weights by exploiting the fact that we have two representations for the arbitrage price of the call option:

$$f(t, S_t) = \beta_t E_Q((S_T - K)_+/\beta_T | \mathcal{F}_t) = \beta_t U_t.$$

EXERCISE 14.5 (Confiming a Martingale Property). Fill in the calculation that was omitted from Proposition 14.1 by showing

$$E_Q[D_t M_t | \mathcal{F}_s] = D_s M_s \quad \text{for all } 0 \le s \le t \le T. \tag{14.87}$$

As a hint, first recall that equation (14.87) really means

$$E_Q[D_t M_t 1_A] = E[D_s M_s 1_A] \quad \text{for all } 0 \le s \le t \le T \text{ and all } A \in \mathcal{F}_s,$$

and complete the proof of this identity by explaining each link in the chain

$$\begin{aligned} E_Q[D_t M_t 1_A] &= E_Q[D_t M_T 1_A] = E_{Q_0}[D_t 1_A] \\ &= E_{Q_0}[D_s 1_A] = E_Q[D_s M_T 1_A] = E_Q[D_s M_s 1_A]. \end{aligned}$$

CHAPTER 15

The Feynman–Kac Connection

The basic, stripped-down, Feynman–Kac representation theorem tells us that for any pair of bounded functions $q : \mathbb{R} \to \mathbb{R}$ and $f : \mathbb{R} \to \mathbb{R}$ and for any bounded solution $u(t,x)$ of the initial-value problem

$$u_t(t,x) = \frac{1}{2}u_{xx}(t,x) + q(x)u(t,x) \qquad u(0,x) = f(x), \tag{15.1}$$

we can represent $u(t,x)$ by the *Feynman–Kac Formula:*

$$u(t,x) = E\left[f(x+B_t)\exp\left(\int_0^t q(x+B_s)\,ds\right)\right]. \tag{15.2}$$

The value of this result (and its generalizations) rests in the link that it provides between important analytic problems, such as the solution of the PDE (15.1), and important probability problems, such as the calculation of the expectation (15.2). Remarkably, the benefits of this connection flow in both ways.

For us, the most immediate benefit of the Feynman–Kac formula (15.2) is that it gives us a way to get information on the *global* behavior of a sample path of Brownian motion. To give a concrete example, suppose we consider the amount of time T_t that Brownian motion spends in the set $[0,\infty)$ during the time period $[0,t]$. Because we can represent T_t as the integral

$$T_t = \int_0^t 1(B_s \geq 0)\,ds,$$

we can take $q(x) = -\lambda 1(x \geq 0)$ and $f(x) = 1$ in the Feynman–Kac formula (15.2) to find that $u(t,0)$ represents the Laplace transform of T_t. By solving the initial-value problem (15.1), we can calculate that transform and subsequently deduce that T_t has the arcsin distribution, a marvelous result due to P. Lévy.

Naturally, there are connections between equations such as (15.1) and (15.2) for processes that are more general than Brownian motion, and we will find that these connections can be used to obtain representations for the solutions of interesting PDEs. In particular, we will find versions of the Feynman–Kac formula that yield new and informative representations for the solutions for the Black–Scholes PDE and its various extensions.

15.1. First Links

We already know from our discussion of the diffusion equation that for a well-behaved function $f : \mathbb{R} \to \mathbb{R}$ there is a unique bounded solution $u(t,x)$ of the initial-value problem

$$u_t(t,x) = \frac{1}{2}u_{xx}(t,x) \qquad u(0,x) = f(x), \tag{15.3}$$

and we also know that this solution may be written as a Gaussian integral

$$u(t,x) = \frac{1}{\sqrt{2\pi t}} \int_{-\infty}^{\infty} f(u) e^{-(u-x)^2/2t}\, du. \tag{15.4}$$

Now, just by inspection, we can see that $u(t,x)$ can also be written as

$$u(t,x) = E[f(x+B_t)], \tag{15.5}$$

and — mirabile dictu — this formula is precisely the Feynman–Kac formula for the special case of $q(x) \equiv 0$.

This modest observation serves to remind us of many tools that might be used to prove the Feynman–Kac formula for general q, and, in particular, we are warned that Itô's formula might prove handy. Still, before we dig into the derivation of the general Feynman–Kac formula, we will first consider another important special case. When $q(x)$ is a constant, the Feynman–Kac formula turns out to have two informative interpretations that make easy work of guessing the general formula.

Brownian Motion: Killed or Discounted

One of the most general ways of building new stochastic processes out of old ones is by a construction called *killing*. The simplest example of this construction is given by *exponentially killed* Brownian motion, and anyone who happens to investigate this process has an excellent chance of discovering the Feynman–Kac formula.

To begin the construction, take any nonnegative random variable T. Brownian motion killed at time T is defined to be the process $\{X_t\}$ with values in the set $\mathbb{R} \cup \{\text{coffin}\}$, which is defined at time t by

$$X_t = \begin{cases} B_t & \text{if } 0 \le t \le T \\ \text{coffin} & \text{if } T < t. \end{cases}$$

Here, of course, the *coffin* state is a special state that we have introduced so that Brownian motion will have a place to go when it gets killed. The terminology is insensitive but traditional.

Now, given any $f : \mathbb{R} \to \mathbb{R}$, we can extend the definition of f to $\mathbb{R} \cup \{\text{coffin}\}$ by defining $f(\text{coffin}) = 0$, so we can introduce an analog to the Brownian motion formula (15.5) that solves the heat equation (15.3). To be completely specific, we specialize T to have the exponential distribution with parameter λ, so we have $P(T \ge t) = e^{-\lambda t}$ for $t \ge 0$, and, for the moment at least, we also assume that T is independent of the Brownian motion $\{B_t\}$. In this case, the process $\{X_t\}$ is called *exponentially killed* Brownian motion with *instantaneous killing rate* $\lambda \ge 0$.

We now ask ourselves whether we can find an initial-value problem that is satisfied by

$$u(t,x) = E[f(x+X_t)].$$

We certainly have $u(0,x) = f(x)$ just from the fact that $X_0 = x$, so we only need to work out u_t. From the definition of $u(t,x)$ together with the independence of the process $\{B_t\}$ and the killing time T, we have the explicit factorization

$$u(t,x) = E[f(x+B_t)1(T>t)] = e^{-\lambda t} E[f(x+B_t)], \tag{15.6}$$

so by direct calculation we have

$$
\begin{aligned}
u_t(t,x) &= e^{-\lambda t}\frac{\partial}{\partial t}E[f(x+B_t)] - \lambda e^{-\lambda t}E[f(x+B_t)] \\
&= e^{-\lambda t}\frac{1}{2}\frac{\partial}{\partial x^2}E[f(x+B_t)] - \lambda E[f(x+B_t)1(T>t)] \\
&= \frac{1}{2}\frac{\partial}{\partial x^2}E[f(x+B_t)1(T>t)] - \lambda E[f(x+B_t)1(T>t)] \\
&= \frac{1}{2}u_{xx}(t,x) - \lambda u(t,x).
\end{aligned}
$$

The net result is quite promising, and, formally at least, it tells us that for the special choice $q(x) \equiv -\lambda$ the formula given by

$$
(15.7) \qquad u(t,x) = E[f(x+B_t)\exp(\int_0^t q(x+B_s)\,ds)] = E[f(x+B_t)e^{-\lambda t}]
$$

will provide us with a representation of the solution of $u_t = \frac{1}{2}u_{xx} + q(x)u$ with $u(0,x) = f(x)$. In other words, the Feynman–Kac formula is valid when $q(x)$ is a nonpositive constant.

This simple observation is bound to present itself to anyone who investigates exponentially killed Brownian motion, and those investigators who are blessed with the youthful desire to generalize will find the discovery of the Feynman–Kac formula to be just a few lines away. The key step is to look for an alternative interpretation of the expectation (15.7), and, given our economic orientation, nothing could be more natural than to regard $E[f(x+B_t)e^{-\lambda t}]$ as a discounted payout where λ is interpreted as an interest rate. With this point of view, there is no reason to stick with just constant interest rates, and, if we simply consider interest rates that depend on $\{x+B_t\}$, then the Feynman–Kac formula leaps before our eyes.

15.2. The Feynman–Kac Connection for Brownian Motion

This story suggests how one might guess the Feynman–Kac formula, but guessing is not the same as proving, and that is our next task. There are many different proofs of the Feynman–Kac formula, but the method of proof that is most in keeping with the ideas that have been developed here is based on Itô's formula. In a nutshell, the key idea is to find a martingale $\{M_s\}$ for which we have

$$
M_0 = u(t,x) \quad \text{and} \quad E(M_t) = E\left[f(x+B_t)\exp\left(\int_0^t q(x+B_s)\,ds\right)\right].
$$

Any martingale trivially satisfies $E(M_0) = E(M_t)$, and, pleasantly enough, this bland identity is good enough to give us the Feynman–Kac formula. One can think of this technique as a kind of interpolation method, and there are many interesting identities that have analogous proofs.

THEOREM 15.1 (Feynman–Kac Representation Theorem for Brownian Motion). *Suppose that the function $q : \mathbb{R} \to \mathbb{R}$ is bounded, and consider the initial-value problem*

$$u_t(t,x) = \frac{1}{2}u_{xx}(t,x) + q(x)u(t,x) \qquad u(0,x) = f(x), \tag{15.8}$$

where $f : \mathbb{R} \to \mathbb{R}$ is also bounded. If $u(t,x)$ is the unique bounded solution of initial-value problem (15.8), then $u(t,x)$ has the representation

$$u(t,x) = E\left[f(x+B_t)\exp\left(\int_0^t q(x+B_s)\,ds\right)\right]. \tag{15.9}$$

PROOF. The most creative step in the interpolation method is the invention of the interpolating martingale. This always requires substantial experimentation, but with some experience and a bit of luck one may reasonably come to consider the process $\{M_s\}$ that is defined for $0 \le s \le t$ by setting

$$M_s = u(t-s, x+B_s)\exp\left(\int_0^s q(x+B_v)\,dv\right). \tag{15.10}$$

By Itô's formula and the assumption that $u(t,x)$ solves the PDE (15.8), we have

$$\begin{aligned} du(t-s,x+B_s) &= u_x(t-s,x+B_s)\,dB_s \\ &\qquad + \frac{1}{2}u_{xx}(t-s,x+B_s)\,ds - u_t(t-s,x+B_s)\,ds \\ &= u_x(t-s,x+B_s)\,dB_s - q(x+B_s)u(t-s,x+B_s)\,ds, \end{aligned}$$

so by the product rule we see dM_s equals

$$\begin{aligned} &\exp\left(\int_0^s q(x+B_r)\,dr\right)\left\{u_x(t-s,x+B_s)\,dB_s - q(x+B_s)u(t-s,x+B_s)\,ds\right\} \\ &\qquad + \exp\left(\int_0^s q(x+B_r)\,dr\right)\left\{q(x+B_s)u(t-s,x+B_s)\,ds\right\} \\ &= \exp\left(\int_0^s q(x+B_r)\,dr\right)\left\{u_x(t-s,x+B_s)\right\}dB_s. \end{aligned}$$

This last formula tells us that $\{M_s : 0 \le s \le t\}$ is a local martingale, and from the defining equation (15.10) of $\{M_s\}$ we find

$$\sup_{0\le s\le t}|M_s| \le ||u||_\infty \exp(t||q||_\infty), \tag{15.11}$$

where $||u||_\infty < \infty$ and $||q||_\infty < \infty$ by the boundedness hypotheses on q and u. The inequality (15.11) tells us the local martingale $\{M_s : 0 \le s \le t\}$ is bounded, and as a consequence it is an honest martingale.

The crudest consequence of the martingale property is the equality of the expectations $E[M_0] = E[M_t]$, but this modest fact is at the epicenter of our plan. In longhand, it tells us that

$$E[M_0] = u(t,x) = E(M_t) = E\left[f(x+B_t)\exp\left(\int_0^t q(x+B_s)\,ds\right)\right], \tag{15.12}$$

just as we wanted to show. □

The pattern demonstrated by this sly proof can be used to establish a large number of representation theorems for functions that satisfy parabolic PDEs. We will soon use the same device to prove a generalization of Theorem 15.1 that will help us understand the Black–Scholes PDE more deeply, but first we will look at a classic application of the basic Feynman–Kac formula for Brownian motion.

15.3. Lévy's Arcsin Law

To see how one can use the Feynman–Kac formula on a concrete problem, there is no better exercise than to work through Mark Kac's derivation of the famous Lévy Arcsin Law for the distribution of the amount of time T_t that Brownian motion spends in the positive half-line $[0, \infty)$ during the time period $[0, t]$. The law adds considerably to our intuition about Brownian motion (or random walk), and it also leads to a clear understanding of many empirical phenomena that might otherwise seem paradoxical.

THEOREM 15.2 (Lévy's Arcsin Law). *For any $0 \leq p \leq 1$ and any $t \geq 0$, we have*

$$P(T_t \leq pt) = \frac{2}{\pi}\arcsin\sqrt{p} = \frac{1}{\pi}\int_0^p \frac{du}{\sqrt{u(1-u)}}. \tag{15.13}$$

To get to the physics behind this law, one should first note that the arcsin density $\pi^{-1}u^{-1/2}(1-u)^{-1/2}$ piles up mass near $u = 0$ and $u = 1$. It also places minimal mass near $u = \frac{1}{2}$. This allocation of mass has some interesting consequences.

For the traditional coin flip game, Lévy's Arcsin Law tells us that during a long playing session you are at least twenty times more likely to be ahead for more than 98% of the time than to be ahead for just $50 \pm 1\%$ of the time. The arcsin law is one of the theoretical facts that shows why even the fairest world can seem unfair to a casual observer.

Translation to a PDE

In the introduction to the chapter, we noted that taking $q(x) = -\lambda 1(x > 0)$ and $f(x) = 1$ leads us to

$$-\lambda T_t = \int_0^t q(B_s)\, ds.$$

The Feynman–Kac connection also tells us that we have

$$E[\exp(\int_0^t q(x + B_s)\, ds)] = u(t, x), \tag{15.14}$$

provided that $u(t, x)$ is the unique solution of the initial-value problem

$$u_t = \frac{1}{2}u_{xx} - \lambda 1(x > 0)u \quad \text{and} \quad u(0, x) = 1 \text{ for all } x \in \mathbb{R}. \tag{15.15}$$

We now have a natural plan. We just need to solve the initial value problem (15.15), find the Laplace transform of T_t, and check that this transform coincides with that of the arcsin density. In essence, this plan works, but there are still a couple of new challenges.

Solving the PDE

Written in longhand, the initial-value problem (15.15) really gives us two equations:

$$u_t(t, x) = \begin{cases} \frac{1}{2}u_{xx}(t, x) - \lambda u(t, x) & x > 0 \\ \frac{1}{2}u_{xx}(t, x) & x \leq 0, \end{cases}$$

and the initial condition can be stated more precisely as

$$u(0^+, x) = 1,$$

where $u(0^+, x)$ denotes the limit of $u(t, x)$ as t approaches 0 through positive values.

We now face an amusing bit of mental gymnastics. The function $u(t, x)$ represents a Laplace transform, but the quickest way to solve the initial-value problem (15.15) is by taking the Laplace transform of both sides to convert it into an ODE. The solution of the resulting ODE will then be a Laplace transform of a Laplace transform, and, although this seems like a one-way ticket without possibility of return, special features of the problem make the round trip easy.

If we take the Laplace transform of $u(t, x)$ in the t variable and write

$$\hat{u}(\alpha, x) \stackrel{\text{def}}{=} \int_0^\infty e^{-\alpha t} u(t, x)\, dt,$$

then integration by parts shows that $u_t(t, x)$ is transformed to $1 + \alpha\hat{u}(\alpha, x)$ while differentiation under the integral sign shows that $u_{xx}(t, x)$ is transformed to $\hat{u}_{xx}(\alpha, x)$. All told, our PDE (15.3) transforms nicely into the more amenable ODE

$$1 + \alpha\hat{u}(\alpha, x) = \begin{cases} \frac{1}{2}\hat{u}_{xx}(\alpha, x) - \lambda\hat{u}(\alpha, x) & x > 0 \\ \frac{1}{2}\hat{u}_{xx}(\alpha, x) & x \leq 0. \end{cases} \tag{15.16}$$

The general solution of this ODE may be obtained by the usual exponential substitution, and, after a little routine work, we find that the only *bounded* solutions of the ODE (15.16) are given by

$$\hat{u}(\alpha, x) = \begin{cases} \frac{1}{\alpha+\lambda} + c_0 \exp(-x\sqrt{2(\alpha+\lambda)}\,) & x > 0 \\ \frac{1}{\alpha} + c_1 \exp(x\sqrt{2\alpha}\,) & x \leq 0, \end{cases} \tag{15.17}$$

where c_0 and c_1 are constants of integration that remain to be determined. Once these constants are found, we will have established the existence of a solution to the initial-value problem (15.15). The uniqueness of the solution then follows from the uniqueness for the corresponding ODE and the uniqueness of the Laplace transform.

Using Smoothness of Fit

The most natural way to find the two constants c_0 and c_1 is to hunt for a system of two linear equations that they must satisfy. We are lucky to have such a system close at hand. Any solution of the PDE $u_t(t, x) = \frac{1}{2}u_{xx}(t, x) + q(x)u(t, x)$ must be twice differentiable in x, and for any $u(x, t)$ that is even continuously differentiable we have

$$u(t, 0^+) = u(t, 0^-) \quad \text{and} \quad u_x(t, 0^+) = u_x(t, 0^-).$$

As a consequence, the Laplace transform $\hat{u}$ must also satisfy

$$\hat{u}(\alpha, 0^+) = \hat{u}(\alpha, 0^-) \quad \text{and} \quad \hat{u}_x(\alpha, 0^+) = \hat{u}_x(\alpha, 0^-), \tag{15.18}$$

and these equations will give us the necessary system.

When we calculate the corresponding limits in formula (15.17), we see that the first equation of (15.18) gives us

$$\frac{1}{\alpha+\lambda} + c_0 = \frac{1}{\alpha} + c_1$$

and the second equation of (15.18) gives us

$$-c_0\sqrt{2(\alpha+\lambda)} = c_1\sqrt{2\alpha}.$$

When we solve this system for c_0 and c_1, we then find

$$c_0 = \frac{\sqrt{\alpha+\lambda}-\sqrt{\alpha}}{\sqrt{\alpha}(\alpha+\lambda)} \quad \text{and} \quad c_1 = \frac{\sqrt{\alpha}-\sqrt{\alpha+\lambda}}{\alpha\sqrt{\alpha+\lambda}}.$$

We can use these constants in (15.17) to complete our formula for the solution of our initial-value problems for $\hat{u}(t,x)$ and $u(t,x)$, but we are really only after $\hat{u}(\alpha, 0^+)$. To make life easy, we just let $x = 0$ in (15.17) and collect terms to find

$$\hat{u}(\alpha, 0^+) = \int_0^\infty e^{-\alpha t}u(t,0)\,dt = \frac{1}{\sqrt{\alpha(\alpha+\lambda)}}. \tag{15.19}$$

This is as simple a formula as we could have hoped to find, and we have every reason to be confident that the proof of the arcsin law will soon be in the bag.

CHECKING THE TRANSFORM — CONFIRMING THE LAW

Lévy's Arcsin Law says that T_t/t has the arcsin density on $[0,1]$, and one line of calculus will confirm that this is equivalent to showing that T_t has the arcsin density on $[0,t]$. In terms of the Laplace transform, this means that the proof of Lévy's theorem will be complete if we show

$$E(e^{-\lambda T_t}) = \int_0^t \frac{1}{\pi}\frac{e^{-\lambda s}}{\sqrt{s(t-s)}}\,ds. \tag{15.20}$$

By the uniqueness of the Laplace transform, the identity (15.20) will follow if we can prove the equality of the transforms

$$\int_0^\infty E(e^{-\lambda T_t})e^{-\alpha t}\,dt = \int_0^\infty \int_0^t e^{-\alpha t}\frac{1}{\pi}\frac{e^{-\lambda s}}{\sqrt{s(t-s)}}\,ds\,dt. \tag{15.21}$$

The left-hand side of the integral (15.21) is just $\hat{u}(t,0)$, for which we have the lovely formula (15.17), so the proof of Lévy's theorem will be complete once we check that

$$\frac{1}{\sqrt{\alpha(\alpha+\lambda)}} = \int_0^\infty \int_0^t e^{-\alpha t}\frac{1}{\pi}\frac{e^{-\lambda s}}{\sqrt{s(t-s)}}\,ds\,dt. \tag{15.22}$$

Fortunately, the right-hand side of equation (15.22) is easy to compute. If we start with Fubini's theorem, we can finish with two applications of the familiar gamma integral

$$\int_0^\infty \frac{e^{-\gamma t}}{\sqrt{t}}\,dt = \sqrt{\frac{\pi}{\gamma}}$$

to find in quick order that

$$\begin{aligned}
\frac{1}{\pi}\int_0^\infty \frac{e^{-\lambda s}}{\sqrt{s}}\int_s^\infty \frac{e^{-\alpha t}}{\sqrt{t-s}}\,dt\,ds &= \frac{1}{\pi}\int_0^\infty \frac{e^{-\lambda s}}{\sqrt{s}}\int_0^\infty \frac{e^{-\alpha(t+s)}}{\sqrt{t}}\,dt\,ds \\
&= \frac{1}{\pi}\int_0^\infty \frac{e^{-(\lambda+\alpha)s}}{\sqrt{s}}\,ds\int_0^\infty \frac{e^{-\alpha t}}{\sqrt{t}}\,dt \\
&= \frac{1}{\pi}\sqrt{\frac{\pi}{\alpha+\lambda}}\sqrt{\frac{\pi}{\alpha}}.
\end{aligned}$$

Now, just cancelling the π's completes our check of equation (15.22) and finishes the proof of Lévy's Arcsin Law.

15.4. The Feynman–Kac Connection for Diffusions

The martingale interpolation technique that helped us prove the Feynman–Kac formula for Brownian motion can also be used to get analogous results for more general processes. The proof of the next theorem closely follows the basic pattern, but the calculations are still instructive.

THEOREM 15.3. *Let $q : \mathbb{R} \to \mathbb{R}$ and $f : \mathbb{R} \to \mathbb{R}$ be bounded and suppose that $u(t, x)$ is the unique solution to the problem defined by the equation*

$$u_t(t,x) = \frac{1}{2}\sigma^2(x)u_{xx}(t,x) + \mu(x)u_x(t,x) + q(x)u(t,x) \tag{15.23}$$

and the initial condition

$$u(0,x) = f(x).$$

If the functions $\mu : \mathbb{R} \to \mathbb{R}$ and $\sigma : \mathbb{R} \to \mathbb{R}$ are Lipschitz and satisfy the growth condition

$$\mu^2(x) + \sigma^2(x) \leq A(1 + x^2) \tag{15.24}$$

for some constant $A > 0$, then the function $u(t, x)$ then has the representation

$$u(t,x) = E\left[f(x + X_t)\exp\left(\int_0^t q(x + X_s)\,ds\right)\right], \tag{15.25}$$

where the process X_t is the unique solution of the SDE

$$dX_t = \mu(X_t)\,dt + \sigma(X_t)\,dB_t \qquad X_0 = 0. \tag{15.26}$$

PROOF. From our earlier experience, we can easily guess that we would do well to consider the process

$$M_s = u(t - s, X_s)\exp\left(\int_0^s q(x + X_v)\,dv\right). \tag{15.27}$$

If we now write $M_s = U_s I_s$, where U_s and I_s provide eponymous shorthand for the two natural factors of M_s, then the next step is to use the product rule to show $\{M_s : 0 \leq s \leq t\}$ is a local martingale.

By Itô's formula applied to U_s, we have

$$\begin{aligned}
dU_s &= u_x(t-s, X_s)\,dX_s + \frac{1}{2}u_{xx}(t-s, X_s)\,dX_s \cdot dX_s - u_t(t-s, X_s)\,ds \\
&= u_x(t-s, X_s)\sigma(X_s)\,dB_s \\
&\qquad + \left\{\frac{1}{2}u_{xx}(t-s, X_s)\sigma^2(X_s) + u_x(t-s, X_s)\mu(X_s) - u_t(t-s, X_s)\right\}ds, \\
&= u_x(t-s, X_s)\sigma(X_s)\,dB_s - q(x + X_s)u(t-s, X_s)\,ds,
\end{aligned}$$

and for I_s Itô's formula reduces to ordinary calculus to give

$$dI_s = \left\{\exp\left(\int_0^s q(x + X_v)\,dv\right)\right\}q(x + X_s)\,ds = I_s q(x + X_s)\,ds. \tag{15.28}$$

The product rule then gives us

$$\begin{aligned} dM_s &= I_s dU_s + U_s dI_s \\ &= I_s\{u_x(t-s, X_s)\sigma(X_s)\, dB_s - q(x+X_s)u(t-s, X_s)\, ds\} \\ &\quad + u(t-s, X_s)I_s q(x+X_s)\, ds \\ &= I_s u_x(t-s, X_s)\sigma(X_s)\, dB_s. \end{aligned}$$

Just like last time, this SDE confirms that $\{M_s\}$ is a local martingale, and the first step of the interpolation argument is complete.

Finally, the defining representation (15.27) for M_s and boundedness hypotheses on u and q imply

$$\sup_{0\le s\le t} |M_s| \le ||u||_\infty \exp(t||q||_\infty) < \infty, \tag{15.29}$$

so again the local martingale $\{M_s : 0 \le s \le t\}$ is bounded, and we see that $\{M_s\}$ is an honest martingale. By the martingale identity, we have $E(M_0) = E(M_t)$, and this identity is just a transcription of equation (15.25) so the proof of the Feynman–Kac representation is complete. □

15.5. Feynman–Kac and the Black–Scholes PDEs

The traditional hunting ground for financial applications of the Feynman–Kac method is the class of stock and bond models that may be written as

$$dS_t = \mu(t, S_t)\, dt + \sigma(t, S_t) dB_t \quad \text{and} \quad d\beta_t = r(t, S_t)\beta_t\, dt, \tag{15.30}$$

where the model coefficients $\mu(t, S_t)$, $\sigma(t, S_t)$, and $r(t, S_t)$ are given by explicit functions of the current time and current stock price. This set of models is less general than those we considered in the previous chapter where the coefficients only needed to be nonanticipating processes, but it is still an exceptionally rich and important class. At the barest minimum, it contains the classic Black–Scholes model where the coefficients take the specific forms

$$\mu(t, S_t) \equiv \mu S_t,\ \sigma(t, S_t) \equiv \sigma S_t, \text{ and } r(t, S_t) \equiv r$$

for constants μ, σ, and r.

The real charm of the set of models defined by the system (15.30) is that it is essentially the largest class for which one can use PDE methods to price contingent claims. In fact, if we simply repeat our original derivation of the classic Black–Scholes PDE, we can show that the time t arbitrage price $u(t, S_t)$ of the time T European claim $X = h(S_T)$ will satisfy the terminal-value problem

$$u_t(t,x) = -\frac{1}{2}\sigma^2(t,x)u_{xx}(t,x) - r(t,x)xu_x(t,x) + r(t,x)u(t,x) \tag{15.31}$$

$$u(T,x) = h(x). \tag{15.32}$$

The calculations that bring us to this important analytical problem are worth revisiting, but we will skip that step so that we can go directly to our main task, which is to modify the Feynman–Kac method to provide a representation for the solution of the terminal-value problem (15.31). Nevertheless, some time is well spent with Exercise 15.1, which invites the reader to discover that the terminal value problem (15.31) can be used to price claims in a class of models that is even a bit more general than the class defined by the system (15.30).

The logical structure of the Feynman–Kac connection is our main concern here, and we are prepared to make strong assumptions to set aside any purely technical concerns. In particular, we will assume that the nonnegative interest rate process $r(t,x)$ is bounded, and we will assume that the stock parameters $\mu(t,x)$ and $\sigma(t,x)$ satisfy both the Lipschitz condition

$$(\mu(t,x)-\mu(t,y))^2+(\sigma(t,x)-\sigma(t,y))^2 \le A(x-y)^2$$

and the linear growth rate condition

$$\mu^2(t,x)+\sigma^2(t,x)\le B(1+x^2).$$

We will even assume that $h(x)$ is bounded, and some may worry for a moment that this assumption is simply too strong. After all, the European call option corresponds to the unbounded function $h(x)=(x-K)_+$. As a practical matter, this concern is groundless. If we replace $h(x)$ by $h_0(x)=\min(h(x),M)$, where M denotes the total of all of the money in the universe, then h_0 is bounded and even sharp-penciled hedge fund partners will be happy to accept the pricing of $h_0(S_T)$ as a satisfactory surrogate for the pricing of $h(S_T)$.

With these ground rules in place, we are ready to state the main result of this section — the Feynman–Kac Formula for the solution of the general Black–Scholes PDE. The formula that we obtain may seem to be a bit more complex than those we have found before, but this complexity is largely due to some notational clutter that is forced on us in order to deal with a *terminal* value problem.

THEOREM 15.4. *If $u(t,x)$ is the unique bounded solution of the terminal-value problem given by equations (15.31) and (15.32), then $u(t,x)$ has the representation*

$$u(t,x)=E\left[h(X_T^{t,x})\exp\left(-\int_t^T r(s,X_s^{t,x})\,ds\right)\right] \tag{15.33}$$

where for $s\in[0,t]$ the process $X_s^{t,x}$ is defined by taking $X_s^{t,x}\equiv x$ and where for $s\in[t,T]$ the process $X_s^{t,x}$ is defined to be the solution of the SDE:

$$dX_s^{t,x}=r(s,X_s^{t,x})X_s^{t,x}\,dt+\sigma(s,X_s^{t,x})\,dB_s \quad \text{and} \quad X_t^{t,x}=x. \tag{15.34}$$

PROOF. We naturally want to exploit the familiar device of interpolating martingales, although this time our interpolation points will be chosen a bit differently. If we set

$$M_s=u(s,X_s^{t,x})\exp\left(-\int_t^s r(v,X_v^{t,x})\,dv\right)=U_sI_s, \qquad \text{for } t\le s\le T, \tag{15.35}$$

then we trivially find that $M_t=u(t,x)$. Also, the expectation of M_T is equal to the right-hand side of the target identity (15.33), so, if we can prove that M_s is a martingale for $s\in[t,T]$, then proof of the theorem will be complete.

As usual, we will extract the martingale property of M_s from a product rule calculation of dM_s. The calculation goes even more smoothly than before, but this time we will use the numerical notation for partial derivatives in order to minimize symbol delirium. First, we calculate dU_s by Itô's formula to find

$$\begin{aligned}
dU_s &= \left\{ u_1(s, X_s^{t,x}) + \frac{1}{2}\sigma^2(s, X_s^{t,x}) u_{22}(s, X_s^{t,x}) \right\} ds + u_2(s, X_s^{t,x})\, dX_s^{t,x} \\
&= \left\{ u_1(s, X_s^{t,x}) + \frac{1}{2}\sigma^2(s, X_s^{t,x}) u_{22}(s, X_s^{t,x}) + u_2(s, X_s^{t,x}) r(s, X_s^{t,x}) X_s^{t,x} \right\} ds \\
&\quad + u_2(s, X_s^{t,x}) \sigma(s, X_s^{t,x})\, dB_s.
\end{aligned}$$

Next we note that $dI_s = -I_s r(s, X_s^{t,x})\, ds$, so the product rule gives us

$$\begin{aligned}
dM_s &= I_s dU_s + U_s dI_s \\
&= I_s \{ dU_s - U_s r(s, X_s^{t,x})\, ds \} \\
&= I_s \Big\{ u_1(s, X_s^{t,x}) + \frac{1}{2}\sigma(s, X_s^{t,x}) u_{22}(s, X_s^{t,x}) \\
&\qquad + u_2(s, X_s^{t,x}) r(s, X_s^{t,x}) X_s^{t,x} - u(s, X_s^{t,x}) r(s, X_s^{t,x}) \Big\} ds \\
&\quad - I_s u_2(s, X_s^{t,x}) \sigma(s, X_s^{t,x})\, dB_s.
\end{aligned}$$

Now, by the key assumption that $u(t,x)$ satisfies the Black–Scholes PDE (15.31), we see that the braced ds term of dM_s must vanish, and as a consequence we see that M_s is a local martingale, as we suspected. Furthermore, because $r(t,x)$ is nonnegative and $u(t,x)$ is bounded, we see from the definition of M_s given by equation (15.35) that M_s is bounded. This implies that M_s is an honest martingale and completes the proof of the theorem. □

Tracking the Progress

At first glance, the formula for $u(t,x)$ given by the integral (15.33) may look almost as abstract as the general arbitrage pricing formula that we found by the martingale method. In fact the integral representation (15.33) is several steps closer to the ground. One big point in its favor is that nothing further remains to be found; in particular, there is no Q to be calculated as in the case of the preceding chapter.

Any difficulties that may reside in the integral (15.33) can be traced to the process $\{X_s^{t,x} : t \leq s \leq T\}$ and its SDE

$$dX_s^{t,x} = r(s, X_s^{t,x}) X_s^{t,x}\, dt + \sigma(s, X_s^{t,x})\, dB_t \quad \text{and } X_t^{t,x} = x. \tag{15.36}$$

When the solutions of this SDE are well understood, there is an excellent chance that the expectation (15.33) may be obtained explicitly, and Exercise 15.2 gives a simple illustration of this lucky circumstance by showing that for the classical Black–Scholes model the Feynman–Kac formula (15.33) reduces precisely to the Black–Scholes formula.

For more difficult problems, the most telling feature of the Feynman–Kac formula (15.33) is that it presents a problem that is directly accessible to simulation. The explicit form of the SDE for the process $X_s^{t,x}$ means that a suitable interpretation Euler method for the numerical solution of ODEs may be used to generate sample paths of $X_s^{t,x}$. As usual, the overall accuracy of these simulations may be difficult to judge, but in this case we have the important reassurance that if the Euler steps are small enough then the simulations will be valid. This observation and the opportunity to reduce the size of the Euler step in a series of independent simulations may be developed into a practical approach to the evaluation of the discounted expectation (15.33).

The field of SDE simulation is large and rapidly growing. Practical, accurate simulation is evolving into one of the most important topics in stochastic calculus, but we will not pursue the development of simulation here. For fans of the Feynman–Kac formula, the key observation is that each advance in simulation technique and each increase in computational speed adds value to the Feynman–Kac representation for the solution of the general Black–Scholes PDE.

15.6. Exercises

The first two exercises deal with the relationship of the Black–Scholes PDE and the Feynman–Kac formula. The first of these explores the most general model that leads to a PDE of the Black–Scholes type, and the second checks that the Feynman–Kac formula essentially contains the classical Black–Scholes formula.

The third exercise returns to the roots of the Feynman–Kac formula and studies the occupation time of an interval. The first two parts of the problem are easily done by hand, but the remaining parts are done most pleasantly with help from Mathematica or Maple.

EXERCISE 15.1 (A General Black–Scholes PDE). The general Black–Scholes PDE given by equation (15.31) does not contain the drift coefficient $\mu(t, S_t)$, and this fact gives us an interesting hint. Could it be that the same equation would continue to hold for a more general stock and bond model than that given by the system (15.30)? Show that this is the case by proving that equation (15.31) continues to hold even where we only assume that

$$dS_t = \mu_t \, dt + \sigma(t, S_t) \, dB_t \quad \text{and} \quad d\beta_t = r(t, S_t)\beta_t \, dt, \tag{15.37}$$

where the only conditions that we place on μ_t are that it be an adapted process which is integrable for almost all ω.

EXERCISE 15.2 (Feynman–Kac and the Black–Scholes Formula). Consider the classic Black–Scholes economy

$$dS_t = \mu S_t \, dt + \sigma S_t \, dB_t \quad \text{and} \quad d\beta_t = r\beta_t \, dt, \tag{15.38}$$

where μ, σ, and $r \geq 0$ are constants. Work out the integral given by equation (15.33) in this case, and show that the Feynman–Kac formula reduces to the Black–Scholes formula. You may save on arithmetic by showing that the formula (15.33) boils down to a representation that we found earlier.

EXERCISE 15.3 (Occupation of an Interval).

Extend the calculations that were done in Kac's proof of the Lévy Arcsin Law to discover the Laplace transform of the amount of time that Brownian motion spends in an interval.

(a) Suppose that $0 \leq a \leq b$ and let $\widetilde{T}_t$ be the amount of time that Brownian motion spends in $[a, b]$ during the time period $[0, t]$. Use the Feynman–Kac formula to find an initial-value problem for a function $u(t, x)$ with the property that $u(t, 0)$ is the Laplace transform of $\widetilde{T}_t$.

(b) Convert your PDE to an ODE by taking Laplace transforms and then solve the resulting ODE.

(c) Check your solution of the ODE by setting $a = 0$ and letting $b \to \infty$. Show that the limit recaptures the result we found for T_t in Section 15.3.

(d) Use the result of part (b) to find a formula for the Laplace transform of the expected value $\mu(t) = E[\widetilde{T}_t]$. If you know a bit about Laplace transforms, you should then be able to determine the asymptotic behavior of $\mu(t)$ as $t \to \infty$.

APPENDIX I

Mathematical Tools

The purpose of this appendix is to provide some quick guidance to the results of probability theory and analysis that are used in the main text. Although the results summarized here suffice to make the text self-contained, a supplemental text such as Jacod and Protter [36] may serve as a wise investment.

Expectation and Integration

The term *expectation* that is used throughout probability theory is just shorthand for the Lebesgue integral that is taken with respect to a probability measure. The calculations of probability theory hardly ever force us to go all the way back to the definition of the Lebesgue integral, and a great many probability calculations require nothing more than the linearity and positivity properties of expectation.

To begin, we say that X is a *simple function* if it can be written as a finite linear combination of indicator functions of measurable sets,

$$X = \sum_{i=1}^{n} c_i 1_{A_i},$$

and we define the expectation of such an X by the natural formula

$$E(X) = \sum_{i=1}^{n} c_i P(A_i).$$

Next, for any nonnegative random variable Y, we define the expectation of Y by

$$E(Y) = \sup\{E(X) \colon X \leq Y \text{ with } X \text{ a simple function}\}.$$

Finally, for general random variables we define the expectation by reduction to the case of nonnegative variables. Specifically, for general Y we introduce the nonnegative random variables

$$Y^+ = Y\,1(Y \geq 0) \text{ and } Y^- = -Y\,1(Y \leq 0),$$

so the two expectations $E(Y^+)$ and $E(Y^-)$ are well defined, and, as a last step, we define the expectation of Y by

$$E(Y) = E(Y^+) - E(Y^-),$$

provided that $E(Y^+)$ and $E(Y^-)$ do not both equal infinity.

In a course on integration theory, one would now need to show that these definitions do indeed give us an expectation that is linear in the sense that for real a and b we have

$$E(aX + bY) = aE(X) + bE(Y).$$

Because the expectation is defined by use of a supremum in the key step, the linearity of the expectation is not immediate. Nevertheless, with a little work one can show that defining properties of a probability measure suffice to establish the required linearity.

The day-to-day workhorses of integration theory are the dominated convergence theorem (DCT), the monotone convergence theorem (MCT), and Fatou's lemma. Each of these results provides us with a circumstance where we can change the order of a limit and an expectation. Here they are as they are most commonly used in probability theory.

DOMINATED CONVERGENCE THEOREM (DCT). *If $P(X_n \to X) = 1$ and $|X_n| \leq Y$ for $1 \leq n < \infty$ where Y satisfies $E(Y) < \infty$, then $E(|X|) < \infty$ and*

$$\lim_{n\to\infty} E(X_n) = E\left(\lim_{n\to\infty} X_n\right) = E(X).$$

MONOTONE CONVERGENCE THEOREM (MCT). *If $0 \leq X_n \leq X_{n+1}$ for all $n \geq 1$, then*

$$\lim_{n\to\infty} E(X_n) = E\left(\lim_{n\to\infty} X_n\right),$$

where the limits may take infinity as a possible value.

FATOU'S LEMMA. *If $0 \leq X_n$ for all $n \geq 1$, then*

$$E\left(\liminf_{n\to\infty} X_n\right) \leq \liminf_{n\to\infty} E(X_n),$$

where the limits may take infinity as a possible value.

Any one of these three integration results can be used as the basis of a proof for the other two, so none of the three has a strong claim on being the most fundamental. Nevertheless, courses in integration theory almost always take the MCT as the basis for the development. As a practical matter, the DCT is the result that is most important for probabilists.

Probabilists use Fatou's lemma less often than the DCT, but there are times when it is the perfect tool. This is commonly the case when we have information about the distribution of the individual terms of the series $\{X_n\}$, and we would like to show that the limit has finite expectation. For example, if we know that $X_n \geq 0$ and that $X_n \to X$ with probability one, then Fatou's lemma tells us that if we can show $E(X_n) \leq 1$, then $E(X)$ exists and $E(X) \leq 1$. On the other hand, if we try to use the DCT to show that $E(X)$ exists, then we would need to show that $Y = \sup_n X_n$ has a finite expectation. Such a bound requires information about the *joint* distributions of the $\{X_n\}$ that may be hard to obtain. This is quite a contrast to Fatou's lemma, where we only need information about the *marginal* distributions to show the existence of $E(X)$.

Conditional Expectations

Conditional expectations are discussed at length in the text, and the only point left for this appendix is to note that the DCT, MCT, and Fatou's Lemma carry over to conditional expectations with largely cosmetic changes. To get the conditional versions of the MCT and Fatou's lemma, there are two steps. First, we naturally want to replace $E(\cdot)$ by $E(\cdot \mid \mathcal{F})$ at each appearance. Second, since the conclusions now refer to random variables instead of numbers, one needs to note that the identities now only hold with probability one. We can illustrate these steps with the DCT, which needs the most care, although the principle is the same in all three cases.

CONDITIONAL DCT. *If $P(X_n \rightarrow X) = 1$ and $|X_n| \leq Y$ for $1 \leq n < \infty$ where $E(Y \mid \mathcal{F}) < \infty$ with probability one, then on a set of probability one we have $E(|X| \mid \mathcal{F}) < \infty$ and*

$$\lim_{n\to\infty} E(X_n \mid \mathcal{F}) = E\Big(\lim_{n\to\infty} X_n \mid \mathcal{F}\Big) = E(X \mid \mathcal{F}).$$

Probability Theory Basics

The text naturally presupposes some familiarity with probability theory, but we should still recall some basic points that must be taken as common ground. Two fundamental facts that we often use without comment are *Markov's inequality,*

$$P(X \geq \lambda) \leq E(X)/\lambda \text{ provided that } X \geq 0 \text{ and } \lambda > 0,$$

and *Chebyshev's inequality,*

$$P(|X-\mu| \geq \lambda) \leq \mathrm{Var}(X)/\lambda^2 \text{ where } \mu = E(X) \text{ and } \lambda > 0.$$

Also, our most trusted tool for proving that some event occurs with probability one is the Borel–Cantelli lemma. This much-used lemma comes in two parts.

BOREL–CANTELLI LEMMA. *If $\{A_i\}$ is any sequence of events, then*

$$\sum_{i=1}^{\infty} P(A_i) < \infty \textit{ implies that } P\Big(\sum_{i=1}^{\infty} 1_{A_i} < \infty\Big) = 1,$$

and if $\{B_i\}$ is any sequence of independent *events, then*

$$\sum_{i=1}^{\infty} P(B_i) = \infty \textit{ implies that } P\Big(\sum_{i=1}^{\infty} 1_{B_i} = \infty\Big) = 1.$$

Finally, we are always guided by the law of large numbers and the central limit theorem, even though they are used explicitly on only a few occasions in the text.

STRONG LAW OF LARGE NUMBERS. *If $\{X_i\}$ is a sequence of independent random variables with the same distribution, $E(|X_i|) < \infty$, and $E(X_i) = 0$ then*

$$P\Big(\frac{1}{n}(X_1 + X_2 + \cdots + X_n) \textit{ converges to } 0\Big) = 1.$$

CENTRAL LIMIT THEOREM. *If $\{X_i\}$ is a sequence of independent random variables with the same distribution, $E(X_i) = 0$, and $\mathrm{Var}(X_i) = 1$, then*

$$\lim_{n\to\infty} P\left(\frac{X_1 + X_2 + \cdots + X_n}{\sqrt{n}} \leq x\right) = \frac{1}{\sqrt{2\pi}} \int_{-\infty}^{x} e^{-u^2/2}\, du.$$

In fact, the text even provides proofs of these two important results, although these proofs are admittedly strange if one does not already know the customary arguments. In the course of our introduction to martingale theory in discrete time, we find a proof of the strong law of large numbers, and much later we find that the central limit theorem can be viewed as one of the corollaries of the Skorohod embedding theorem.

Hilbert Space, Completeness, and L^2

In a metric space S with metric ρ, we say that a sequence $\{x_n\} \subset S$ is a Cauchy sequence if

$$\rho(x_n, x_m) \to 0 \text{ as } n, m \to \infty,$$

and we say that the metric space S is *complete* if every Cauchy sequence converges to an element of S. One of the most useful complete metric spaces is $C[0,1]$, the set of continuous functions on $[0,1]$ with the metric given by

$$\rho(f,g) = \sup_{x\in[0,1]} |f(x) - g(x)|.$$

The completeness of $C[0,1]$ is a consequence of the fundamental fact that if a sequence of continuous functions on a compact set converges uniformly, then the limit is also a continuous function.

One of the reasons that Lebesgue's integral was an immediate hit when it was introduced in 1904 is that the new integral made it possible to view the inner product space $L^2(dP)$ as a complete metric space. In the language of probability theory, this means that if $\{X_n\}$ is a sequence of random variables for which we have

$$\lim_{n,m\to\infty} ||X_n - X_m||_2 = 0,$$

then there is an $X \in L^2(dP)$ such that

$$||X_n - X||_2 \to 0 \text{ as } n \to \infty.$$

The usual tools of probability theory make it easy for us to give a proof of this important fact, and the proof even offers good practice with the material we have reviewed thus far. In outline, one first shows the existence of X by means of subsequence and Borel–Cantelli arguments, and, once a candidate in hand, one shows the required L^2 convergence by a simple application of the triangle inequality.

To flush out the details, we first note that the sequence

$$r(N) = \sup_{\{m,n\geq N\}} ||X_m - X_n||_2$$

decreases monotonically to zero, so we can choose a subsequence n_k such that $r(n_k) \leq 2^{-k}$. Next, we note by Markov's inequality that

$$P(|X_{n_{i+1}} - X_{n_i}| \geq 1/i^2) \leq i^4 2^{-2i},$$

so the Borel-Cantelli lemma guarantees there is a set Ω_0 of probability one such that for each $\omega \in \Omega_0$ we have

$$|X_{n_{i+1}}(\omega) - X_{n_i}(\omega)| \leq 1/i^2$$

for all but a finitely many i. We can therefore define a random variable X for all $\omega \in \Omega_0$ by the converging sum

$$X(\omega) = X_{n_1}(\omega) + \sum_{n=1}^{\infty}\{X_{n_{i+1}}(\omega) - X_{n_i}(\omega)\}.$$

Finally, for any n there is a k such that $n_k \le n < n_{k+1}$, so we have

$$\begin{aligned}||X_n - X||_2 &\le ||X_n - X_{n_k}||_2 + ||X_{n_k} - X||_2 \\ &\le ||X_n - X||_2 + \sum_{j=k}^{\infty} ||X_{n_{j+1}} - X_{n_j}|| \le 2^{-k} + \sum_{j=k}^{\infty} 2^{-j} = 3 \cdot 2^{-k}.\end{aligned}$$

Since $k \to \infty$ as $n \to \infty$, the last inequality shows that $||X_n - X||_2 \to 0$, as required to show that $L^2(dP)$ is complete.

Notational Matters

The notation that one uses to describe function spaces has natural variation, though less than one finds in urban pigeons. For example, L^p always denotes a set of functions for which the absolute p'th power is integrable, but there are many different measure spaces where these functions may live, and the plumage adorning $L^p(\mu)$ or $L^p(dP)$ just reflects this variation.

For us, the basic case is given when the measure space is the familiar probability space $(\Omega, \mathcal{F}, P)$, although even here there is one special case that should be singled out. When Ω is $[0,1]$, $\mathcal{F}$ is the Borel σ-field, and P is the uniform measure on $[0,1]$ (so for $A = [0,x]$ we have $P(A) = x$), then we write $L^2[0,1]$ instead of just $L^2[dP]$. There is no defense to the charge that this notation is inconsistent, except the Emmersonian standby that a foolish consistency is the hobgoblin of little minds. In the construction of the Itô integral that is given in Chapter 6, we will work at once with several different function spaces, and we will need to introduce several variations on the basic notation to make the distinctions clear.

Bessel and Parseval

A sequence $\{\phi_n\}$ in $L^2(dP)$ is called an *orthonormal sequence* if it satisfies

$$E(\phi_k^2) = 1 \text{ and } E(\phi_j \phi_k) = 0 \text{ when } j \neq k.$$

Such sequences owe much of their importance to the fact that for any $X \in L^2$ the generalized Fourier coefficients

$$a_n = E(\phi_n X)$$

can tell us a lot about X, and vice versa. For example, one of the simplest but most useful connections is given by *Bessel's inequality*:

$$\sum_{n=1}^{\infty} a_n^2 \le E(X^2).$$

Despite its importance, the proof of Bessel's inequality is as easy as pie. We begin with the trivial inequality

$$0 \le E\Big[\big(X - \sum_{n=1}^{m} a_n \phi_n\big)^2\Big],$$

and then we do honest arithmetic using the definition of orthonormality to close the loop:

$$E\left[\left(X-\sum_{n=1}^{m}a_n\phi_n\right)^2\right] = E(X^2)-2\sum_{n=1}^{m}a_nE(\phi_nX)+E\left[\left(\sum_{n=1}^{m}a_n\phi_n\right)^2\right]$$
$$= E(X^2)-\sum_{n=1}^{m}a_n^2.$$

We now face a natural question. When does Bessel's inequality reduce to a genuine equality? This is an extremely fruitful direction for investigation, and we are fortunate that there is a definitive answer.

PARSEVAL'S THEOREM. *For any orthonormal sequence $\{\phi_n : 1 \le n < \infty\}$ in $L^2(dP)$, the three following conditions are equivalent:*

(A) the finite linear combinations of the functions $\{\phi_n\}$ form a dense subset of $L^2(dP)$,

(B) the only $X \in L^2(dP)$ that satisfies $E(\phi_nX)=0$ for all n is $X=0$, and

(C) for any $X \in L^2(dP)$, we have Parseval's identity,

$$E(X^2)=\sum_{n=1}^{\infty}a_n^2 \quad \textit{where} \quad a_n=E(\phi_nX).$$

PROOF. To show that condition A implies condition B, we first suppose that we have $E(\phi_nX)=0$ for all n, and then note that condition A tells us for any $\epsilon>0$ that we have an integer N and real numbers $\{\alpha_n\}$ such that

$$X=\sum_{n=1}^{N}\alpha_n\phi_n+R_N \quad \text{and} \quad E(R_N^2)\le\epsilon^2.$$

Now, if we compute $E(X^2)$ by orthogonality, we find

$$E(X^2)=\sum_{n=1}^{N}\alpha_n^2+2\sum_{n=1}^{N}\alpha_nE(\phi_nR_N)+E(R_N^2),$$

and, when we recall our assumption that $E(\phi_kX)=0$, we also find

$$E(\phi_kR_N)=E\left[\phi_k(X-\sum_{n=1}^{N}\alpha_n\phi_n)\right]=-\alpha_k.$$

The two preceding formulas tell us that

$$E(X^2)=\sum_{n=1}^{N}\alpha_n^2-2\sum_{n=1}^{N}\alpha_n^2+E(R_N^2)\le\epsilon^2,$$

and, because $\epsilon>0$ was arbitrary, we see that $X=0$. This completes the proof of the first implication.

Next, to show that condition B implies condition C, we begin by noting that for $a_n=E(X\phi_n)$ Bessel's inequality tells us

$$\sum_{n=1}^{\infty}a_n^2<\infty,$$

so the sum defined by

$$\sum_{n=1}^{\infty} a_n \phi_n$$

converges in $L^2(dP)$ to a some random variable which we may denote as Y. To complete the proof of condition C, we just need to show that $X = Y$. This is in fact quite easy since we have

$$E[(X - Y)\phi_n] = a_n - a_n = 0,$$

and condition B then guarantees that $X - Y = 0$.

Now, to close the loop through all three conditions, we only need to show that condition C implies condition A, and this is the simplest implication of the lot. We have

$$||X - \sum_{n=1}^{N} a_n \phi_n||_2^2 = \sum_{n=N+1}^{\infty} a_n^2,$$

and the last sum goes to zero as $N \to \infty$ because Bessel's inequality provides the convergence of the infinite sum. □

Traditionally, an orthonormal sequence that satisfies the first condition of Parseval's theorem is called a *complete orthonormal sequence,* and the main consequence of Parseval's theorem is that for any such sequence $\{\phi_n\}$ and any $X \in L^2$, we may represent X as a weighted sum of the elements of the sequence.

Parallelograms and Projections

The complete inner product spaces are so useful they have a special name — Hilbert spaces. The finite-dimensional Hilbert spaces include $\mathbb{R}^n$ for all $n < \infty$, but even the infinite-dimensional Hilbert spaces have a geometry that continues to echo familiar features of the finite dimensional Euclidean spaces. For example, if u and v are elements of a Hilbert space, then simple algebraic expansion will prove that

$$||u - v||^2 + ||u + v||^2 = 2||u||^2 + 2||v||^2.$$

This identity already contains an important law of Euclidean geometry; it tells us that the sum of the squares of the diagonals of a parallelogram equals the sum of the squares of the sides.

HILBERT SPACE PROJECTION THEOREM. *If H_0 is a closed linear subspace of the Hilbert space H, then for every $X \in H$ there is a unique Y in H_0 such that*

(A1.1) $$||X - Y|| = \inf\{ \, ||X - Z|| : Z \in H_0 \, \}.$$

This Y can also be characterized as the unique element of H_0 such that

(A1.2) $$\langle X - Y, Z\rangle = 0 \text{ for all } Z \in H_0.$$

In particular, any $X \in H$ may be written uniquely as $X = Y + W$ where $Y \in H_0$ and

$$W \in H_0^{\perp} \stackrel{\text{def}}{=} \{W : E(WZ) = 0 \quad \forall Z \in H_0\}.$$

PROOF. To show the existence of Y, we let α denote the value of the infimum in (A1.1), and choose a sequence of elements $Y_n \in H_0$ such that

$$||X - Y_n||^2 \leq \alpha + 1/n.$$

Now, if we let $u = X - Y_n$ and $v = X - Y_m$, the parallelogram identity tells us

$$||Y_m - Y_n||^2 + ||(X - Y_m) - (X - Y_n)||^2 = ||X - Y_m||^2 + ||X - Y_n||^2,$$

or, more informatively,

$$||Y_m - Y_n||^2 \leq 2||X - Y_m||^2 + 2||X - Y_n||^2 - 4||X - (Y_m + Y_n)/2||^2.$$

But we have $||X - (Y_m + Y_n)/2||^2 \geq \alpha^2$ by definition of α and the fact that the average $(Y_m + Y_n)/2 \in H_0$ since H_0 is a linear space, so we find

$$||Y_m - Y_n||^2 \leq 2(\alpha + 1/m) + 2(\alpha + 1/n) - 4\alpha^2 = 2/m + 2/n.$$

This tells us that $\{Y_n\}$ is a Cauchy sequence in H, so Y_n converges to some $Y \in H$. But H_0 is closed and $Y_n \in H_0$ so in fact we have $Y \in H_0$.

To prove the uniqueness of the minimizer Y, we can also use the parallelogram law. For example, suppose that $W \in H_0$ also satisfies $||X - W|| = \alpha$, then we also have $||Y + W||^2 = 4||(Y + W)/2||^2 \geq 4\alpha^2$ so

$$\begin{aligned} ||Y - W||^2 &= 2||Y||^2 + 2||W||^2 - ||Y + W||^2 \\ &\leq 2\alpha^2 + 2\alpha^2 = 0. \end{aligned}$$

Next, to check the orthogonality condition (A1.2), we first note that by the definition of Y we have for all $Z \in H_0$ that

$$\inf_t ||(Y - X) + tZ||^2 = ||X - Y||^2.$$

But by calculus, we know that the quadratic function of t given by

$$||(Y - X) + tZ||^2 = ||Y - X||^2 - 2t\langle X - Y, Z\rangle + t^2||Z||^2$$

has a minimum at $t = \langle X - Y, Z\rangle/||Z||^2$ with associated minimum value

$$||X - Y||^2 = \langle X - Y, Z\rangle^2/||Z||^2.$$

Since this minimum cannot be less than $||X - Y||^2$, we must have $\langle X - Y, Z\rangle^2 = 0$.

Finally, we need to show that the orthogonality condition (A1.2) characterizes Y. But, if there were a $W \in H_0$ with

$$\langle X - W, Z\rangle = 0 \text{ for all } Z \in H_0,$$

then the fact that Y satisfies the same condition would let us take the difference to find

$$\langle Y - W, Z\rangle = 0 \text{ for all } Z \in H_0.$$

But $Y - W \in H_0$, so we can take $Z = Y - W$ in the last equation to deduce that $W = Y$. □

There is a great deal more that one might discuss in this Appendix such as the solution of elementary ODEs, the basics of power series, and the use of transforms, such as those of Laplace and Fourier. We only use the simplest aspects of these topics, and, with luck, the brief introductions that are given in the text will provide all the necessary background.

APPENDIX II

Comments and Credits

In his preface to *Fourier Analysis*, Körner says " ... a glance at the average history of mathematics shows that mathematicians are remarkably incompetent historians." Körner might have claimed exception for himself yet he did not, and neither can this author. Nevertheless, history is the thread that binds us all, and the acknowledgement of our teachers gives recognition to everyone who would teach.

Only a few of the ideas in this volume are new, but, with luck, not too many are stale (and even fewer false). The main purpose of this appendix is to acknowledge the sources that have been drawn upon. Also, at times, it points out a place one might go for further information, or it shares an observation that did not fit comfortably into the main text.

Chapter 1: Random Walk and First Step Analysis

Virtually all probability books deal with simple random walk, but Feller [27] takes the elementary treatment to remarkable heights. Even as the book begins to show its age, anyone interested in probability needs a copy within easy reach.

Epstein [23] gives a careful description of casino games, and any excursion on the Internet will reveal that there is a huge specialized literature on gambling. Much of the specialized material is pap, but some is reasonably serious, such as Griffin [28]. Gambling possibly deserves more attention from probabilists and statisticians. After all, it is the cradle of our craft, and it is also finance of a remarkably pristine sort.

Wilf [60] gives a deep but easy-to-read development of the modern theory of generating functions, the superset for our probability generating functions. Also, Lin and Segel ([41], p. 234) offer modest but pointed coaching about the use of Newton series versus the Taylor expansion: " ... experience teaches us that in constructing power series one should use the binomial expansion whenever possible."

Chapter 2: First Martingale Steps

This chapter puts more weight on L^2 martingales than is common. The motivation for this allocation of effort was the desire to give the reader additional familiarity with L^2 arguments before digging into the construction of Brownian motion in Chapter 3 and the L^2 construction of the Itô integral in Chapter 6. By beginning with the L^2 theory, we also have the chance to meet localization arguments before they are faced at gale force during the development of the Itô integral in $\mathcal{L}^2_{\text{LOC}}$.

Nevertheless, one really cannot do without the usual L^1 theory and the upcrossing inequality; so, after soul searching and foot dragging, this material was also added. For further discussion of the theory of discrete-time martingales, one cannot do better than consult Neveu [50] or Williams [66].

The notion of a *self-improving inequality* is based on a comment from Pólya and Szegö ([54], Vol.II, p. 30), where they note that Landau's proof of the maximum modulus principle shows how "... a rough estimate may sometimes be transformed into a sharper estimate by making use of the generality for which the original estimate is valid."

The elegant argument of Lemma 2.1, which shows how to convert an "inner inequality" to an L^p inequality, can be traced back at least to Wiener's work on the ergodic theorem and the work of Hardy and Littlewood on the maximal functions of differentiation theory.

Chapter 3: Brownian Motion

The decision to use the language of wavelets in the construction of Brownian motion was motivated by the discussion of Brownian motion in Y. Meyer's inspirational book *Wavelets: Algorithms and Applications*. True enough, the essence of the construction goes back to P. Lévy, who simplified Wiener's construction through the introduction of the Schauder functions $\{\Delta_n\}$, but there is still value in using the new language. Dilation and translation properties are the defining elements of wavelet theory, and they are also at the heart of the symmetry and fractal qualities of Brownian motion.

For a quick, well-motivated introduction to the application of wavelets, one does well to begin with the essay of Strang [59]. This paper gives a concise introduction to the core of wavelet theory, and it also reports on the interesting contests between wavelets and Fourier methods in the development of high-definition television. One now finds a deluge of book-length treatments of wavelets. To choose among these is a matter of taste, but one collection that seems to offer a good balance between theory and practice is Benedito and Frazier [5].

The original construction of Wiener used the beautiful, but more complex, representation:

$$B_t = t\,Z_0 + \sum_{n=1}^{\infty} \sum_{k=2^{n-1}}^{2^n-1} Z_k \sqrt{2}\, \frac{\sin \pi k t}{\pi k}.$$

Such sums over geometric blocks were fundamental to much of Wiener's work in harmonic analysis, and, in the case of Brownian motion, the trick of using blocks is essential; other arrangements of the same summands may fail to converge uniformly.

For many years, the undisputed masterwork of Brownian motion was that of Itô and McKean [35]. This is still a remarkable book, but now there are many rich sources that are more accessible. The two works that have been most useful here are those of Karatzas and Shreve [39] and Revuz and Yor [55]. The collection of formulas in the handbook of Borodin and Salminen [8] is a treasure trove, but the user needs to bring a pick and shovel. The handbook's gems are not easy to unearth.

Chapter 4: Martingales: The Next Steps

The idea of using Jensen's inequality to prove Hölder's inequality is a bit ironic; after all, the usual proof goes via the elementary bound $xy \leq x^p/p + y^q/q$, yet one of the easiest ways to prove this bound is by Jensen. *Plus ça change ...*

Dudley ([18], p. 140) gives an interesting historical account of Hölder's inequality, including the surprising observation that the name *Roger's inequality* may be more appropriate.

Exercise 21 of Bass ([3], p. 80) motivated the argument that we used to obtain the uniform integrability needed in the proof of Doob's stopping time theorem. Williams ([66] p. 128) provides an alternative method to establish the required uniform integrability that is brief and elegant, though less quantitative. One of the interesting consequences of Lemma 4.4 is that it yields a systematic way to sharpen some common estimates. In particular, the lemma may be used to convert almost any O-estimate that one obtains from Markov's inequality into a o-estimate, although in most cases there are several ways to get the same improvement.

Chapter 5: Richness of Paths

Our development of the Skorohod embedding begins with a traditional idea, but the linear algebra connections are new. Friedman [27] gives a motivated discussion of the applications of dyads to integral equations and ODEs; it fits nicely with our development of embedding theory.

Dubins [15] provides an important refinement of the Skorohod embedding that is more consonant with a martingale view of the world. Dubins shows that essentially any discrete-time martingale can be embedded in essentially any continuous martingale without having to call on any exogeneous randomization. When the square integrable martingale $\{M_n\}$ is embedded in Brownian motion, Dubins' method again provides stopping times that satisfy the critical expectation identity $E[\tau_n] = E[M_n^2]$.

Exercise 5.3 was motivated by the discussion of Lévy's construction of Brownian motion given by Y. Meyer ([48] p. 18).

Chapter 6: Itô Integration

This reasonably barehanded approach to the Itô integral is distilled from many sources, the most direct being lectures given by David Freedman one summer at Stanford in the early 1980s. These lectures may in turn owe a debt to the amazingly terse volume *Stochastic Integrals* by H. P. McKean. The treatments of stochastic integration given in Chung and Williams [12], Durrett [21], and Karatzas and Shreve [39] are all directed to a somewhat more sophisticated audience. The treatments of Ikeda and Watanabe [33], Protter [51], Revuz and Yor [55], and Rogers and Williams [56] are even more demanding, yet always worth the effort.

The Latin quote from Leibniz is taken from Eriksson et al. ([24], p. 60), where one can also see an interesting photograph of the appropriate page from the Leibniz manuscript. To judge from *The Cambridge Companion to Leibniz* [38], mathematics only had a walk-on role in this philosopher's life.

CHAPTER 7: LOCALIZATION AND ITÔ'S INTEGRAL

Localization may seem technical, but it is a technicality that makes life easier — not harder. If one's goal were to give the most parsimonious development of continuous time martingale theory, the efficient plan would be to take local martingales as the primitive objects and to consider proper martingales just as a special case. Naturally, such a plan does not make sense if one's goal is rather to gain more members for the club of people who understand both martingales and local martingales. For many potential members, the initiation fee would be too high.

The formulation and proof of Proposition 7.13 follow the discussion given by Revuz and Yor ([55], p. 180).

CHAPTER 8: ITÔ'S FORMULA

The basic analysis-synthesis approach to Itô's formula is present in some form in almost any development of the formula, but here the aim has been to make the steps absolutely explicit. The chain rule approach to the Itô formula for geometric Brownian motion seems to be a novel twist, but this is not much to crow about since the chain rule method is really quite crude. It must be set aside for the more powerful tools of the box calculus.

The development of quadratic variation given here is based in part on that of Karatzas and Shreve ([39], pp. 32–35).

The example in Exercise 8.5 of a local martingale that is not a martingale is from Revuz and Yor ([55], p. 194).

Axler et al. [1] is a delightful place to read further about harmonic functions.

CHAPTER 9: STOCHASTIC DIFFERENTIAL EQUATIONS

The existence and uniqueness theorems developed here follow a natural pattern based on the Picard iteration scheme of ODEs. The application of Picard's method to stochastic integrals goes back to Itô [34]. It is used almost universally and has the benefit of applying equally well to systems of SDEs.

For one-dimensional SDEs there is an interesting alternative approach to existence theory due to Engelbert and Schmidt. A valuable development of this approach is given in Karatzas and Shreve ([39], pp. 329–353).

The intriguing "squared" Brownian bridge of Exercise 9.2 was introduced by Li and Pritchard [43], where it was found as the solution to a sock sorting problem!

The text hardly touched on the important topic of the numerical solution of SDEs. A useful introduction to this rapidly developing field can be found in Kloeden and Platen [40] together with the instructive experimental companion volume of Kloeden, Platen, and Schurz [41].

Exercise 9.6 is the text's only brush with the important topic of the estimation of parameters of financial models. Even here we stop short of developing the empirically essential idea of *implied volatility*. Fortunately, the well known text of Campbell, Lo, and MacKinlay [11] provides easy and accessible coverage of this topic, as well as many other topics of importance in empirical finance.

Chapter 10: Arbitrage and SDEs

This chapter owes a special debt to the founding fathers, Fisher Black and Myron Scholes. The line-by-line dissection given here of their classic paper [7] does not seem to have any direct antecedent, though there is a loose parallel in Pólya's discussion of Euler's memoir "Discovery of a Most Extraordinary Law of the Numbers Concerning the Sum of Their Divisors." In *Induction and Analogy in Mathematics*, Pólya ([52], pp. 90–107) translates almost all of Euler's memoir in order to comment on what it has to teach about the psychology of invention and inductive reasoning.

The discussion of the put-call parity formula is based on the original article of Stoll [60]. Exercise 10.3 and the minor observation that the Black–Scholes PDE can be used to prove the put-call parity formula seem to be new.

The translation of the *Mumonkan* by Sekida [57] never fails to inspire.

Chapter 11: The Diffusion Equation

There are enjoyable elementary discussions of the diffusion equation in Lin and Segel [44] and Feynman et al. [26]. In the first of these, a diffusion equation is found for the population of slime mold amoebae that is only a bit more involved than the basic diffusion equation of our mice, yet it quickly leads to novel questions in biology.

Körner [42] also presents a lively discussion of the diffusion (or heat) equation, and in particular, he gives an entertaining synopsis of Kelvin's use of temperature to estimate the age of the Earth. Burchfield [9] provides an extensive treatment of Kelvin's investigation of this problem.

For the mathematician, the most focused account of the heat equation is probably the elegant text of Widder [62]. An especially instructive part of Widder's book is the discussion of his uniqueness theorem for nonnegative solutions for the heat equation. Karatzas and Shreve ([39], pp. 256–261) also give an interesting probabilistic discussion of Widder's uniqueness theorem.

The parabolic maximum principle is treated in almost all texts on partial differential equations. The discussion in John ([37], pp. 215–218) is particularly clear, and it helped form the development followed here. John ([37], pp. 211–213) also gives a detailed discussion of an example showing nonuniqueness of solutions of the heat equation.

Zwillinger ([68], pp. 235–237) gives a right-to-the-point discussion of Euler's equidimensional equation. The classic text of Carslaw and Jaeger [10] contains the explicit solution of many boundary value problems for diffusion equations, although the boundary constraints studied by Carslaw and Jaeger are better focused for applications to physics than to finance.

Wilmott, Howison, and Dewynne [65] provide a valuable introduction to option pricing from the point of view of of partial differential equations, and their monograph [64] possibly provides the most extensive treatment of the PDE approach to option pricing that is available.

Shaw's monograph [58] gives a fascinating, individualistic development of derivative pricing from a computational point of view that is not quite mainline PDE. This book also contains much practical advice that cannot be found in other sources.

CHAPTER 12: REPRESENTATION THEOREMS

The proof of Dudley's Theorem follows his beautiful paper [17] with the small variation that a conditioning argument in the original has been side-stepped.

The idea of exploiting alternative σ-fields when calculating the density of the hitting time of a sloping line is based on Karatzas and Shreve ([39], p. 198).

The proof of the martingale representation theorem given here is based in part on the development in Bass ([3], pp. 51–53). The time change representation (Theorem 12.4) is due to Dubins and Schwarz [16] and K.E. Dambis [14].

The π–λ theorem is a challenge to anyone who searches for intuitive, memorable proofs. The treatment given by Edgar ([23], pp. 5–7) is a model of clarity, and it formed the basis of our discussion. Still, the search continues.

CHAPTER 13: GIRSANOV THEORY

The idea of importance sampling occurs in many parts of statistics and simulation theory, but the connection to Girsanov theory does not seem to have been made explicit in earlier expositions, even though the such a development is prefectly natural and presumably well-understood by experts.

The parsimonious proof of Theorem 13.2 was suggested by Marc Yor.

The treatment of Novikov's condition is based on the original work of R.S. Liptser and A.N. Shiryayev [45] and the exposition in their very instructive book [46]. Exercise 13.3 modifies a problem from Liptser and Shiryayev [45], which for some reason avoids calling on the Lévy–Bachelier density formula.

CHAPTER 14: ARBITRAGE AND MARTINGALES

The sources that had the most influence on this chapter are the original articles of Cox, Ross, and Rubinstein [13], Harrison and Kreps [31], Harrison and Pliska [32] and the expositions of Baxter and Rennie [4], Duffie [20], and Musiela and Rutkowski [49].

The pedagogical artifice of the martingale pricing formula as the best guess of a streetwise gambler evolved from discussions with Mike Harrison, although no reliable memory remains of who was talking and who was listening.

Proposition 14.1 and its proof were kindly provided by Marc Yor.

The discussion of the American option and the condition for no early exercise was influenced by ideas that were learned from Steve Shreve.

CHAPTER 15: FEYNMAN–KAC CONNECTION

The Feynman–Kac connection is really a part of Markov process theory, and it is most often developed in conjunction with the theory of semi-groups and the theory of Kolmogorov's backward and forward equations. This chapter aimed for a development that avoided these tools, and the price we pay is that we get only part of the story.

The pleasant development of martingale theory solutions of PDEs given by Durrett [21] formed the basis of our proof of Theorem 15.1, and the discussion in Duffie [20] informed our development of the application of the Feynman–Kac formula to the Black–Scholes model. Duffie [19] was the first to show how the Feynman–Kac connection could be applied to Black–Scholes model with stochastic dividends and interest rates.

Appendix I: Mathematical Tools

There are many sources for the theory of Lebesgue integration. Both Billingsley [6] and Fristedt and Gray [28] give enjoyable developments in a probability context. Bartle [2] gives a clean and quick development without any detours.

Young [67] provides a very readable introduction to Hilbert space. Even the first four chapters provide more than one ever needs in the basic theory of stochastic integration.

Three Bonus Observations

- George Pólya seems to have changed his name! In his papers and collected works we find Pólya with an accent, but the accent is dropped in his popular books, *How to Solve It* and *Mathematics and Plausible Reasoning.*
- There was indeed a *General Case*, a graduate of West Point. The General gave long but unexceptional service to the U.S. Army.
- Michael Harrison has a useful phrase that everyone should have at hand if challenged over some bold (or bald) assumption, such as using geometric Brownian motion to model a stock price. The swiftest and least reproachable defense of such a long-standing assumption is simply to acknowledge that it is a *custom of our tribe.*

Bibliography

[1] S. Axler, P. Bourdon, and W. Ramey, *Harmonic Function Theory*, Springer-Verlag, New York, 1995.

[2] R.G. Bartle, *The Elements of Integration*, Wiley, New York, 1966.

[3] R.F. Bass, *Probabilistic Techniques in Analysis*, Springer-Verlag, New York, 1992.

[4] M. Baxter and A. Rennie, *Financial Calculus: An Introduction to Derivative Pricing*, Cambridge University Press, New York, 1996.

[5] J.J. Benedito and M.W. Frazier, eds., *Wavelets: Mathematics and Applications*, CRC Press, Boca Raton, FL, 1993.

[6] P. Billingsley, *Probability and Measure*, 3rd Ed., Wiley, New York, 1995.

[7] F. Black and M. Scholes, Pricing of Options and Corporate Liabilities, *J. Political Econ.*, **81**, 637–654, 1973.

[8] A.N. Borodin and P. Salminen, *Handbook of Brownian Motion — Facts and Formulae*, Birkhäuser, Boston, 1996.

[9] J.D. Burchfield, *Lord Kelvin and the Age of the Earth*, Macmillan, New York, 1975.

[10] H.S. Carslaw and J.C. Jaeger, *Conduction of Heat in Solids*, 2nd Ed., Oxford University Press, Oxford, UK, 1959.

[11] J.Y. Campbell, A.W. Lo, and A.C. MacKinlay, *The Econometrics of Financial Markets*, Princeton University Press, Princeton, NJ, 1997.

[12] K.L. Chung and R. J. Williams, *Introduction to Stochastic Integration*, 2nd. Ed., Birkhäuser, Boston.

[13] J. Cox, S. Ross, and M. Rubinstein, Option pricing:a Simplified approach, *J. of Finan. Econ.* **7**, 229–263, 1979.

[14] K.E. Dambis, On the decomposition of continuous martingales, *Theor. Prob. Appl.* **10**, 401–410, 1965.

[15] L.E. Dubins, On a Theorem of Skorohod, *Ann. Math. Statist.*, **30**(6), 2094–2097, 1968.

[16] L.E. Dubins and G. Schwarz, On continuous martingales, *Proc. Nat. Acad. Sci. USA* **53**, 913–916, 1965.

[17] R.M. Dudley, Wiener Functionals as Itô Integrals, *Ann. Probab.*, **5**(1), 140–141, 1977.

[18] R.M. Dudley, *Real Analysis and Probability*, Wadsworth and Brooks/Cole, Belmont, CA, 1989.

[19] D. Duffie, An Extension of the Black-Scholes Model of Security Valuation, *J. Econ. Theory*, **46**, 194–204, 1988.

[20] D. Duffie, *Dynamic Asset Pricing Theory*, 2nd Ed., Princeton University Press, Princeton, NJ, 1996.

[21] R. Durrett, *Stochastic Calculus: A Practical Introduction*, CRC Press, New York, 1996.

[22] R. Durrett, *Probability: Theory and Examples*, 2nd Ed., Duxbury Press, New York, 1996.

[23] G.A. Edgar, *Integral, Probability, and Fractal Measures*, Springer-Verlag, New York, 1998.

[24] K. Eriksson, D. Estep, P. Hansbo, and C. Johnson, *Computational Differential Equations* Cambridge University Press, Cambridge, UK, 1996.

[25] R.A. Epstein, *The Theory of Gambling and Statistical Logic*, revised edition, Academic Press, San Diego, CA, 1977.

[26] R.P. Feynman, R.B. Leighton, and M. Sands, *The Feynman Lectures on Physics*, Addison–Wesley, Reading, MA, 1963.

[27] B. Friedman, *Principles and Techniques of Applied Mathematics*, Dover Publications, Mineola, NY, 1990.

[28] B. Fristedt and L. Gray, *A Modern Approach to Probability Theory*, Birkhäuser, Boston, 1997.

[29] W. Feller, *Introduction to Probability, Vol.I*, 3rd Ed., Wiley, New York, 1968.

[30] P.A. Griffin, *The Theory of Blackjack*, 6th Ed., Huntington Press, Las Vegas, NV, 1999.

[31] J.M. Harrison and D. Kreps, Martingales and Arbitrage in Multiperiod Securities Markets, *J. Econ. Theory*, **20**, 381–408, 1979.

[32] J.M. Harrison and S.R. Pliska, Martingales and Stochastic Integrals in the Theory of Continuous Trading, *Stoch. Proc. and Appl.*, **11**, 215–260, 1981.

[33] N. Ikeda and S. Watanabe, *Stochastic Differential Equations and Diffusion Processes*, North-Holland, New York, 1981.

[34] K. Itô, On a Stochastic Integral Equation, *Proc. Imperial Acad. Tokyo*, **22**, 32-35, 1946.

[35] K. Itô and H. McKean, *Diffusion Processes and their Sample Paths*, 2nd printing, corrected, Springer-Verlag, New York, 1974.

[36] J. Jacod and P. Protter, *Probability Essentials*, Springer Verlag, New York, 2000.

[37] F. John, *Partial Differential Equations*, 4th Ed., Springer-Verlag, New York, 1982.

[38] N. Jolley, *The Cambridge Companion to Leibniz*, Cambridge University Press, Cambridge, UK, 1995.

[39] I. Karatzas and S.E. Shreve, *Brownian Motion and Stochastic Calculus*, 2nd Ed., Springer-Verlag, New York, 1991.

[40] P.E. Kloeden and E. Platen, *Numerical Solution of Stochastic Differential Equations*, Springer-Verlag, New York, 1995.

[41] P. E. Kloeden, E. Platen, and H. Schurz, *Numerical Solution of SDE Through Computer Experiments*, Springer-Verlag, New York, 1994.

[42] T.W. Körner, *Fourier Analysis*, Cambridge University Press, Cambridge, UK, 1990.

[43] W.V. Li and G.P. Pritchard, A Central Limit Theorem for the Sock-sorting Problem, in *Progress in Probability*, **43**, E. Eberlein, M. Hahn, M. Talagrand, eds., Birkhäuser, Boston, 1998.

[44] C.C. Lin and L.A. Segel, *Mathematics Applied to Deterministic Problems in the Natural Sciences*, Society for Industrial and Applied Mathematics, Philadelphia, 1988.

[45] R.S. Liptser and A.N. Shiryayev, On Absolute Continuity of Measures Corresponding to Diffusion Type Processes with Respect to a Wiener Measure, *Izv. Akad. Nauk SSSR, Ser. Matem.*, **36**(4), 874–889, 1972.

[46] R. S. Liptser and A. N. Shiryayev, *Statistics of Random Processes I: General Theory* (2nd Edition), Springer Verlag, New York, 2000.

[47] H.P. McKean, Jr., *Stochastic Integrals*, Academic Press, New York, 1969.

[48] Y. Meyer, *Wavelets: Algorithms and Applications* (Translated and revised by R.D. Ryan), Society for Industrial and Applied Mathematics, Philadelphia, 1993.

[49] M. Musiela and M. Rutkowski, *Martingale Methods in Financial Modelling*, Springer-Verlag, New York, 1997.

[50] J. Neveu, *Discrete-Parameter Martingales*, (Translated by T. Speed), North-Holland Publishing, New York, 1975.

[51] P. Protter, *Stochastic Integration and Differential Equations: a New Approach*, Springer-Verlag, 1990.

[52] G. Polya, *Induction and Analogy in Mathematics: Vol. I of Mathematics and Plausible Reasoning*, Princeton University Press, Princeton, NJ, 1954.

[53] G. Polya, *How to Solve It: A New Aspect of Mathematical Method*, 2nd Ed., Princeton University Press, Princeton NJ, 1957.

[54] G. Pólya and G. Szegö, *Problems and Theorems in Analysis, Vols. I and II*, Springer-Verlag, New York, 1970.

[55] D. Revuz and M. Yor, *Continuous Martingales and Brownian Motion*, (3rd Edition) Springer Verlag, New York, 1999.

[56] L.C.G. Rogers and D. Williams, *Diffusions, Markov Processes, and Martingales. Volume One: Foundations*, 2nd Ed. Cambridge University Press, 2000.

[57] K. Sekida (translator and commentator), *Two Zen Classics: Mumonkan and Kekiganroku*, Weatherhill, New York, 1977.

[58] W.T. Shaw, *Modelling Financial Derivatives with Mathematica: Mathematical Models and Benchmark Algorithms*, Cambridge University Press, Cambridge, UK, 1998.

[59] G. Strang, Wavelet Transforms versus Fourier Transforms, *Bull. Amer. Math. Soc.*, **28**(2), 288–305, 1993.

[60] H.R. Stoll, The Relationship Between Put and Call Option Prices, *J. Finance*, **24**, 802–824, 1969.

[61] E.O. Thorp and S.T. Kassouf, *Beat the Market*, Random House, New York, 1967.

[62] D.V. Widder, *The Heat Equation*, Academic Press, New York, 1975.

[63] H.S. Wilf, *Generatingfunctionology*, Academic Press, Boston, 1990.

[64] P. Wilmott, S. Howison, and J. Dewynne, *Option Pricing: Mathematical Models and Computations*, Cambridge Financial Press, Cambridge, UK, 1995.

[65] P. Wilmott, S. Howison, and J. Dewynne, *The Mathematics of Financial Derivatives: A Student Introduction*, Cambridge University Press, Cambridge, UK, 1995.

[66] D. Williams, *Probability with Martingales*, Cambridge University Press, New York, 1991.

[67] N. Young, *An Introduction to Hilbert Space*, Cambridge University Press, New York, 1988.

[68] D. Zwillinger, *Handbook of Differential Equations*, 2nd Ed., Academic Press, New York, 1992.

Index

Applications of Mathematics

(continued from page ii)

35 Kushner/Yin, **Stochastic Approximation and Recursive Algorithms and Applications,** Second Ed. (2003)
36 Musiela/Rutkowski, **Martingale Methods in Financial Modeling: Theory and Application** (1997)
37 Yin/Zhang, **Continuous-Time Markov Chains and Applications** (1998)
38 Dembo/Zeitouni, **Large Deviations Techniques and Applications**, Second Ed. (1998)
39 Karatzas/Shreve, **Methods of Mathematical Finance** (1998)
40 Fayolle/Iasnogorodski/Malyshev, **Random Walks in the Quarter Plane** (1999)
41 Aven/Jensen, **Stochastic Models in Reliability** (1999)
42 Hernández-Lerma/Lasserre, **Further Topics on Discrete-Time Markov Control Processes** (1999)
43 Yong/Zhou, **Stochastic Controls: Hamiltonian Systems and HJB Equations** (1999)
44 Serfozo, **Introduction to Stochastic Networks** (1999)
45 Steele, **Stochastic Calculus and Financial Applications** (2000)
46 Chen/Yao, **Fundamentals of Queueing Networks: Performance, Asymptotics, and Optimization** (2001)
47 Kushner, **Heavy Traffic Analysis of Controlled Queueing and Communication Networks** (2001)
48 Fernholz, **Stochastic Portfolio Theory** (2002)
49 Kabanov/Pergamenshchikov, **Two-Scale Stochastic Systems** (2003)
50 Han, **Information-Spectrum Methods in Information Theory** (2003)
51 Asmussen, **Applied Probability and Queues** (2003)
52 Robert, **Stochastic Networks and Queues** (2003)
53 Glasserman, **Monte Carlo Methods in Financial Engineering** (2003)

